Conserving Nature
in Greater Yellowstone

CONSERVING NATURE IN GREATER YELLOWSTONE

Controversy and Change in
an Iconic Ecosystem

ROBERT B. KEITER

The University of Chicago Press
Chicago and London

The University of Chicago Press, Chicago 60637
The University of Chicago Press, Ltd., London
© 2025 by Robert B. Keiter
All rights reserved. No part of this book may be used or reproduced in any manner whatsoever without written permission, except in the case of brief quotations in critical articles and reviews. For more information, contact the University of Chicago Press, 1427 E. 60th St., Chicago, IL 60637.
Published 2025
Printed in the United States of America

34 33 32 31 30 29 28 27 26 25   1 2 3 4 5

ISBN-13: 978-0-226-84124-3 (cloth)
ISBN-13: 978-0-226-84126-7 (paper)
ISBN-13: 978-0-226-84125-0 (e-book)
DOI: https://doi.org/10.7208/chicago/9780226841250.001.0001

Library of Congress Cataloging-in-Publication Data

Names: Keiter, Robert B., 1946– author
Title: Conserving nature in Greater Yellowstone : controversy and change in an iconic ecosystem / Robert B. Keiter.
Description: Chicago : The University of Chicago Press, 2025. | Includes bibliographical references and index.
Identifiers: LCCN 2024046651 | ISBN 9780226841243 cloth | ISBN 9780226841267 paperback | ISBN 9780226841250 ebook
Subjects: LCSH: Nature conservation—Yellowstone National Park Region | Yellowstone National Park—Environmental conditions | Ecosystem management—Yellowstone National Park Region
Classification: LCC QH76.5.Y45 K45 2025 | DDC 333.750978752—dc23/eng
LC record available at https://lccn.loc.gov/2024046651

♾ This paper meets the requirements of ANSI/NISO Z39.48-1992 (Permanence of Paper).

In memory of Joe Sax,
for his profound contributions to nature conservation
and
to Don Sharp,
stalwart friend and hiking companion

CONTENTS

Preface   ix

Introduction   1

1. Greater Yellowstone and Ecosystem Conservation   13

2. Nature Conservation in the GYE National Parks   35

3. Restoring Endangered Species:
Grizzly Bears and Wolves   65

4. Maintaining Migratory Wildlife:
Bison, Elk, and Pronghorn   92

5. The Multiple-Use Lands:
From Clearcuts to Playgrounds   126

6. A Fragmenting Landscape:
The Private Land Conservation Challenge   165

7. Ecosystem Conservation Revisited   190

Conclusion   227

Acknowledgments   241
Notes   245
Index   299

PREFACE

It was the summer of 1967, driving through Yellowstone on an extended western road trip. I was a passenger in my friend's sports car with the top down when he pulled into a turnout amid several cars, with both of us wondering what the attraction is. Moments later, the bear ambled over to our car, which was now trapped in the middle of a growing cluster of curious onlookers all safely inside their vehicles. Unable to move or to exit the car, I ended up in a face-to-face encounter with the black bear whose slathering mouth almost immediately alerted me it was looking for food—not an uncommon occurrence those days in Yellowstone. Once I got a good look at the large claws resting only inches away on the car's doorframe, the potential seriousness of the situation was evident. The encounter ended without incident, thankfully, when the bear was distracted away. But the event left an indelible impression—one that has evolved over time, just as the park and the surrounding landscape has changed.

Little did I then know or contemplate that my professional career would take me back to Yellowstone repeatedly, and that I would witness the profound transformation that has occurred throughout the region, including revised bear-management policies. I grew up amid a national park controversy that played out a few hundred feet from our backyard, namely the 1950s battle to preserve the C&O Canal and the adjacent Potomac River corridor from a new highway designed to connect Washington, DC's burgeoning suburbs with the nation's capital city. Though unable to stop what is now the Clara Barton Parkway, energized conservation groups succeeded in convincing the president and then Congress to protect the canal and river corridor, establishing eventually the Chesapeake and Ohio National Historic Park. Years later, I was a new law professor at the University of Wyoming when the opportunity arose to examine how Glacier National Park, named in 1980 as the nation's most threatened park due to nearby development pressures, might address those impending external threats. It was soon evident that Yellowstone confronted similar problems in the form of clearcutting, energy leasing, and geothermal drilling on adjacent national forest lands, as well as a dwindling grizzly bear population. The ecological integrity of both parks was at serious risk.

A couple new ideas, however, offered hope these problems might be brought under control: the Greater Yellowstone Ecosystem (GYE) concept and the related ecosystem-management strategy. These seminal ideas viewed the expansive, jurisdictionally fragmented Yellowstone region as an integrated ecological system that should be managed as an interconnected entity with a focus on preserving its ecological integrity. A new, assertive conservation group, the Greater Yellowstone Coalition, had seized these ideas and was promoting the need for a regional perspective in order to protect the area's charismatic wildlife and stunning natural features that had given rise to the park and surrounding wilderness areas in the first place. These ideas soon proved powerful enough to take hold even beyond the Yellowstone country, injecting a new ecological mentality into the federal land-management agencies, though not without controversy aplenty. Since then I have continued to observe and study how these ideas have served to reshape long-standing natural resource-management policies and capture the public imagination.

Just as it was the genesis for the national park idea, the Yellowstone country incubated these ideas, serving as a critical testing ground for ecosystem conservation. Given the region's jurisdictional complexities and the ongoing changes afoot in the region's communities, the quest to preserve the GYE's extraordinary natural character and free roaming wildlife has proven quite challenging at times, as evidenced by the strident local resistance to the Yellowstone wolf reintroduction initiative. Indeed, seemingly endless political battles as well as court cases have been fought and refought over grizzly bears, wolves, bison, wildfire, logging, mining, drilling, grazing, subdivisions, snowmobiling, property rights, state sovereignty, and the list goes on. Importantly, these conflicts have been—and are being—addressed within the context of the GYE idea and a mounting consensus supporting the need to preserve this far-flung region's unique natural character.

This book seeks to recount how this widespread commitment to nature conservation at the ecosystem scale has taken hold, what that has meant on the ground, and implications for the future. It is a story about ideas, controversy, and change, abetted by science, politics, economics, law, and cultural evolution. It begins in the post–World War II era, with an emerging commitment to ecological science within the Park Service and Forest Service, the beginnings of conservation biology as a dimension of ecology, validation of the wilderness concept with passage of the Wilderness Act, the emergence of an assertive environmental movement, and dramatic changes in the nation's social fabric and values. These developments soon coalesced in the Yellowstone country, bringing meaningful and sometimes painful change to the region's public lands as well as the communities dotted around them—a process that continues unabated today.

To tell the story of nature conservation in the GYE, the book unfolds topically with a focus on the primary issues that have confronted the region during the past sixty years. The book begins, following a short introductory chapter, by describing the GYE, reviewing the relevant science underlying an ecosystem-based management approach, and outlining the political-legal framework governing GYE conservation policy, including a brief sketch of how these policies evolved during the latter half of the twentieth century. It then focuses on the GYE's two national parks and critical resource-management challenges that Yellowstone and Grand Teton have faced. Next, it reviews GYE wildlife issues involving grizzly bears, wolves, bison, and elk that span the region's jurisdictional boundaries and have regularly proven contentious. Then it focuses on the GYE national forests, where significant policy changes regarding timber, energy, mining, and livestock are evident, while wildlife, wilderness, and recreation concerns have risen in importance. It next explains the importance of the region's privately owned lands, describes the emerging conservation initiatives taking hold on these lands, and examines the options available to communities and property owners, along with promising new landscape-scale initiatives. The ensuing penultimate chapter examines the impact ecosystem-based management has had on the GYE, the forces that have brought nature conservation to the fore as a regional priority, the conservation challenges that lie ahead, and the impact of the GYE ecosystem conservation effort elsewhere. The book concludes by extracting instructive lessons from the GYE experience with an ecological conservation approach, and with observations on the next-generation conservation challenges facing the region and its natural attributes.

While researching and writing this book, I regularly found myself returning to the fact that more than fifty years had elapsed since my initial encounter with Yellowstone and its bears. Though then awed by the park's raw beauty and wildness, I had scant knowledge about its wildlife, ecology, or management, or its relationship to the surrounding landscape. And I was not alone. That has changed dramatically, however, throughout the region. Today, we better understand this unique, complex, and extraordinary place widely known as the Greater Yellowstone Ecosystem where wild nature still prevails. We appreciate more fully the interconnections with the surrounding region, as well as what is required to conserve the GYE's wildlife, natural processes, and wilderness character. But change—warming temperatures, migration obstacles, wildlife diseases, and rampant subdivision activity, for example—is ever present, posing increasingly urgent challenges in the enduring quest to conserve this special place. Drawing upon our accumulated experience in pursuit of ecosystem conservation, as captured in the ensuing pages, we have the fundamental knowledge required to meet these impending challenges. The looming question is whether we have the will to employ this knowledge and the requisite commitment to carry these nature conservation efforts forward to preserve this iconic landscape.

Introduction

Number 10 likely did not suspect any danger as it wandered the remote mountainside following its own instincts. The male wolf and its mate, Number 9, had earlier crossed out of Yellowstone National Park and made their way to Mount Maurice in Montana's Custer National Forest. They were preparing to den so Number 9 could bear her pups, some of the first wolves to be born in the Yellowstone country in nearly seventy years. A few months earlier, at the dawn of 1995, fourteen Canadian wolves had been transported to the park under a federal reintroduction initiative designed to restore this legendary predator to lands it once occupied. Returning the long-absent wolf to the park not only restored dormant predator-prey relationships, but also set in motion ecological changes that would, in time, reshape the Yellowstone ecosystem.

Number 10 would not witness any of this, however, once Chad McKittrick spotted him. In an outlaw act driven by apparent bloodlust and wolf hatred, McKittrick took quick aim and killed Number 10 with a single bullet from 140 yards away. Federal law enforcement agents soon identified McKittrick as the culprit and charged him with violating the Endangered Species Act (ESA). Because they were part of an experimental population, Number 10 and the other reintroduced wolves were protected from being hunted or killed, whether within the park or outside it—a legal status that would continue until the population was deemed recovered.[1]

Of course the 1995 Yellowstone wolf reintroduction was quite controversial and long resisted by local politicians, the ranching community, and others in Montana, Wyoming, and Idaho. But the National Park Service, US Fish and Wildlife Service, conservation groups, and allies had persisted and eventually persuaded Congress that wolves belonged in the nation's first national park. That did not mollify those local residents who resented the federal intrusion as well as the wolves, reflecting the same sentiments underlying the original all-out eradication campaign that rid the West of wolves by the 1930s.[2]

Nonetheless, a Montana jury that included several ranchers and farmers convicted McKittrick of violating the ESA by shooting the federally protected wolf. The jury did not believe McKittrick's mistaken identity defense, rendering a verdict that suggested a surprising degree of local acceptance for the Yellowstone wolf restoration project. On appeal, the Ninth Circuit Court of Appeals affirmed not only the conviction but also the wolf reintroduction project itself, finding the park lacked an established wolf pack, which would have precluded use of the ESA's experimental population provision. In a related Wyoming case, the Tenth Circuit Court of Appeals similarly confirmed that the Yellowstone wolf reintroduction was legal under the ESA.[3]

These court decisions, however, did not end the controversy over the wolves. With the wolf population growing rapidly, the three Northern Rockies states soon pushed to have wolf-management responsibility returned to them, which would end federal protection and expose the wolves to hunting and trapping. A series of court challenges ensued as well as congressional intervention that disaggregated Wyoming from the other two states, a manifestation of the Greater Yellowstone region's jurisdictional complexity. Eventually, the three states assumed legal responsibility for managing the wolves, but when Montana and Idaho drastically loosened their wolf hunting and trapping laws, conservation groups initiated another round of litigation and a federal status review that was ultimately rejected. Although wolf restoration is regarded as an unalloyed ecological success, there is little reason to believe the controversy ends there.[4]

The Yellowstone wolf reintroduction illuminates the vital role nature conservation plays across the region, now widely known as the Greater Yellowstone Ecosystem (GYE), and the conflicts inherent in such a conservation agenda. By any standard, the restored wolves are symbolic of wildness; their presence constitutes powerful evidence of a societal commitment to preserving the region's wildlife and wilderness character. With no regard for the jurisdictional boundary lines we have imposed on the landscape wolves now range—as Number 10 did—across the region, confirming it as an interconnected ecosystem and the need to manage it as such. Meanwhile, controversy persists over the wolves and extends to other regional wildlife-management issues, reflecting the challenges involved in pursuing an ecosystem conservation agenda in even the expansive and relatively undeveloped Yellowstone country.

Those challenges are equally evident in the case of migratory wildlife and the historic pathways they follow across the GYE and onto the broader landscape. Just as the Yellowstone wolf restoration initiative broke new ground, the same is true for ongoing efforts to identify and protect wildlife migration corridors to maintain a seasonal ecological process vital to the GYE's ungulate populations. Like the wolf restoration effort, wildlife corridor conservation extends across existing boundary lines, requiring multiagency collaboration as well as private landowner cooperation. In several instances, the region's migration pathways extend beyond the GYE's conventionally regarded boundaries, effectively expanding the conservation effort to a larger landscape scale—which also increases the possibility for controversy.[5]

But that did not occur, at least not initially, when federal and state officials joined with conservation organizations and a private landowner during the early 2000s to preserve a historic 150-mile-long migration corridor christened "Path of the Pronghorn." The nation's first federally designated wildlife migration corridor, the Path of the Pronghorn runs from Grand Teton National Park through the Bridger-Teton National Forest, then crosses private ranchlands as well as a state highway before reaching winter range on more southerly Bureau of Land Management (BLM) lands that host major natural-gas fields. To safeguard the corridor and the annual antelope herd migration, the Park Service acquired a strategic state-owned parcel, the Forest Service added protective provisions to its governing plan, a local land trust secured a conservation easement on critical ranchlands, the state of Wyoming constructed a wildlife overpass spanning a busy state highway, and the BLM established a protective "area of critical environmental concern" at the terminus.[6]

Though a widely acclaimed example of collaborative conservation, the Path of the Pronghorn has not escaped controversy even as it has spawned additional corridor conservation efforts in Wyoming and surrounding states. The conservation easement was tested when the ranch's new owner began to build illegally in the protected pathway contrary to the easement's terms. When an energy company sought to expand its drilling operations into the pathway, conservation groups sued in an effort to block the project, but were unsuccessful, highlighting the challenges involved when even a high-profile wildlife conservation achievement collides with energy development in Wyoming. And despite its acclaim, the Path of the Pronghorn has yet to receive designation as an official state wildlife corridor, largely because of the potential ramifications for future energy activity. Nonetheless, the corridor serves as an innovative model for next-generation conservation efforts in the GYE, which are effectively expanding ecosystem conservation approaches onto the broader landscape with attendant jurisdictional complexities.[7]

Following her hereditary instincts, mule deer Number 19 played the next major pioneering role in GYE corridor conservation efforts. Biologist Hall Sawyer collared Number 19 during winter 2011 in southern Wyoming's Red Desert area to track her movements, expecting she would not stray more than thirty miles. When Sawyer recovered her collar, to his surprise, she had seasonally moved 150 miles northward to lush summer range in the Hoback area a bit south of Jackson Hole, validating another long-range migration that overlapped the GYE. Since then, biologists have confirmed the herd's migration pattern, conservation groups have acquired key private lands in the pathway, and the state has officially designated the Red Desert to Hoback Migration Corridor. Yet work remains to enable the herd to navigate safely a mosaic of federal, state, and private lands, as well as drill rigs, highways, fences, and subdivision pressures evident along the migration route.[8]

In short, nature conservation in the GYE is complex and often contentious. The region's myriad jurisdictional boundaries complicate essential ecosystem-based conservation efforts necessary to knit together the landscape, the responsible agencies, and landowners in order to safeguard GYE wildlife populations and pathways. Despite the best intentions and coordination efforts, controversy lurks both outside and inside the region's two national parks, embracing the region's diverse federal, state, and private lands. These cross-boundary conservation challenges are not new to the GYE region, having initially surfaced with the establishment of Yellowstone National Park and persisting since then.

* * *

Beginning with its inception in 1872 Yellowstone National Park has been synonymous with nature conservation, and remains so today. As the world's first national park, Yellowstone has served as a model for natural heritage preservation efforts across the nation and beyond. The park boasts the full complement of wildlife that the early Euro-American explorers originally encountered in their westward journeys as well as the stunning features and natural processes that have long shaped the area landscape and wilderness character. Few places in the world can say the same or offer a similar experience. But the expansive 2.2-million-acre park—large as it is—does not stand alone. It is surrounded by extensive national forests, sprawling ranchlands, and fast-growing communities that are inextricably connected with the park, its wildlife, and natural features. This extended area—widely known as the Greater Yellowstone Ecosystem (GYE)—is today synonymous with nature conservation and related ecological management efforts. Yet even in the world-renowned Yellowstone region, generally regarded as the cradle of conservation, the transition to thinking in ecosystem terms and adjusting conservation efforts to this scale has not come easily.

Since the mid-twentieth century the GYE has spawned a succession of conservation controversies that have redefined federal resource-management priorities as well as the region itself. During the 1960s, when protests erupted over Yellowstone's annual elk culling to limit herd sizes, the National Park Service's resource-management policies came under intense scrutiny. The upshot was a new, science-oriented policy predicated on minimal intervention in the ecological processes that defined the national park setting. The new policy, however, just fueled more controversy. Elk numbers soon spiked, provoking an intense decades-long debate over the animals' impact on the park's range ecosystems. Similarly, the park's free-ranging bison herd grew in numbers, prompting more bison to leave the park in winter for food, where they presented a disease threat to domestic cattle with potentially dire economic implications. Meanwhile, bear numbers plummeted after the Park Service abruptly stopped permitting them to be human-fed. The park's declining grizzly bear population soon prompted an Endangered Species Act listing that has kept the bear under federal management for the past fifty years. Yellowstone's related experiment allowing wildfires to burn came under harsh criticism when the 1988 fires rampaged across the park, precipitating a far-reaching review of federal wildfire policy. The ultimately successful decades-long effort to restore wolves to Yellowstone similarly provoked fierce political resistance that continues today, though now focused on state wolf-management policy. Moreover, the Park Service's seemingly harmless decision during the 1960s to open Yellowstone to snowmobiles during wintertime eventually sparked a prolonged skirmish over appropriate recreational activity in a national park setting. In every respect, these Yellowstone-related resource conflicts have profoundly altered national park and wildlife-management policy across the region.[9]

Federal forest policy has likewise prompted controversy and change in the GYE. As World War II receded in the nation's mind, the Forest Service shifted into high gear, adopting an industrial production management model that soon brought controversy to the GYE's national forests. During the 1970s the agency consummated the largest timber sale outside Alaska on Idaho's Targhee National Forest, seeking to salvage beetle-damaged timber and invigorate the area's flagging economic fortunes. The resulting sales soon came under attack by a conservation community alarmed at the loss of wildlife habitat and the scarred landscapes left by the agency's manic clearcutting. Faced with repeat international oil embargoes during the 1970s, the United States turned to the Overthrust Belt running through the GYE, widely thought to contain rich oil deposits, in an effort to develop domestic energy supplies. Conservation groups resisted this incursion, challenging federal leasing and drilling proposals to protect the GYE's national park viewsheds, critical wildlife habitat, and wilderness values. Similar battles have been waged over mining, which has historic roots in the GYE but today threatens the region's environmental qualities and natural appearance. Livestock grazing, historically widespread across the GYE's public lands, practically ensures conflict with bears, wolves, and other predators—animals that the Park Service and others are committed to protecting. And when Congress adopted the Wilderness Act in 1964, the stage was set for extended battles over the fate of the GYE's undeveloped national forest lands. In each instance, the Forest Service has confronted a choice between ecological conservation and intrusive development activity, and its decisions have reverberated well beyond forest boundaries.[10]

The GYE's human communities have also found themselves entangled in changes that have contributed to regional controversy. The 1950s battle over the expansion of Grand Teton National Park foretold future changes in the region's rural, extraction-oriented character. Since then a growing number of GYE communities have gradually transformed into amenity-based towns dependent on the region's natural qualities for economic sustenance. Major ski areas at Jackson and Big Sky are emblematic of the industrial-scale recreation economy that has taken hold along with a soaring tourism economy, driven by visitation to the region's national parks. The GYE's attractions have drawn legions of new residents, as reflected in population growth figures placing several local counties among the fastest growing nationally. With their arrival, traditional ranchlands are being subdivided, changing the landscape's open appearance and fragmenting critical wildlife habitat. The fallout also features growing wealth disparities in several communities, enhanced wildfire concerns as new homes sprout adjacent to national forest lands, and newly arrived ranch owners with little knowledge about wildlife needs and few community connections. By several accounts, these transformative changes herald arrival of the New West across much of the GYE, reflected in clashes between newer arrivals intent on protecting the region's natural values and longtime residents wary of these cultural, economic, and other changes.[11]

Although still embroiled in complex resource-management controversies, the GYE is a quite different place today than a half century ago. The grizzly bear population has rebounded, wolves have been restored to Yellowstone, and both animals have now established residence well beyond the national parks. Elk numbers have decreased, while ungulate migration patterns are much better understood, prompting growing efforts to safeguard critical migration corridors. The bitter bison-management conflict tied to a tenuous domestic livestock disease threat persists, having sent thousands of park bison to slaughter, though bison transfers to Native American reservations are helping to mitigate this unseemly situation. More intense wildfires pose a growing seasonal problem manifested in smoky skies and unhealthy air, while also threatening new residences situated next to flammable forestlands. Commercial timber sales have mostly disappeared from the GYE national forests, while oil, gas, and mining activity is confined largely to the GYE's periphery. Domestic sheep grazing has been dramatically reduced, but predator conflicts with cattle remain a troublesome recurrent problem. The GYE's private ranchlands are now seen as critical to sustaining the region's ecological integrity, giving rise to robust land trust activity and conservation easement transactions, along with an infusion of federal dollars to incentivize landowner conservation. Significant growth in tourism and recreation is evident nearly everywhere, putting additional stresses on the region's wildlife and its backcountry lands. Overshadowing all of these resource-management challenges is climate change, lending a real sense of urgency to ongoing GYE conservation efforts.[12]

* * *

The "Greater Yellowstone Ecosystem" concept has assumed a central role in regional nature conservation efforts. Drawing upon emergent science, conservation organizations and others seized the concept during the 1980s in an effort to curtail the sweep of environmental problems besetting the region. The GYE concept, although still controversial in some quarters, is nonetheless deeply rooted in the region's history. The term

"Greater Yellowstone" first surfaced more than a century ago, just ten years after Yellowstone National Park was established. It is attributed to General Philip Sheridan, a legendary Civil War hero, who took an early interest in the new park. The general is credited with bringing the US Cavalry to the park to stop local poaching and other incursions, and for resisting efforts to open the park to the railroads. During an 1882 visit to Yellowstone, upon observing that the park's wildlife seasonally depended on lower-elevation habitat beyond the boundary line, Sheridan coined the "Greater Yellowstone" term in an effort to expand the new park. Although Sheridan's park expansion idea never gained a political foothold, President Benjamin Harrison accomplished nearly the same in 1891 when he invoked a new law empowering the president to establish forest reserves on the public domain. Harrison's creation of the Yellowstone Timber Land and Reserve—the nation's first federal forest reserve—buffered the park on its eastern and southern borders by removing these lands from further settlement. Subsequent presidents soon added other forest reserves adjacent to the park, creating an extensive federal land complex in this remote mountainous region.[13]

Upon passage of the 1916 National Parks Organic Act, the new National Park Service assumed responsibility for Yellowstone and renewed the park expansion idea. Superintendent Horace Albright proposed dramatically expanding the park southward toward Jackson Hole in an effort to protect the spectacular Teton mountain range and important wildlife habitat. Local dude ranchers, fearing new roads would disturb the natural setting and upset their Eastern clientele attracted to the area's wilderness qualities, joined with the Forest Service, whose lands were at risk, to oppose Albright's proposal. Opposition crystallized when Emerson Hough, a well-known *Field and Stream* correspondent, penned an article endorsing the expansion proposal by asserting "Give her Greater Yellowstone and she will inevitably become Greater Wyoming." Hough's misinterpreted statement galvanized the Wyoming legislature to oppose the proposal, sealing its fate. Nonetheless, the episode introduced the "Greater Yellowstone" phrase into public discourse and laid the groundwork for the eventual establishment of Grand Teton National Park.[14]

Less than forty years later, the term "Greater Yellowstone Ecosystem" entered the regional lexicon, grounded in grizzly bear biology. During the 1950s biologists John and Frank Craighead initiated a pioneering study of Yellowstone's grizzly bears that employed new technologies enabling them to collar and electronically track the bears to better understand their habits and range. They discovered that the bears roamed widely through and beyond the park in their quest for food and mates. To describe the bears' expansive range, they coined the term "Greater Yellowstone Ecosystem." Although the Craigheads eventually found themselves at odds with the Park Service over its revised bear-management policies, the GYE term was soon adopted by scientists and environmental groups to describe the region, its ecological connections and importance. It has become the central concept driving regional nature conservation efforts—efforts that are being expanded to an even larger landscape scale to protect vital wildlife migration routes and to connect the GYE with other, more distant ecosystems.[15]

The related ecosystem-management concept is grounded in late twentieth-century insights about the natural world. Ecosystem management draws heavily on the ecological sciences and conservation biology principles in an effort to protect biodiversity and ecosystem integrity for long-term human benefit. After sputtering in the GYE, ecosystem management emerged as a federal resource-management policy during the early 1990s as a response to unbridled logging in the Pacific Northwest, growing species extinction concerns, and external activities threatening national parks. Simply put, ecosystem management promotes science-driven, coordinated, and adaptive resource-management efforts to conserve biodiversity and natural processes at an ecologically appropriate scale, while also supporting economically and socially sustainable communities. It endeavors, through a commitment to interagency coordination and planning, to break down the jurisdictional boundaries that have long separated the responsible agencies in order to better align resource-management strategies with ecological realities on the ground. Laws like the Endangered Species Act, National Environmental Policy Act, and National Forest Management Act provide a legal foundation for ecosystem-management policy.[16]

As the twenty-first century has unfolded, the landscape conservation idea has taken hold, incorporated into federal policy by the Obama administration and now strongly endorsed in scientific, academic, and conservation circles. Recognizing the true impacts of impending climate-related and other human-driven environmental changes, landscape conservation expands the ecosystem-management approach to a larger landscape scale. It seeks to enlist federal public land managers along with private and other nonfederal landowners to prioritize nature conservation by protecting species, migratory passageways, ecosystem services, and natural processes. It is also connected to the emergent 30×30 campaign, which calls for protecting 30 percent of the earth's lands and waters by 2030 as a necessary means to preserve biodiversity and stave off likely extinctions in the face of a warming climate and unrelenting development pressures. These shifts in resource-management policy are currently playing out across the GYE, though more often through informal rather than formal coordination efforts. Consequently, the expansive GYE provides a vital testing ground—albeit an evolving one—for related efforts seeking to institutionalize a landscape conservation approach in ecologically critical venues at home and abroad.[17]

* * *

What follows examines the emergence of nature conservation as a priority concern across the GYE, how it has taken root, the conflicts it has spawned, and the changes it has engendered, as well as the challenges lying ahead. The book endeavors to answer several interrelated questions regarding these developments. One is whether, despite the region's ecological and jurisdictional complexities, we are managing to conserve (and restore) the GYE as a functional ecological entity. A second question explores how regional conservation advances have occurred, focusing on the role played by the GYE concept, related ecosystem-management approaches, and the diverse forces undergirding these efforts. Yet another goes to the implications arising from GYE conservation efforts, both in the ecosystem and beyond it. Finally, are the strategies that have brought us this far in the GYE relevant to tomorrow's next-generation conservation challenges?

To answer these questions, the book takes a thematic approach centered on the GYE national parks, national forests, wildlife policies, and the region's ecologically important private lands. It identifies and analyzes the dynamic scientific, political, social, economic, and legal forces propelling the profound changes that have evolved during the past sixty years. In many respects the policy changes are remarkable, largely reflecting a decided shift in priorities toward preserving the GYE's natural attributes, wildlife populations, and unique character. The results on the ground are notable, too, as measured by the region's wildlife numbers, habitat conditions, and extant ecological processes. How this has occurred offers critical lessons in the evolution of nature conservation policy and the interrelated role of science, politics, and law in bringing about the evident changes. As the GYE story continues to unfold, much can be learned from these efforts and accomplishments, while much also remains to be done.

As we shall see, the GYE's jurisdictional complexity has complicated regional nature conservation efforts. No grand vision or plan has yet emerged, so the progress achieved thus far has largely occurred issue by issue, including some notable intergovernmental coordination accomplishments, as in the case of the grizzly bear. In the GYE national forests resource-management priorities have shifted dramatically as timber, mining, and energy activities have dwindled during the past thirty years due in part to federal law and related policy shifts. Meantime, wildlife conservation and recreation concerns have assumed greater importance, but new wilderness designations have proven elusive during the past forty years. Federal-state relations remain tense over such matters as wolf and bison management, while tensions also persist among the three GYE states, most notably in the case of Wyoming's elk feedground policy.

As nature conservation has gained a firmer foothold on the GYE federal lands, attention has turned to the region's private lands, which face escalating development pressures and related habitat fragmentation problems. These private land development issues as well as the GYE's persistent wildlife-management challenges are governed primarily by state laws that—unlike related federal laws—are notoriously weak. Powerful local political and economic forces are also at work, sometimes in harmony with and other times in conflict with the region's emerging conservation agenda. Nonetheless, the GYE concept and ecosystem-management policies have clearly risen to the fore across the region, while the scale of conservation efforts is expanding to embrace the larger landscape.

Of course, considerable uncertainty surrounds the GYE's future amid the embryonic Anthropocene era with the human imprint now evident virtually everywhere on the earth. In fact, we are still comprehending the full effect climate change will have on the GYE, its wildlife, forests, rangelands, water, and communities. We are witnessing, with trepidation, the impacts that population growth and private land development are visiting on the region. We perceive the region's national parks may be overcrowded, diminishing the visitor experience and adversely affecting wildlife and other resources. We see the impacts escalating recreational activity has on backcountry trails and sensitive wildlife. We are anxious about the potentially devastating effects chronic wasting disease could have on the region's elk herds, while the long-standing brucellosis disease problem persists unresolved. And we remain at odds over the GYE's grizzly bears and wolves—still regarded by some as fearsome predators—that now widely roam the region, sparking controversy outside the national parks.

How then, drawing upon past conservation achievements and lessons, can we shape the future to preserve the GYE's ecological integrity and special attributes? Can we maintain the GYE national park experience? Can we preserve robust populations of grizzlies, wolves, bison, and elk for the enjoyment of future generations? Can we sustain—and perhaps even expand—the region's wilderness acreage as well as the open spaces that render the area so unique? Can we establish and protect wildlife migration corridors that stretch across public as well as private lands? Can we do all of this in the face of warming temperatures and ever-mounting development pressures? Moreover, how might these lessons apply elsewhere? Our forbearers managed to preserve much of the Yellowstone region's natural attributes, begging the question whether we have the will and wisdom to sustain while also improving upon their conservation achievements. If we cannot protect nature in the expansive and relatively intact GYE, then where can we hope to preserve our natural heritage?

CHAPTER ONE

Greater Yellowstone and Ecosystem Conservation

Nature conservation in the Greater Yellowstone Ecosystem is as complex and challenging as the region itself. The GYE's remarkable natural features and systems extend across a diverse landscape overseen by an array of federal, state, tribal, and local governmental entities, creating a complicated jurisdictional setting. While the agencies responsible for the region's public lands and wildlife have remained constant, the GYE's human communities are enmeshed in changes that have spawned a growing commitment to preserving the area's natural character. The region's evident preservation impulses are linked to our better understanding of natural systems and our evolving relationship with nature. Ecosystem science and conservation biology, having come of age during the late twentieth century, are shaping wildlife conservation efforts across the region. A greater appreciation for the natural world as well as changing economic and social conditions underlie an evident regional conservation consciousness. Other forces have been at work, too. A maturing body of law generally supports GYE ecological preservation efforts and brought the courts squarely into several conservation conflicts. With much of the GYE in federal ownership, national and local political considerations feature prominently in the ongoing regional conservation dialogue. All of these factors are at work in GYE ecosystem-management efforts, not only accounting for the regional transformations evident today but also setting the stage for future changes.

A Complex and Contested Landscape

The GYE is an interconnected yet complicated landscape when viewed in ecological and legal terms for nature conservation purposes. The GYE is generally understood to encompass roughly 23 million acres, up from roughly 14 million acres when the concept was initially conceived for conservation purposes during the 1980s. The natural setting—regularly characterized as "the largest, nearly intact, temperate ecosystem in the world"—can only be described as extraordinary, a wonderland of unique features, charismatic wildlife, and spectacular scenery that still conjures a sense of the early West.

Much of the GYE rests on a high plateau formed by volcanic activity, which accounts for the region's numerous geothermal features. Several mountain ranges intertwine across the area, including the renowned Teton, Absaroka, Wind River, and Beartooth ranges. These mountain ranges are largely forested, blanketed by thick stands of pine, fir, and aspen trees. High elevation, sagebrush-studded plains and basins separate the mountain ranges and provide critical wildlife habitat. Three continental river systems—the Colorado, Columbia, and Mississippi—originate in the GYE. The area boasts all of the wildlife species present when settlers first arrived, including grizzly bears, wolves, elk, bison, pronghorn, moose, bighorn sheep, bald eagles, trumpeter swans, cutthroat trout, and the list goes on. Following age-old instincts, the region's wildlife routinely cross the manmade boundary lines we have imposed on the landscape as they migrate seasonally and otherwise go about making a living for themselves. These superlative natural attributes and the GYE's sheer size help account for its ecological significance, both in terms of the diverse species making the region home and the natural processes that still prevail across the landscape.[1]

Not a recognized formal legal entity, the GYE contains an amalgam of federal, state, and private lands that bridge three states—Idaho, Montana, and Wyoming. Roughly 15 million acres (or two-thirds of the GYE) is in federal ownership, giving Congress final say over how these lands are managed and putting the four responsible federal land-management agencies in an elevated regional role. Two world-famous national parks—Yellowstone and Grand Teton—constitute the core of the ecosystem and extend across 2.5 million acres. Moving outward, the GYE includes five national forests, extensive wilderness areas, three national wildlife refuges, three protected wild and scenic river segments, and lower-elevation public lands. The three GYE states have scattered school trust landholdings in the area, while privately owned lands, including many large ranches, cover roughly 30 percent of the region. Still largely rural in appearance, the GYE spans twenty counties and embraces numerous small towns, each with its own economic, demographic, and cultural characteristics. While federal law largely dictates management of GYE's parks, forests, and other federal lands, state and local law dictates what occurs on the region's privately owned lands, many of which are ecologically significant given their location at lower elevations and near water sources. State law also governs wildlife management outside the region's national parks and wildlife refuges. In short, the GYE manifests extreme jurisdictional complexity, making it difficult to coordinate resource-management priorities and practices at an ecosystem or regional level.[2]

Yellowstone and Grand Teton national parks represent the centerpieces of the GYE, encompassing an area larger than Delaware and Rhode Island combined. The parks are managed by the National Park Service, which is charged with protecting their scenery, wildlife, and other natural features in an unimpaired condition while accommodating visitors. More than two million acres in the parks have been recommended for wilderness designation and are managed accordingly. Both parks are known for their scenic wonders: Old Faithful, Grand Canyon of the Yellowstone, Mammoth Terrace, the Teton spires, and Jenny Lake to name a few. They also host an extraordinary array of wildlife, including a growing number of grizzly bears, free-ranging bison, and a restored wolf population. Due to the harsh winter weather that prevails in the elevated parks, many animals exit during winter months in search of more suitable seasonal habitat. Both parks attract legions of visitors—now exceeding four million annually at each—whose presence supports a thriving regional tourism and recreation economy tied to nearby gateway communities. As visitation has mounted, the Park Service has found itself confronting difficult questions pitting its conservation responsibilities against visitor pressures.[3]

Five national forests adjoin the GYE national parks, covering about 11 million acres. Overseen by the US Forest Service, the Bridger-Teton, Shoshone, Caribou-Targhee, Custer-Gallatin, and Beaverhead-Deerlodge national forests are managed under a multiple-use mandate that includes timber, mining, energy development, livestock grazing, recreation, fish, wildlife, and wilderness. The Forest Service, though long imbued with a commodity production ethic, has recently been shifting its resource-management emphasis from extractive uses toward conservation and recreation. In fact, recreational activity on the GYE forests has accelerated in recent years, reflected in several popular downhill ski resorts,

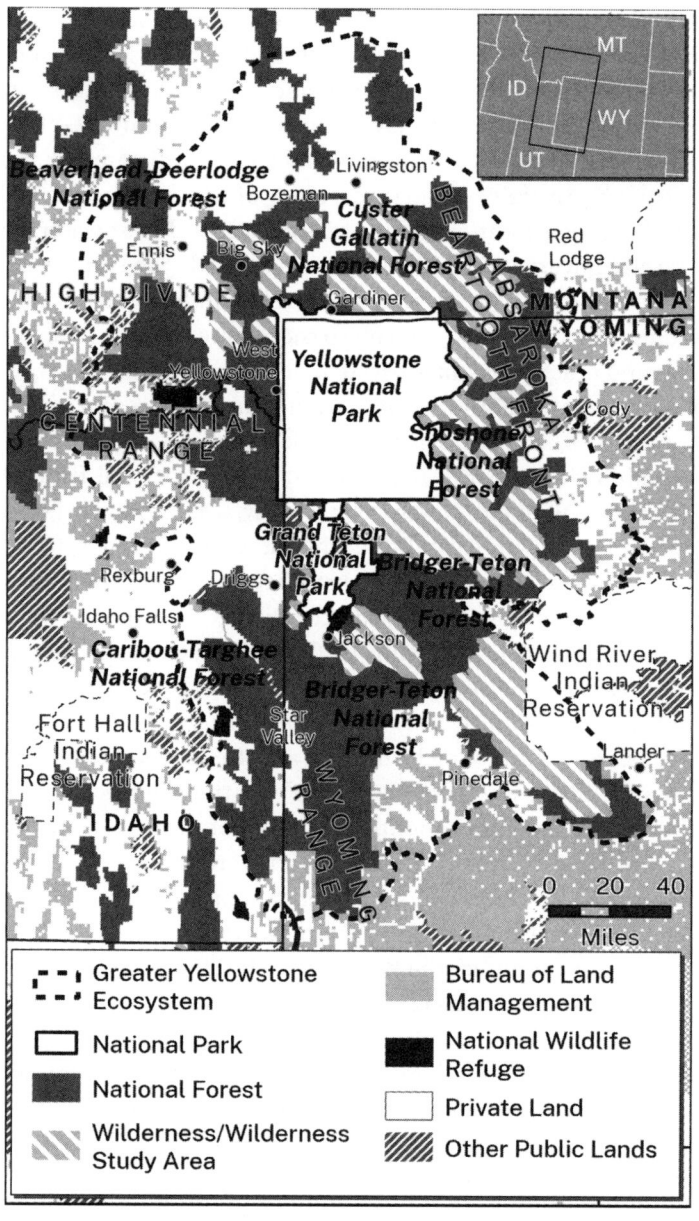

The Greater Yellowstone Ecosystem. The Greater Yellowstone Ecosystem embraces roughly 23 million acres in and around Yellowstone and Grand Teton national parks, including more than four million acres of designated wilderness and wilderness study areas in the five surrounding national forests. The map depicts the GYE's various political boundaries, land ownerships, and legal designations, illuminating the region's jurisdictional complexity. Map by the University of Utah DIGIT Lab.

off-road vehicle (ORV) traffic, and mountain-biking activity. The GYE national forests also boast more than four million acres of officially designated wilderness and wilderness study areas, where roads, ORVs, mountain bikes, and the like are prohibited. Several of these wilderness areas, including the Absaroka-Beartooth, Jedediah Smith, Teton, and Washakie, are situated adjacent to the two GYE national parks, creating an unbroken expanse of undisturbed lands that offer important habitat for the region's wildlife and backcountry recreational experiences. Although national forest management practices have long sparked controversy across the region, recent years have seen a notable shift in the issues generating conflicts.[4]

The GYE also encompasses other federal and tribal lands occupying important ecological niches. The US Fish and Wildlife Service manages three national wildlife refuges: the National Elk Refuge, situated north of Jackson, Wyoming; the Red Rock Lakes refuge in Montana's Centennial Valley; and the Grays Lake refuge in southeast Idaho. The National Elk Refuge provides migratory elk seasonal winter habitat where the animals are fed daily, while the other two refuges protect wetlands vital to migratory waterfowl. The Bureau of Land Management (BLM) oversees lower-elevation public lands located mostly on the outer portions of the GYE. The BLM operates under a multiple-use standard that includes oil and gas production, mining, livestock grazing, and wildlife habitat. It manages major oil and gas fields south of Pinedale, Wyoming, which also overlays important winter wildlife and sage grouse habitat. The GYE boasts three federally protected wild and scenic river segments, namely the Clarks Fork and Upper Snake River in Wyoming and East Rosebud Creek that flows eastward from Montana's Beartooth mountain range. In addition, the Wind River Indian Reservation—home to the Arapahoe and Shoshone tribes—occupies the GYE's southeastern flank, embraces portions of the Wind River mountain range (part of which the tribes designated the Wind River Roadless Area), and boasts the same array of species found in the two national parks.[5]

The federal agencies overseeing these extensive GYE public lands early on recognized the region's interconnectedness and the need to coordinate their management efforts. In 1964 Park Service and Forest Service officials established the Greater Yellowstone Coordinating Committee (GYCC), which today consists of managers from the GYE's two national parks, five national forests, the US Fish and Wildlife Service, the BLM, and, recently added, officials from the three states' wildlife agencies, plus an executive coordinator who provides staff support. Though lacking any legal mandate or authority, the GYCC meets periodically, providing the region's federal land managers with a forum for discussing common resource-management problems, building personal relationships, and supporting scientific research projects.

During the 1980s, facing congressional pressure over Yellowstone's declining grizzly bear population and lack of coordination among the responsible agencies, the GYCC initiated a high-profile "Vision" exercise designed to establish coordinated, region-wide resource-management goals and planning protocols. Although the GYCC's Vision process produced an initial draft document calling for a "world-class model" of ecosystem management on "a landscape where natural processes are operating with little hindrance on a grand scale," the effort ultimately stalled in the face of intense local political opposition. Since then the GYCC has proceeded cautiously, generally shying away from conflict and focusing instead on less contentious matters, such as invasive species, fisheries, wildfire management, and climate change research. Outside the GYCC, however, the GYE federal agencies are engaged in issue-specific committees overseeing the grizzly bear recovery effort, bison management, and other conflict-laden issues. But for the most part, change on the ground in the GYE has proceeded in a less centralized, less structured manner.[6]

The nonfederal GYE acreage is largely in private ownership, covering roughly seven million acres or about 30 percent of the ecosystem. The region's private lands contain scenic ranches, critical wildlife habitat, and various towns. Large intact ranches, generally situated in valley locations near flowing water, have historically afforded wildlife vital habitat during winter months and helped to maintain scenically attractive open space. That is changing, however, as ranches change hands. Many ranchers—a powerful political force in the three GYE states—still maintain traditional livestock operations, dependent on nearby public lands for seasonal grazing access where they face increasing conflicts with predators and recreationists. Other ranches are being sold to wealthy new buyers seeking a "hobby" ranch or vacation retreat. With regional population pressures mounting, developers are also purchasing available ranches and then subdividing them to provide new housing stock. The net effect is an increasingly fragmented and cluttered landscape that is deeply problematic from an ecological perspective.[7]

Because the GYE extends across parts of Wyoming, Idaho, and Montana, these three states are prominent players in regional conservation matters. Politically conservative, the states each contain extensive federal lands and, absent contrary congressional legislation, they exercise concurrent jurisdiction over the region's national forest and BLM lands, which includes wildlife-management responsibility. The three states have a long history of dependence on the extractive resource industries, including mining, timber, oil and gas, and livestock, for tax revenues and local employment opportunities. During the past several decades, however, the economic profile of each state has changed, discernably shifting away from the natural resource industries and toward the service sector. In fact, tourism and recreation now constitute a major segment of their separate economic portfolios, a fact that has prompted each of them to create tourism and outdoor recreation offices. Nonetheless, the states regularly challenge federal public land and conservation policies, generally concerned that too much of the public domain is being put off-limits to commodity production activity. All three states also have a strong tradition supporting private property rights, as well as relatively weak zoning and land-use planning laws. With two-thirds of the GYE situated in Wyoming, the "Cowboy State" plays an outsized role in the region, one that has been characterized as "forceful and strident" over resource-management issues. According to one knowledgeable longtime resident, "the [GYE] states will pursue their own interests and preserve their own authority."[8]

Numerous towns dot the GYE landscape, each with its own history, profile, and connections to the nearby public lands. Several towns serve as gateway communities to Yellowstone and Grand Teton national parks. Long reliant on tourism, Jackson, Wyoming, stands out among the region's gateway communities. Home to a major ski area and situated within sight of the spectacular Teton mountain range, Jackson has become a major tourist destination with a soaring real estate market. In fact, Teton County, Wyoming, ranks as the wealthiest county per capita in the nation. Drawing upon the state's tax haven reputation, the county has attracted droves of high-wealth individuals enticed by the area's scenic beauty and outdoor lifestyle. These new residents generally support local conservation organizations and efforts, making the community more progressive on these issues than the rest of the state. But their presence has also transformed the community into an expensive—some would say elite—enclave, where service workers can no longer afford to live. Many Jackson workers now live over the mountain in Teton Valley, Idaho, or forty miles south in Wyoming's Star Valley, where they are daily bused by Jackson employers to their local workplaces.[9]

Most of the GYE region's principal communities have likewise experienced profound change during the past several decades. In Teton Valley,

Idaho, the towns of Driggs and Victor are shedding their agricultural heritage in the face of a burgeoning real estate market, catering to priced-out Jackson workers and new arrivals seeking affordable property and outdoor recreation opportunities. To capitalize on this transition, the local Grand Targhee Ski Area, long a modest operation catering to local skiers, has petitioned the Forest Service to greatly expand its facilities. In Wyoming, the town of Pinedale was transformed during the early 2000s with the development of the Pinedale Anticline gas field on federal lands south of the town, which attracted legions of oilfield workers who soon altered the quiet town's character. Meantime, wealthy newcomers have been purchasing nearby ranches, sometimes subdividing them to accommodate more newcomers attracted to this scenic rural area at the foot of the Wind River mountain range. To the north, the town of Cody, Wyoming, represents the eastern gateway to Yellowstone; it has long relied on tourism, featuring its celebrated Buffalo Bill Center of the West museum and Western-themed art galleries to attract visitors. Other Wyoming communities situated in the GYE include Lander, Dubois, and the Wind River Indian Reservation, all of which have ties to the region's public lands and are affected by its evolving resource-management policies.[10]

In Montana change is likewise evident in communities like Bozeman, Big Sky, and West Yellowstone. Home to a major university and regional airport, Bozeman is one of the nation's fastest-growing mid-sized cities; its population has more than doubled since 2000, and now stands at 58,250 residents. At the present growth rate, the Bozeman area is projected to host a population the size of Salt Lake City in less than twenty years. Surrounding Gallatin County has likewise seen explosive growth and accompanying sprawl that is transforming former ranchlands into a series of subdivisions and malls. The pandemic did not slow the growth rate; rather, new residential sales accelerated as people fled the nation's crowded urban areas for the relative security and outdoor amenities available in Montana. South of Bozeman, the resort community of Big Sky fills an entire mountain valley with ski runs, condos, and second homes, along with the exclusive, millionaires-only Yellowstone Club. Further south, West Yellowstone serves as a small gateway community of 1.300 residents at the park's western entrance, but summertime visitors more than double that number daily while winter recreationists keep the town busy during the off-season.[11]

Similar transformations are afoot elsewhere in the Montana portion of the GYE. Once a railroad and mill town, Livingston is in transition following closure of the lumber mill, with new residents arriving in search of outdoor opportunities on nearby public lands. Red Lodge was once a vibrant mining town, but today relies on recreation and tourism for its sustenance, offering visitors the opportunity to enter Yellowstone over the scenic Beartooth Highway. Gardiner, another gateway community at Yellowstone's northern entrance, may appear largely unchanged, but the town's limited housing stock is being converted to Airbnbs for visitors, which is beginning to price local residents out of the housing market. Paradise Valley, stretching along the Yellowstone River from Livingston to Gardiner, is seeing increased subdivision activity and new construction on its scenic ranchlands. And in the Madison Valley, a historic ranching area one mountain range west of Yellowstone, large ranches are changing hands, conservation easements are being placed on others, and new outdoor shops, guide services, and other businesses linked to tourism and recreation are appearing with regularity.[12]

These dramatic changes to the GYE's human landscape are imposing troubling new pressures on the region's public and private lands. Population growth is prompting ranch sales, land subdivision, and sprawl, altering the region's open-space appearance and available wildlife habitat. New businesses linked to technology and the service industries are proliferating, shifting the region's economic character away from resource extraction. New arrivals to the GYE are bringing new perspectives and values with them, often an environmental ethic committed to preserving the area's appearance and wildlife—an ethic at odds with many old-timers who have long relied on logging, mining, ranching, and other resource development activities for their well-being. Both old and new residents, as well as visitors, are pursuing ever more recreational activities on the GYE public lands and waters, creating environmental problems as well as conflicts with other users. In short, the GYE has come to embody the "New West," reflecting "a shift away from the 'old west' heritage of utilitarianism extraction toward a new culture of natural amenity-minded transplants, influenced by new ideas about nature and motivated to both enjoy and protect its natural amenities."[13]

One aspect of the changing GYE is the proliferation of nonprofit conservation groups interested in preserving the area. More than 180 environmental organizations are now working in the region, two-thirds of whom have appeared since 1986. They include national groups like the Wilderness Society, Sierra Club, and National Parks Conservation Association, along with local groups that often have a more narrowly focused agenda. Among these groups, the Greater Yellowstone Coalition has pursued a region-wide conservation agenda since its founding in the early 1980s; it not only pioneered the GYE concept but early on aggressively sought to curtail logging, mining, grazing and other extractive activities on the region's national forests and other public lands. In 1993 Earthjustice, a nationally prominent nonprofit environmental law firm, established an office in Bozeman and has since provided experienced legal counsel to the region's conservation groups, enabling them to challenge troublesome federal and local resource-management decisions. As the region's conservation groups have grown in number, differences in priorities and strategies are evident, sometimes putting the groups at odds among themselves. These differences can also confront the responsible federal and state agencies with irreconcilable demands, making common ground difficult to discern and litigation virtually inevitable.[14]

Notwithstanding these changes, one powerful pillar of the "Old West"—the ranching community—remains very evident and relevant in the GYE today. Since Euro-Americans settled the region, ranchers have been a mainstay on the GYE landscape and quite influential in state and local politics. With ranchlands accounting for nearly a quarter of the GYE acreage, the ranching community's interests have often collided with nature conservation initiatives, stirring conflict as well as occasional compromise over such matters as wolf restoration, grizzly predation, livestock management, winter habitat, and the like. Regional ranch sales, which are occurring more frequently, likewise have conservation implications, especially when these lands are then subdivided and developed. Conservation easements and similar arrangements have gained some but not universal acceptance within the GYE ranching community, reflecting its growing interest in and commitment to wildlife conservation efforts. In short, ranchers have long occupied a prominent role in regional conservation efforts and policy.[15]

The region's original Indigenous occupants—another distinctive aspect of the West's historical past—have assumed an increasingly relevant role in the GYE today. As sovereign nations, the GYE's Native American tribes have begun to assert themselves more forcefully in national park, public land, and wildlife conservation issues. The tribes and tribal members have joined in cooperative-management arrangements with the federal agencies for conservation purposes, as reflected in the expanding Yellowstone bison-transfer program. And they are affirming treaty hunting and other rights on GYE lands. A critical part of the GYE landscape, the Wind River Reservation in Wyoming provides important habitat for regional wildlife, including a restored buffalo herd. In the GYE as well as elsewhere across the West, the undeniable truth is that Native tribes are aggressively promoting their interests, treaty rights, and traditional Indigenous knowledge, making them an important constituency and a frequent ally in nature conservation matters.[16]

Taking Conservation Science Seriously

As the twentieth century unfolded, science gradually assumed a more important role in natural resource management among the responsible federal agencies, with the Yellowstone region a frequent testing ground for new policies. During its early years, the Park Service frequently ignored science in an effort to preserve scenery and safeguard "good" animals for visitors to enjoy. Wildfires were extinguished to protect scenic green landscapes while predators like wolves were exterminated to maintain elk and deer populations. Although the early Forest Service endorsed Gifford Pinchot's "scientific utilitarianism" resource-management policy, agency officials understood the term to mean ensuring a steady supply of lumber, minerals, and livestock forage; wildlife, fish, and aesthetic concerns were essentially ignored. With the ecological sciences still in their infancy, the long-standing "balance of nature" paradigm held sway as scientists perceived the natural world as a static entity prone to an equilibrium state. Besides, the western states remained thinly populated, so human pressures on the region's public lands and resources were relatively light, which was also true of national park visitation.[17]

Following World War II, however, the sciences were awash in change, precipitating a new understanding of the natural world and the processes shaping it. As the field of ecology matured, scientists realized that the natural world was in a perpetual state of flux. This critical ecological insight soon displaced the prevailing static view of nature, giving rise to a new disequilibrium model, one holding that the natural world was continuously changing in often unpredictable ways due to floods, earthquakes, wildfires, and other disturbances, including the growing human presence on the landscape. A new concept—the ecosystem—linked biological species with their physical environment and thus provided a useful means for scientists to study the natural world and for the public to better understand that world. At the same time, scientists were progressively becoming more concerned about species extinction rates, which they cast as a pending biodiversity crisis. As these insights revealed nature's interconnectedness, the general public gradually developed a deeper understanding of the natural world along with a new environmental awareness that prompted serious questions about heavy-handed resource production policies. In the GYE these advances in knowledge and perception sparked growing concern over the Park Service's wildlife and wildfire policies, as well as the Forest Service's timber-dominated agenda.[18]

It became increasingly evident that Yellowstone was essentially an island, one that could not be preserved intact over the long term as a mere enclave. Island biogeography, derived from the study of species persistence in island settings, conclusively demonstrated that isolated animal populations with limited available habitat trended toward extinction over time. A new discipline coined "conservation biology" made clear that an ecosystem perspective was necessary to sustain the park, its wildlife, and the other natural features that made the GYE area special. A widely circulated scientific paper documenting species' extinctions in North American national parks due to limited available habitat affirmed these insights. Scientists accordingly endorsed three interrelated strategies essential to meet the impending biodiversity crisis: 1) establish and maintain large, connected nature reserves to ensure secure habitat; 2) plan and manage at an ecosystem or landscape scale to coordinate conservation efforts across land ownerships; and 3) employ adaptive management strategies to identify, address, and respond to changing natural conditions. The expansive GYE wildland complex presented an ideal setting for applying such an integrated natural resource-management approach, even though the region's diffuse manmade boundaries presented potential legal and political problems.[19]

The Park Service's early disregard of science changed irrevocably during the 1960s when Secretary of the Interior Stewart Udall, faced with mounting criticism of Yellowstone's policies, commissioned two independent studies. One examined the agency's wildlife-management policies, and the other addressed the agency's general approach to science. The resulting reports—one of which is now widely known as the Leopold Report, named after its lead author—reached similar conclusions: The Park Service must not only incorporate science into its managerial approach, but should "seek to preserve or restore a natural biotic scene" and "maintain[] natural conditions" across the National Park system. The reports noted that national parks were impacted by land uses occurring outside their borders, that most parks were too small to meet the essential habitat needs of resident species, and that past human actions had notably altered ecological processes. While acknowledging the need for occasional human intervention to restore missing species and natural processes, the reports urged the agency to allow nature much freer rein in the national parks. In response, the Park Service began according more credence to science in its decisions, while shifting its management philosophy toward maintaining and restoring natural conditions, and to paying more attention to matters outside park boundary lines.[20]

During this same period, the Forest Service found itself under intense scrutiny over its dominant postwar commitment to timber production and clearcutting. Following adverse court decisions enjoining its clearcutting practices and a series of critical reports, Congress intervened in 1976 by passing the National Forest Management Act (NFMA), which significantly revised the agency's managerial priorities and practices. The NFMA instructed the agency to develop forest plans using "a systematic interdisciplinary approach to achieve integrated consideration of physical biological, economic, and other sciences." It imposed limitations on logging practices, required the agency to address biodiversity in its plans, and created a Committee of Scientists to help develop regulations implementing this new science-based regime. Although the Forest Service continued to press its timber-first agenda, the legislation plainly incorporated new ecological considerations into forest policy—a fact that the courts would soon make clear.[21]

Toward an Ecological Imperative

Beyond these notable advances in scientific knowledge, other forces were also pushing federal resource-management law and policy in the same direction. Having split the atom, developed the nuclear bomb, and traveled to the moon, we began to understand just how fragile the earth and its natural systems were—a realization that helped trigger a shift in societal values toward a new environmental consciousness. Industrial accidents like the 1969 Santa Barbara oil spill and Cuyahoga River chemical fire heightened awareness of how badly we were treating the places where we lived and played. Significant population growth and improved economic conditions, coupled with technological advances, helped to bring the outdoors closer to home for Americans who found themselves with greater financial security and leisure time. These societal changes, in turn, nurtured corresponding attitudinal shifts in how the nation viewed the western public lands and the policies governing them, establishing a foundation for injecting new, legally mandated ecological and wilderness protection considerations into resource-management policy in the GYE and elsewhere.[22]

During the postwar period, societal values were evolving in response to mounting public concerns over industrial pollution and destructive resource-management policies. Two important books, both written by scientists who masterfully connected ecological and ethical principles, helped to instill a new public environmental awareness. Rachel Carson's *Silent Spring* chronicled the devastating impact of pesticides on the "web of life," arguing that humankind must alter its relationship with nature away from a domineering control mentality. Aldo Leopold's *A Sand County Almanac* explained how interconnected the natural and human worlds are, but lamented that humanity was on the verge of irreparably altering the natural world. To address the problem, Leopold proposed his now-famous land ethic: "A thing is right when it tends to preserve the integrity, stability, and beauty of the biotic community. It is wrong when it tends otherwise." A receptive public, taking these informed messages to heart, soon responded by attaching much greater value to the natural world, seeing it as more than a commodity to be harvested or mined. In short, Carson and Leopold linked ecological science with morality, presaging the emergence of an ecological consciousness that was soon translated into Earth Day, the modern environmental movement, and more enlightened laws and policies.[23]

Pervasive postwar social and economic changes also helped foster a new environmental ethos manifested in an interest in the outdoors and in protecting special places. Following demobilization the nation was soon transformed, the result of a dramatic population spurt, unparalleled economic growth, escalating urbanization, increased leisure time, unimaginable technological advancements, and expanded educational opportunities—all of which contributed to greater public interest in visiting, protecting, and recreating in special places like the national parks and other public lands. People also began moving westward, growing the region's cities while also filling up its "empty" spaces, bringing new residents to the region with different values and views about the environment than prevailed among long-term residents. The new interstate highway system opened the West to tourism, while new lightweight camping equipment made the region's backcountry accessible to a generation seeking outdoor experiences. As the years passed the GYE was increasingly enmeshed in these changes, setting the stage for a protracted struggle over the region's public lands and resource-management priorities.[24]

These interrelated scientific, social, and economic changes were soon felt in the nation's political institutions, prompting Congress to adopt an extraordinary array of new environmental laws with major ramifications for the GYE and other public lands. First came the Multiple Use–Sustained Yield Act of 1960, which expressly added outdoor recreation, fish, and wildlife to the Forest Service's resource-management responsibilities. Next, following publication of Wallace Stegner's influential *Wilderness Letter*, was the Wilderness Act of 1964, empowering Congress to set aside, as a protected wilderness area, undeveloped portions of national forest, park, and wildlife refuge lands to maintain their "primeval character [and] . . . natural conditions." Since then, Congress has preserved more than 111.5 million acres as protected wilderness, creating outstanding outdoor recreational opportunities as well as vital ecological strongholds for various wildlife species. Other related laws shortly followed. In 1965 Congress adopted the Land and Water Conservation Act, which provided federal funding to acquire private lands for recreational and conservation purposes. A year later, Congress adopted the National Wildlife Refuge System Administration Act, establishing uniform standards governing the growing refuge system, including an explicit conservation mandate. Two more laws—the Wild and Scenic Rivers Act of 1968 and the National Trails Act of 1968—represented a further national commitment to conservation and recreation on the nation's lands and waters. Each of these statutes, most notably the Wilderness Act and the Land and Water Conservation Act, has been instrumental in protecting and shaping the GYE setting.[25]

Following the initial Earth Day in 1970, Congress continued to expand the nation's portfolio of critical environmental legislation, much of it profoundly impacting public land management as well as private land use and industrial practices. The National Environmental Policy Act of 1969 (NEPA) is widely regarded as the Magna Carta of the nation's environmental laws; it has reshaped federal agency decision-making processes by mandating preparation of a detailed environmental assessment with public involvement before an agency can approve any major project proposals. The federal judiciary has vigorously enforced NEPA to ensure agencies follow the statute's rigorous environmental analysis requirements, frequently enjoining logging, drilling, and other activities on public lands for lack of compliance.[26] In 1973 Congress revised a mostly toothless Endangered Species Act (ESA), elevating the protection of species verging on extinction to a pivotal priority position in federal agency decisions, including those involving resource extraction projects on public lands. As revised, the ESA also extends to private lands, limiting land-use activities that can harm a protected species or its habitat.[27] In addition, Congress adopted two key pollution control laws during the early 1970s, namely the Clean Water Act and the Clean Air Act, which each impose constraints on mining, logging, drilling, and other industrial activities that routinely occur on national forest and BLM public lands.[28] A few years later Congress added landmark hazardous waste control and cleanup legislation.[29] These new environmental laws inserted the federal government into matters traditionally the domain of the states, but that were now seen to transcend state boundaries and deemed to be nationally important.

The 1970s also saw important congressional additions to the legislation governing the public lands that has affected the GYE national forests and other lands. In 1976, with the public and the courts growing increasingly disenchanted over the Forest Service's widespread clear-cut logging practices, Congress adopted the National Forest Management Act (NFMA), essentially remaking the agency's approach to resource management on its forests. The NFMA introduced a detailed, interdisciplinary forest-planning process designed to ensure the agency considered both economic and environmental values in its decision processes, including a new diversity requirement, clearcutting limitations, and public participation obligations.[30] In the GYE and elsewhere, however, the Forest Service's initial round of forest plans revealed little change in the agency's traditional focus on commodity production.

At the same time, Congress passed the Federal Land Policy and Management Act (FLPMA), handing the BLM a new organic charter based on multiple-use principles. The FLPMA established resource inventory and planning requirements, and also extended the wilderness concept to BLM lands, giving the agency explicit preservation obligations.[31] In 1978 Congress added the so-called Redwood Amendment to the National Park Service's organic legislation, reinforcing the agency's duty to protect its lands and resources from potentially damaging activities, including those occurring outside park boundaries.[32] In 1980 Congress passed the Alaska National Interest Lands Conservation Act (ANILCA), marking a dramatic expansion in the nation's commitment to preserving public lands by establishing more than 157 million acres of new national park, refuge, and wilderness lands in the forty-ninth state. Remarkably, Congress was convinced to draw the boundaries for these protected Alaskan lands along ecosystem lines, further injecting an ecological perspective into public land law and policy.[33]

Collectively, these laws provided a legal foundation for significant federal ecosystem-management initiatives during the 1990s. Confronted with a prolonged battle over the impact of rampant timber harvesting on the northern spotted owl in the Pacific Northwest's old-growth forests, the federal courts blocked all logging on the region's federal forests, citing violations of the principal 1970s environmental and forest-management laws. The newly inaugurated Clinton administration responded by directing the Forest Service and the BLM to jointly prepare a regional, science-based forest-management plan that would safeguard the spotted owl and other species while enabling commercial timber harvesting to continue. The resulting Northwest Forest Plan covered 24 million acres and extended across three states, nineteen national forests, and seven BLM districts, representing an unprecedented interagency coordination effort geared to preserving the region's old-growth ecosystems and dependent species.[34] The federal judge whose rulings triggered the plan affirmed it by observing: "Given the condition of the forests, there is no way the agencies could comply with the environmental laws *without* planning on an ecosystem basis."[35] Building upon this legal endorsement, the Clinton administration embraced ecosystem management as federal agency policy and undertook an array of ecosystem-based initiatives across the country, including in California's Sierra Nevada mountain range, the Interior Columbia River Basin, the California Desert, and Florida's Everglades. Although these initiatives did not all survive the ensuing political crosswinds, they nonetheless legitimized ecosystem management as a viable federal policy, one that rested upon a sound scientific and legal foundation.[36]

Since then, federal law and policy has continued to evolve toward protecting the public lands for nature conservation purposes, while incorporating ecological science into the agencies' planning and management decisions. Under the Wilderness Act, Congress has designated nearly 112 million acres as protected wilderness and another 33 million acres as wilderness study areas, which provides nearly the same level of legal protection.[37] Presidents Clinton, Obama, and Biden all aggressively employed the 1906 Antiquities Act to establish more than a dozen large-scale national monuments, several of which were designed to protect ecosystems intact. Congress blessed these executive actions when it established a new National Landscape Conservation System under the BLM's aegis.[38] The National Wildlife Refuge System Improvement Act of 1997 established a new comprehensive conservation planning process for the nation's wildlife refuges as well as a new "biological integrity, diversity, and environmental health" management standard. In the National Parks Omnibus Management Act of 1998, Congress gave the Park Service scientific research and related management responsibilities while also acknowledging that parks are part of larger ecosystems. In 2020 Congress passed the Great American Outdoors Act, making $900 million annually available to the Land and Water Conservation Fund to acquire sensitive lands for nature conservation purposes.[39] Moreover, all fifty states—including Montana, Wyoming, and Idaho—have adopted conservation easement laws that enable state agencies and land trusts to acquire interests in private lands with conservation value, which can often be conjoined with nearby federal lands to promote an ecosystem-based conservation agenda.[40]

Drawing upon this legal foundation, the federal agencies have incorporated ecological conservation policies and practices into their resource-management responsibilities. In 2000 the Forest Service employed its rulemaking authority to adopt the roadless area rule that places 58.5 million national forest acres off-limits for new road building or timber harvesting—a designation that emphasizes the conservation and recreational value of this acreage. In 2012 the Forest Service revised its NFMA planning regulations, establishing "ecological sustainability" as a primary forest-management goal. The regulations committed the agency to using the "best available science" and preparing landscape assessments in its forest-planning process, while enumerating climate change, ecosystem integrity, wildfire, ecosystem services, and connectivity as factors to address in the plans.[41] In 2023 the Forest Service proposed amending all forest plans to put old-growth trees off-limits for timber-harvesting purposes.[42] In 2006 the Park Service revised its *Management Policies*, which establish system-wide resource-management standards and practices. The revisions instruct park officials to "use all available tools to protect park resources and values from unacceptable impacts"; to "consider the park in its full ecological . . . contexts . . . as part of a surrounding region"; and to cooperate with neighbors to mitigate potentially harmful external activities.[43] The US Fish and Wildlife Service has translated its "ecological integrity" and comprehensive conservation planning statutory mandates into policy documents instructing refuge managers to maintain "genetically viable" breeding populations, to maximize habitat blocks, and to ensure connectivity between blocks.[44] Moreover, in 2024 the Bureau of Land Management adopted new regulations incorporating "conservation" use, landscape inventories, and an "ecosystem resilience" standard into its multiple-use management responsibilities.[45]

Similar ecological conservation developments are evident at the ground level. The Obama administration established multiagency Landscape Conservation Cooperatives to promote research on climate change, its impact on federal resources, and landscape-scale conservation strategies to address these impacts. Confronted with an alarming decline in sage grouse numbers due to loss of habitat, the Obama administration implemented a multistate sage grouse management plan that covers 165 million acres, representing a stunning landscape-scale wildlife conservation initiative. However, the Trump administration, eager to open acreage for mining and energy activity, proceeded to defund the LCCs and to significantly modify the initial sage grouse plan—a move that was blocked by the courts. For its part, the Biden administration has embraced a vigorous conservation agenda, reflected in its America the Beautiful campaign to preserve biodiversity, its various climate change initiatives, and a new BLM conservation rule.[46]

In the Age of the Anthropocene, the question is whether these laws and related nature conservation policies—many established nearly a half century ago to address that era's environmental concerns—are sufficient to meet the challenges ahead. The Anthropocene, according to the scientific community, embodies an unprecedented world where humankind's imprint is ubiquitous across the globe. Human activities—cultivated agriculture, industrialization, mining, dam building, fossil-fuel usage, atomic explosions, and the like—have dramatically altered the natural world with far-reaching environmental consequences, one of which is anthropomorphic climate change. In this human-dominated world, does the legal concept of wilderness as an untrammeled natural setting still ring true, or is wilderness merely a quaint cultural construct, one where more (not less) human manipulation will be necessary to preserve climate-imperiled species and existing ecological processes? Do today's nature reserves, such as Yellowstone National Park at 2.2 million acres, still represent an effective nature-conservation strategy in a warming world? And does a legal regime that exalts private property rights ultimately undermine rather than advance ecological conservation efforts by promoting rampant development and landscape fragmentation? While yesterday's laws and policies endorse an ecological approach to nature conservation that has helped to safeguard special places like the GYE, it remains to be seen whether these same laws and policies will prove adequate to meet our looming conservation challenges.[47]

Change Comes to the GYE

Midway through the twentieth century, amid the profound scientific, social, and legal changes chronicled above, Yellowstone National Park assumed a prominent role in what amounted to a major reassessment of our relationship with nature. As dramatic changes in conservation policy began taking hold in the 1970s, the park became a testing ground for new ecologically oriented management strategies. The Park Service—long charged with conserving wildlife and the natural setting—had by then stopped feeding bears and culling elk herds, and was allowing wildfires to burn in the backcountry. The fundamental idea was to permit nature to take its course with minimal human intervention. The Forest Service, despite its quite different utilitarian management philosophy, had also decided to let some fires burn unchecked, while it assumed responsibility for recently decreed wilderness areas where active management was prohibited. At the same time, the Forest Service was ramping up its timber harvests and welcoming oil, gas, and mining operations on its lands. It was soon apparent, however, that these industrial activities threatened local wildlife and were altering the area's natural appearance. Meanwhile, the Craighead brothers had introduced the "Greater Yellowstone Ecosystem" term to describe the habitat required to sustain the region's grizzly bears. The notion that the Yellowstone National Park was part of a larger ecosystem that included the adjoining national forests was taking shape, along with a nascent interest in preserving nature in this expansive setting.[48]

Regional change accelerated during the 1980s, bringing the GYE concept into popular discourse as well as aggressive efforts to protect the region's ecological integrity. A dramatic drop in the grizzly bear population catalyzed Congress and the agencies into action, confirming that the region's public lands were interconnected and the need for coordinated management. An aggressive new conservation organization—the Greater Yellowstone Coalition—emerged with a commitment to employing science and law to protect the region's wildlife and wild places. The dramatic 1988 Yellowstone wildfires further affirmed the region's park and forest lands were ecologically connected despite existing boundary lines, which required a new level of coordinated management among the responsible agencies. The ravages of unrestrained clearcutting on the region's national forest lands along with new oil and gas leases abutting the two national parks stirred public concern over potential wildlife and aesthetic impacts. As the decade wound down, the Park Service and Forest Service—acting through the Greater Yellowstone Coordinating Committee—responded to these concerns with a high-profile interagency Vision initiative intended to bring a new ecological perspective to federal management policy in the region. Though well intentioned at the outset, the effort ultimately fizzled in the face of local resistance.[49]

During the 1990s the federal land-management agencies embraced the ecosystem-management concept, and priorities began to shift on the GYE national forests. The Clinton administration endorsed ecological science and ecosystem management as the lodestar for managing the public lands, driven in part by its efforts to resolve the Pacific Northwest's spotted owl controversy. In the GYE a combination of litigation, administrative appeals, and public pressure prompted notable changes in the Forest Service's timber-management practices on the national forests. The industry's interest in oil and gas leasing faded due to international market conditions. An interagency grizzly bear recovery effort showed progress as bear numbers slowly began to rebound. In a remarkable development wolves were returned to Yellowstone National Park, restoring a long-absent apex predator and sense of wildness, as well as confirming the important role ecological processes played in the region. Efforts to reestablish a large gold mine just outside Yellowstone met sustained resistance that succeeded, with presidential intervention, in blocking the project. Controversy mounted within the region's ranching communities over the disease threat posed by Yellowstone's growing bison population. Yet the Park Service and Forest Service, chastened by their earlier Vision initiative experience, avoided any similar regional coordination effort, focusing instead on discrete, issue-based efforts, such as the joint federal-state Interagency Grizzly Bear Committee.[50]

Since then, change has only accelerated in the GYE, impacting both the region's public and private lands. By all accounts the Forest Service is a very different agency than it was a few decades ago, one that has embraced ecological science as a guiding principle while deemphasizing commodity production. Grizzly bear and wolf numbers have continued to grow, such that wolves have been removed from federal protection and the GYE states are pressing to remove the grizzly as well. Despite the Bush and Trump administrations' deep commitment to energy production, oil and gas activity is confined to the periphery of the GYE. New mine proposals have been blocked at both political and judicial levels. Visitation to the GYE parks and recreational activity on the adjoining national forests is rising precipitously, presenting both environmental and user-conflict concerns. New scientific research has documented expansive wildlife migration patterns across the GYE and beyond, effectively expanding our understanding of the ecosystem and what is required to preserve it. People have discovered Greater Yellowstone and are moving to the area in record numbers, putting extreme pressure on the region's private lands with adverse impacts on migratory wildlife. Moreover, the region's original Native American inhabitants have begun asserting themselves, bringing a new voice and force into wildlife and other resource-management matters.[51]

This profound evolution in the GYE and related nature conservation efforts has not come without considerable conflict, which endures today. Bison management remains unresolved, as Yellowstone's bison continue to exit the park in winter onto adjoining national forest and private lands where they are not welcome due to inflated disease concerns. Extensive litigation by conservation groups has forestalled efforts to remove grizzly bears from federal oversight, but the GYE states continue demanding to take over bear management. Climate change is altering the GYE forests, bringing with it more intense wildfires that destroy wildlife habitat and threaten nearby homes being built at an alarming rate next to the forest boundary. Once-plentiful ranchlands are disappearing as new subdivisions and twenty-acre ranchettes spring up to accommodate the region's growing populace. Wildlife habitat keeps disappearing, migration corridors are severed, and growth pressures are evident throughout the region. Along with more people come more recreational users penetrating further into the backcountry and disrupting wildlife patterns. As the pace of change only accelerates, the story of the GYE and ecosystem management holds vital lessons for the region as it wrestles with new challenges that are refocusing nature conservation efforts to an even larger landscape scale.

The ensuing chapters explore in depth the quest to preserve the GYE's superlative natural attributes through an ecologically based management approach. Focusing on the region's national parks, wildlife, national forests, and private lands, each chapter examines the role of science, law, politics, and policy in promoting nature conservation efforts in this jurisdictionally fragmented environment, where federal, state, tribal, and local governments are each involved in determining the GYE's future. The region's rapidly growing populace as well as its diverse advocacy organizations occupy central roles in this unfolding nature conservation drama as do a warming climate and the changes it portends. The stakes are high given the GYE's history and prominence—not only for the region's future but for nature conservation efforts elsewhere across the nation's public and private lands. The goal is to put the GYE lessons in perspective and to understand them given present challenges and those surely awaiting in the years ahead.

CHAPTER TWO

Nature Conservation in the GYE National Parks

Situated at the core of the Greater Yellowstone Ecosystem, Yellowstone and Grand Teton national parks have provided a principal impetus for the movement toward ecosystem management in the region. Indeed, preservation of these two world-famous national parks, including their iconic wildlife populations and stunning scenic features, initially drove park managers, allied conservation groups, and others to embrace the GYE concept and related ecological management approaches. Following the 1963 Leopold Report and its admonition to give nature freer rein in national park resource-management policy, Park Service officials confronted the reality that the prevailing boundary line did not insulate the park from the surrounding world. Elk, bison, and bears freely roaming beyond the boundary inevitably stirred controversy with park neighbors, as did the emergent wildfire policy that allowed some fires to burn unchecked. It was also impossible to ignore the pervasive human presence both inside and outside the parks with attendant impacts on wildlife and other park resources. With hotels, campgrounds, and other facilities scattered throughout the parks, each summer brought even more visitors and automobiles than the summer before, and the nearby gateway communities were ever more dependent economically on these seasonal visitors. To meet their conservation responsibilities in a scientifically responsible manner, park officials would have to think in ecosystem terms and engage with neighboring agencies and communities.

By any measure, the GYE national parks faced major policy challenges related to this evolving approach to nature conservation. The emergent goal of managing park wildlife and ecological processes to maintain naturalness was impossible to achieve without taking account of the landscape beyond park boundaries and the different management objectives that prevailed there. Not all wildlife species and biota found within the parks were welcome; nonnative (or exotic) species were regarded as intruders, requiring active intervention, which seemed at odds with the agency's new hands-off approach. The interests of visitors were also paramount, but their needs and demands were not always compatible with naturalness goals—a problem that has grown as visitor numbers have steadily mounted and new recreational activities have surfaced. Because the parks essentially served as "cash cows" for the nearby gateway communities, any effort to curtail or alter visitation patterns was bound to meet with resistance. All of these concerns were present as the GYE parks proceeded to implement the Leopold Report's new, ecological approach to resource management.

Nature, Science, and Park Management

The Yellowstone country has long been linked to science, though that linkage was not incorporated into Park Service management policies until long after the park was established. The 1870 Washburn expedition to the Yellowstone region, whose reports helped to persuade Congress to create the nation's first national park, extolled the area's extraordinary natural features and also highlighted its potential for scientific research, describing it as "probably the greatest laboratory that nature furnishes on the surface of the globe." Soon after Congress acted on the park proposal, Dr. Ferdinand Hayden, leader of another early Yellowstone expedition, described the decision "as a tribute from our legislators to science," prophesizing that the nation and world would be forever indebted. But in 1916, when Congress adopted the National Parks Organic Act, it did not include any explicit reference to science, following the lead of the act's proponents who were primarily focused on protecting scenery and promoting tourism. Two years later, when Secretary of the Interior Franklin Lane released the "Lane Letter"—intended to be the foundational interpretation of this new organic legislation—he directed the Park Service "for assistance in the solution of administrative problems in the parks relating both to their protection and use . . . [to enlist] the scientific bureaus of the Government."[1]

During the 1930s, the Park Service took initial steps toward a science-based mission, prodded by ranger-naturalist George Melendez Wright. At Wright's urging, the Park Service undertook a wildlife survey to help park managers better understand and oversee the animals they were responsible for protecting. Although Wright's tragic death stalled the agency's early foray into science-based management, his *Faunal Survey* reports set forth key concepts that were eventually embraced by agency officials. Wright's reports pressed park managers to preserve native animals in their primitive state with minimal human interference, to restore extirpated predators, to eliminate exotic species, and to employ scientific principles for resource-management purposes. Echoing the views of Yellowstone's early managers, Wright recognized "the failure of parks as biological entities . . . [due to] their size and boundary location," and advocated for establishing buffers zones adjacent to them or expanding park boundaries to meet the year-round needs of park wildlife. Once World War II ended, the drumbeat for a more science-based resource-management approach intensified. A few Park Service naturalists from Wright's time remained, pursuing field studies designed to better understand nature's workings and wildlife behavior. Among them, Adolf Murie's research on wolves at Mount McKinley (now Denali) and coyotes at Yellowstone garnered considerable attention.[2]

Science finally gained a firm foothold in the Park Service during the 1960s in response to Yellowstone's increasingly unpopular wildlife-management policies and growing ecological knowledge. Since the 1930s Yellowstone had managed its elk and bison much as ranchers did their cattle by limiting herd numbers to safeguard the range grasses they depended on. Park Service marksmen annually culled—simply shot— large numbers of congregating elk, whose carcasses were then donated to nearby Native American reservations. At the same time, bears were on nightly display for visitors—much like in a zoo—while being artificially fed at the park's garbage dumps, and they regularly frequented roadsides and campgrounds to beg visitors for food. As public concern mounted over these practices, Secretary of the Interior Stewart Udall appointed a team of scientists led by A. Starker Leopold, Aldo Leopold's eldest son and himself a respected zoologist, to review the Park Service's wildlife-management policies. By then, the scientific community had begun to question the science underlying the agency's management policies, including wildlife culling, wildfire exclusion, and predator eradication practices. At the same time, Udall enlisted the National Academy of Sciences to review the agency's "natural history and research needs."[3]

The resulting reports recommended significant changes to the Park

Black Bear Begging at Car in Traffic Jam (circa 1960). Credit: R. Robinson, Yellowstone National Park Photo Collection.

Service's science and management policies. The so-called Leopold Report had the most immediate and enduring impact in Yellowstone, where major shifts in the agency's resource-management policies soon ensued. Starker Leopold and his colleagues—in memorable language—proposed that "as a primary goal . . . the biotic associations within each park be maintained or where necessary recreated, as nearly as possible in the condition that prevailed when the area was first visited by white man. A national park should represent a vignette of primitive America." Recognizing it may not be possible to achieve this goal everywhere, they urged recreating "[a] reasonable illusion of primitive America . . . using the utmost in skill, judgment, and ecological sensitivity." The report envisioned a widespread, professionally overseen ecological restoration effort in the national parks that included restoring extirpated predators, controlled burning to mimic natural fire regimes, and eliminating exotic species. Moreover, the report observed that "few of the world's national parks are large enough to be in fact self-regulatory ecological units; rather, most are ecological islands subject to direct or indirect modification by activities and conditions in the surrounding areas." The parallel National Academy report similarly concluded that the Park Service's science program was inadequate and not integrated into resource-management decision processes. Together the reports presaged a dramatic shift in Yellowstone's management practices, one that elevated science to a prominent new role, while simultaneously provoking a new series of controversies over the agency's scientific credibility and the proposed "vignette of primitive America" management goal.[4]

The Leopold Report's "primitive America" goal soon prompted a vigorous debate over the role of Native Americans on the landscape before white settlement, as well as the treatment of these original inhabitants during the establishment of individual national parks. In fact, several Native American tribes originally utilized the Yellowstone country, where they "hunted, fished, gathered plants, quarried obsidian, and used the thermal waters for religious and medicinal purposes."[5] Due to the area's high elevation and harsh winter conditions, however, only one small band of Shoshone people—referred to as Sheepeaters for their dietary habits—actually lived in the park. By the mid-1800s the Native presence in the Yellowstone country began to wane owing to growing white settlement in the surrounding area, though the park lands remained an attractive tribal hunting ground as bison and other wildlife were decimated elsewhere by early settlers. Although some tribes retained ownership of park lands under the 1851 Fort Laramie treaty, these rights were eliminated by the 1868 Fort Laramie treaty, when area tribes were relegated to reservations. And while several tribes retained "the right to hunt on the unoccupied lands of the United States so long as game may be found thereon," these rights were early construed not to extend to the park. By 1879, seven years after the park was established, the Sheepeaters, whose presence had prompted increasing concern among park officials, were gone from the new park.[6]

Throughout Yellowstone's early years, its civilian and military managers sought to deter the Indians from hunting or burning or otherwise damaging park resources during their regular summer forays into the park's more remote areas. By the late 1880s they had succeeded, and the park's original inhabitants were no longer making summer hunting trips into the park. Gradually, Yellowstone officials have come to acknowledge the early Native American presence and impact on the park landscape. Today, the Park Service identifies twenty-seven different tribes that have a "traditional association" with the park and acknowledges their sovereign nation status. Park officials have committed to consulting with the tribes on projects or actions that could affect their spiritual and historical interests. As these relationships have evolved, the tribes are playing an increasingly important role in Yellowstone's wildlife issues, particularly by expanding the park's bison-management options. Discussions are also ongoing about a role for traditional ecological knowledge in park resource management as well as co-stewardship opportunities.[7]

Since the late 1960s controversy has dogged Yellowstone over its revised resource-management policies, which have largely tracked the Leopold Report's recommendations. One persistent issue is whether it is possible to recreate point-in-time presettlement conditions in the park setting, given the dynamic and often unpredictable nature of the ecological processes that define the park environment. Another is whether the agency should allow nature to take its own course without human intervention—often referred to as the "natural regulation" policy—or whether more intensive (or active) management is required to sustain park resources. These issues are complicated by the fact that the park is essentially an island surrounded by lands where quite different management goals and practices prevail. Regardless, Yellowstone officials altered their policies in the wake of the Leopold Report and stopped culling the park's elk and bison herds, letting the animals roam freely across the landscape in search of food. Park officials also stopped feeding the bears, leaving them to fend for themselves from natural foods sources. They reintroduced fire to the park, allowing some wildfires to burn unchecked, and began to promote the idea of wolf restoration. These decisions were inevitably contested, often with the argument that active management is necessary in the constricted confines of the park and to maintain functional ecological systems.[8]

One undeniable reality of Yellowstone's shift toward a more laissez faire management approach was the impact of these changes outside park boundaries. Growing elk and bison herds were not sustainable within the park, so they began migrating in larger numbers onto neighboring lands, where they intermingled with domestic cattle (soon sparking fears of disease), damaged fencing, and prompted safety concerns. The migrating elk also spawned new hunting and guiding opportunities, which meant more local spending and jobs. As grizzly bear numbers gradually rebounded, they posed other problems; the bears sometimes preyed on domestic cattle and sheep and presented very real human safety concerns. The wolf reintroduction debate stirred similar concerns and an even stronger emotional response from livestock operators and park neighbors. The idea that wildfires might be allowed to burn, reversing more than fifty years of federal fire suppression policy, was likewise questioned given the local fuel-loaded forests—a concern that was soon validated when the 1988 fires burned uncontrolled for several months across the park and nearby national forests. Simply put, the revised Yellowstone policies vividly confirmed that the park and surrounding lands were inherently interconnected. What transpired inside the park affected lands, communities, people, and their livelihoods outside the park, and what happened on those lands likewise impacted the park, its animals and appearance. Yellowstone, in short, was part of a larger ecosystem.[9]

These realities were not confined to Yellowstone. As the 1970s drew to a close, the Park Service confronted an array of incompatible external activities that posed a serious threat to the ecological integrity of other individual parks. The problem originally gained national attention at Redwood National Park, where heavy logging on upstream private lands precipitated seasonal flooding in the park, eroding soils and uprooting the storied redwood trees. Following extensive litigation by the Sierra Club to force the Park Service to take action, Congress intervened and, at enormous cost, expanded the park to eliminate the threat. Congress also took the opportunity to amend the Park Service's organic legislation with the so-called Redwood Amendment, clarifying the agency's responsibility to protect park resources from such destructive activities.[10]

Concerned that other parks may face similar problems, Congress directed the Park Service to investigate further. The agency's ensuing 1980 *State of the Parks* report identified an array of "external threats" to individual park units, including air pollution, timber harvesting, energy exploration, and residential development. The report observed that two seemingly isolated parks—Glacier and Yellowstone—faced the most threats within the system. Although Congress entertained several legislative proposals addressing the issue, none passed due to strong political opposition from potentially affected communities, businesses, and landowners as well as other federal agencies. With a legislative solution beyond reach, park protection advocates seized upon the ecological dimensions of the problem and began advocating for an agency-driven ecosystem-management approach to better coordinate resource-management goals and priorities among the affected parties. The Yellowstone region quickly became the primary testing ground for this new approach, which rested upon the same emergent science that park officials had embraced a decade earlier following the Leopold Report.[11]

Since then, the Park Service has formally adopted a science-based management goal of maintaining and restoring natural conditions. Drawing upon the Organic Act's injunction to conserve park resources in an unimpaired condition, the agency's revered *Management Policies* direct park officials "to preserve fundamental physical and biological processes, as well as individual species, features, and plant and animal communities" in order "to preserve naturally evolving ecosystems." Managers are instructed "not [to] intervene in natural biological or physical processes" except in limited circumstances, including when necessary "to restore natural ecosystem functioning that has been disrupted by past or ongoing human activities." These basic policy prescriptions have largely guided Yellowstone's wildlife-management practices, which no longer involve heavy-handed intervention, though the park's bison and lake trout control programs are notable exceptions.

Moreover, recognizing that national parks are part of larger ecosystems, the *Management Policies* direct park managers to pursue "cooperative conservation" efforts with neighbors to meet resource preservation goals. In response, Yellowstone officials, unable to ignore threatening external activities, have actively sought to protect the park from nearby mining project proposals as well as a geothermal project that could have damaged the park's renowned thermal features. And they participate in the Greater Yellowstone Coordinating Committee—a long-standing regional federal land managers group that was convened to improve relations among the agencies and better coordinate management decisions.[12]

In recent years the Park Service's resource-management policies have come under renewed scrutiny as we enter the Anthropocene, confronting a destabilized climate and escalating development pressures. The Park Service has responded by undertaking a scientific reassessment of the agency's resource-management policies. In a 2012 report entitled *Revisiting Leopold: Resource Stewardship in the National Parks*, a committee of respected scientists acknowledged the undeniable: "Environmental changes confronting the National Park System are widespread, complex, accelerating and volatile. These include biodiversity loss, climate change, habitat fragmentation, land-use change, groundwater removal, invasive species, overdevelopment, and air, noise, and light pollution." Noting several key Leopold Report conclusions—the important role of science and the need for active intervention in some instances—the report emphasized that the national parks "are embedded in larger regional and continental landscapes" and endorsed collaborative management approaches. The agency's goal, it proposed, should be to manage for continuous change "in order to preserve ecological integrity" and ensure "ecosystems that are largely self-sustaining and self-regulating." It did not endorse an unalloyed "hands-off" management approach. But it did promote the idea of a "coherent and sustainable national conservation land- and seascape," envisioning a network that tied parks and other conservation areas together to promote resiliency. While lacking the memorable phraseology of the Leopold Report, the reassessment plainly captured the current sense of the scientific community regarding the challenges and responses required to preserve our natural heritage in this time of unprecedented environmental change.[13]

As the Obama administration was winding down, the Park Service incorporated the *Revisiting Leopold* report's recommendations into agency policy. Issued in late 2016, Director's Order 100—subtitled "Resource Stewardship for the 21st Century"—set a resource-management goal to conserve, restore, and protect park resources by "emphasiz[ing] resilience, connectivity at land/seascape scales, and life-cycle stewardship as guiding strategies for resource management." The order called for collaborative approaches to large landscape conservation, while endorsing the precautionary principle, adaptive management, and a commitment to science, law, and the long-term public interest. Along with other Obama-era climate-related initiatives, including the interagency Landscape Conservation Cooperatives, which were established to improve our scientific understanding of climate change and promote landscape-level resource-management strategies, the Director's Order put the agency on a clear path to confront the increasing uncertainties it faced.[14]

The Trump administration, however, rejected these policies. It not only overturned Director's Order 100 and defunded the LCCs, but prioritized fossil fuel development on the public lands with little regard for the surrounding impacts of such activity. The ensuing Biden administration has taken a quite different view, emphasizing the need to adjust resource-management policy to meet the climate change challenge and protect against biodiversity loss. Moving forward, it remains apparent that national parks are embedded in larger landscapes, and that their resource-management policies are infused with an inherent tension between science and politics—a lesson that has been learned in the Yellowstone setting, repeatedly.[15]

Wildfire Management

The image of Yellowstone on fire dominated the national news during the summer of 1988. By several accounts, the park was tragically being destroyed by the very agency responsible for overseeing it, the result of the Park Service's misguided fire-management policy, which allowed wildfires to burn unchecked. Local politicians, business owners, and nearby communities all put the blame on the park, deriding its ill-fated experiment with a natural regulation policy that placed nature ahead of human concerns. Most scientists, however, viewed the agency's wildfire policy quite differently, as an effort to restore fire—a natural ecological process—to its historical role on the landscape. Once the flames were out and recriminations eased, two separate inquiries clarified the facts surrounding the inferno as well as the Park Service's ultimately fruitless efforts to control the blazes, while affirming its basic fire-management policy. As it has throughout history, fire continues to shape and reshape Yellowstone, though now subject to managerial constraints designed to avoid another uncontrollable inferno that burns beyond the park and adjacent federal lands.

Although Native Americans were known to ignite fires to improve hunting and forest conditions, Yellowstone's early military caretakers regularly suppressed wildfires in the new park. From its inception in 1916 the Park Service did the same, adopting the same aggressive fire-control policy that the Forest Service had implemented following the destructive 1910 fires—known as the "Big Burn"—that swept across the interior northwest. Though a few scientists questioned this early suppression policy, it remained in place throughout the National Park system until the 1960s, when the Leopold Report called it squarely into question by endorsing a new naturalness policy. In 1968, with scientific support mounting to reintroduce fire into the parks, the Park Service revised its resource-management policies to allow some natural fires to burn unchecked and to use prescribed burning for vegetative- or wildlife-management purposes. The agency's first significant experiments with controlled fire took place in the Sierra Nevada parks during the late 1960s and early 1970s, where lightning- and Indian-ignited fires had periodically burned these fire-adapted ecosystems. (Coincidentally, the early Yosemite fire program was overseen by a young resource manager named Bob Barbee, who would later preside over the 1988 Yellowstone fires as the park's superintendent.) As the 1980s unfolded, Yellowstone and other large national parks were operating under a policy that viewed fire as a natural ecological process and allowed most backcountry fires to burn, generally without adverse effects.[16]

This natural-process fire policy was put to the test in 1988, when a series of mid-summer fires roared out of control in Yellowstone and the adjacent national forests, ultimately scorching nearly half of the park before an early fall snowstorm snuffed out the flames. Working from a 1972 fire-management plan Yellowstone officials, along with their Forest Service counterparts, originally allowed several remote lightning-ignited

Continental Divide Fire, 1976, Yellowstone National Park.
Credit: Yellowstone National Park Photo Collection.

fires to burn until mid-July when, due to the continued lack of rainfall, they decided to suppress them. It was too late, though. Unexpected high winds combined with unnaturally hot weather fanned the flames, causing several fires to converge while burning embers were blown miles ahead of the flames, making it impossible for the legions of firefighters to maintain fire breaks. Walls of fire raced through the park's dry lodge-pole pine forests, imperiling the Old Faithful Lodge and the historic Mammoth complex as well as nearby communities, and casting an unwelcome pall of smoke over the entire region. The accompanying media and political spectacle made it nearly impossible for park officials to explain that several of the fires originated outside the park, that others were fought from the outset, or that the underlying weather conditions were nearly unprecedented. By then, critics were referring to the firestorm as "Barbee's Barbeque" and placing blame squarely on the Park Service's natural regulation policy.[17]

Once the fires were out and tempers cooled, however, two official reviews essentially exonerated Yellowstone's basic wildfire policy. Both a high-level federal Interagency Fire Management Policy Review team and a panel of scientists convened by the Greater Yellowstone Coordinating Committee found that such fires were an essential natural process in the Greater Yellowstone Ecosystem, but concluded that the park's fire-management policies required significant tightening to avoid future runaway fire events.[18] As predicted by ecologists, most blackened portions of the park were soon awash in new growth, few animals were lost during the fires, and silted streams were returning to normal. Though an extraordinary event, the 1988 Yellowstone fires largely tracked the park's ecological history, which reveals that large-scale fire events regularly swept through its dense lodge-pole pine forests at 200- to 400-year intervals.[19]

Since then, the Park Service has consistently reaffirmed its commitment to a natural fire policy that allows some wildfires to burn and employs controlled burns to emulate natural fire regimes. The National Parks Organic Act's "conserve unimpaired" language gives the agency legal authority to pursue such a naturalness policy. To that end, the agency's *Management Policies* acknowledge that "naturally ignited fire . . . is part of many of the natural systems that are being sustained in parks" and instruct individual parks to prepare fire-management plans "allowing fire to perform its natural role as much as practicable" while protecting firefighter and public safety. The policy, tested once more in 2000 when a Park Service–initiated prescribed burn at New Mexico's Bandelier National Monument raced out of control and destroyed nearly half the town of Los Alamos, was again deemed fundamentally sound with additional sideboards to guide future controlled-burning events.

In 2012, following a series of bad fire seasons across the West, Yellowstone revised its fire-management plan, once more "allowing fire to play its ecological role in the Park to the greatest extent possible." The revised plan also established "suppression strategy zones" designed to enable park officials to respond promptly to fire events and better protect people, property, and park resources. Although the park has not since experienced any serious fire years akin to the 1988 event, climate-induced changes to weather patterns, including warmer, drier summers in the GYE, are increasing the risk of future catastrophic fires, which some believe will require more active management of the park's forests to preserve its ecological integrity.[20]

Wildfire in Yellowstone has tested the role of science in resource-management policy while highlighting the GYE as an integrated ecological entity. Despite the heavy political pressures accompanying the 1988 fires, the notion of fire as an important natural process essential to maintain ecological integrity within national parks was validated and remains a vital part of national park resource-management policy at Yellowstone and elsewhere. The rampaging fires burned across the manmade boundary lines separating the park from the surrounding national forests, thereby validating the interconnectedness of the region's public lands and confirming the importance of interagency coordination for resource-management purposes. The fact that nearby communities were endangered by the spreading flames also brought home the need for better collaboration with the surrounding states, towns, and landowners—a need that extends beyond wildfire to include wildlife, residential development, and other resource-related policy concerns, especially in the face of a rapidly growing regional population. In fact, the need for greater coordination among the park and its neighbors will only expand in the years ahead as a warming atmosphere begins to reshape the GYE and strengthen the calls for landscape-scale planning to sustain this wild place so many people revere.

Nonnative Species: Trout, Goats, and Ecological Integrity

The GYE—regularly touted to be "the largest, nearly intact temperate ecosystem in the world"—is only as intact as the national parks at its core. Although the GYE now embraces the full array of wildlife species present when Euro-Americans arrived, it also contains nonnative—or invasive—species that threaten its ecological integrity. The Park Service has spent twenty-five years seeking to restore Yellowstone Lake and its cutthroat-trout population due to the unwelcome presence of lake trout, an invasive species that has disrupted the watershed's delicate ecology, impacting grizzly bears and other species long dependent on the cutthroat trout that historically populated the lake. In Grand Teton National Park, nonnative mountain goats arrived several decades ago and took up residence on the park's high mountain ridges, where they represent a serious disease threat to the park's declining native bighorn sheep population. The Park Service, consistent with its commitment to a "naturalness" resource-management policy, has embarked on vigorous campaigns to reduce these invasive intruders and restore the native ecology.

The seminal Leopold Report was explicit that "exotic vertebrates, insects, plants, and plant diseases" should be subject to active management in order to create a "reasonable illusion of primitive America."[21] To achieve this goal, the Park Service's *Management Policies* emphasize the maintenance and restoration of native plant and animal species, requiring park managers to "prevent the introduction of exotic species into units of the national park system, and remove, when possible, or otherwise contain individuals or populations of these species that have already become established in parks." The *Policies* explain that "because an exotic species did not evolve in concert with the species native to the place, the exotic species is not a natural component of the natural ecosystem at that place." As a result, "exotic species will not be allowed to displace native species if displacement can be prevented." Although the Organic Act instructs the Park Service to conserve wildlife in the parks, it also empowers the agency to "provide for the destruction of such animals and plant life as may be detrimental to the use of any System unit." Confronted with litigation challenging Park Service decisions to shoot excess deer and elk populations in order to maintain ecosystems, the courts have consistently sustained intentional culling as a viable means to reduce undesirable wildlife numbers. In short, to preserve or restore ecosystem integrity the Park Service is authorized by law and policy to eliminate nonnative species from the parks.[22]

Lake trout were first detected in Yellowstone Lake in 1994, setting off an ecological alarm followed by a sustained effort to substantially reduce this apex predator, which was decimating the native cutthroat trout. Although park managers introduced lake trout and other fish species into various Yellowstone waters during the late nineteenth century to promote sport-fishing opportunities, they were not found in Yellowstone Lake, home to a robust, ecologically important cutthroat-trout population. Not only did the cutthroat trout transport vital nutrients into the sixty-eight spawning streams that fed the lake, they also represented a critical food source for grizzly bears, black bears, osprey, and various other avian species when spawning in these feeder streams during the springtime. Once established in the lake, the larger and more aggressive lake trout, which only spawned in the deep lake waters, predated ferociously on the native cutthroat, dramatically reducing that population. In 1998, by one estimate, "the 125,000 lake trout . . . likely consumed 3–4 million cutthroat trout that year." Other studies showed a similar "precipitous, lake-wide decline in cutthroat trout."[23]

As the cutthroat trout population declined, scientists observed a series of cascading ecological impacts to the lake's food web and beyond. Lake-trout predation on the cutthroat trout affected the flow of nutrients and energy in tributary streams, likely "contributing to a decline in the overall productivity of those waters." As fewer cutthroat trout survived to spawn in Yellowstone Lake's tributary streams, grizzly and black bears lost an important seasonal food source. Scientists estimated that grizzly bears consumed nearly 21,000 spawning cutthroat trout in the late 1980s, but that number declined to just 302 trout in the late 2000s, forcing the bears to seek other food sources. Not only did the number of grizzly bears visiting these spawning streams decline by 63 percent between 1997 and 2009, but no bears were observed on traditional spawning streams in 2008, 2009, and 2011. Because river otters depended heavily on cutthroat trout as a food source, scientists feared that they may likewise be affected by the decline in the annual spawn. Scientists also observed a notable decline in bald eagle and osprey nests on Yellowstone Lake coincident with the decline in the lake's cutthroat-trout population. Where osprey nests averaged thirty-eight nests from 1987 to 1991, only two nests were observed from 2015 to 2019. Put simply, scientists perceived a troubling trophic cascade due to the unnatural presence of lake trout: "it is likely that bears, otters, eagles, ospreys, and other species have been widely displaced throughout the upper Yellowstone River drainage."[24]

In an effort to reverse these ecological trends, Yellowstone mounted an aggressive suppression campaign, designed to dramatically reduce the lake trout population in order to restore cutthroat trout numbers. Building upon recommendations from an independent panel of scientists, Yellowstone finalized a park-wide Native Fish Conservation Plan, along with an environmental assessment, in 2011. The primary goal for Yellowstone Lake is to "restore the natural ecological role of native cutthroat trout."[25] The chosen strategy concentrated on lake-trout suppression through large-scale gillnetting, contracted to an experienced freshwater commercial fishing outfit that would catch and kill adult lake trout with nets deployed from large fishing vessels. Using advanced technology, including telemetry-rigged "Judas fish" to track lake trout, the scientists have learned where lake trout congregate, enabling the gill netters to concentrate their efforts and increase their catch. Following adoption of the plan and acquisition of additional funding, the Park Service was able to secure more vessels and larger nets, and thus ramp up the lake-trout catch.

Beginning in 2012, the effort turned the corner to where the number of lake trout caught annually began to decrease, indicating that the population had peaked. While this represented a milestone event in the restoration effort, the netters have not relented in the suppression effort; they took an annual average of 249,466 lake trout between 2010 and 2014, which rose to an annual average of 331,783 lake trout between 2015 and 2019. By the end of the 2022 season, the gillnetting program had succeeded in killing 4.3 million lake trout. Over time, biologists have also learned where lake trout spawn and are directing their efforts toward killing the eggs before they hatch. Buoyed by this sustained effort, cutthroat trout appear to be rebounding, reflected in the number of large cutthroats being seen and caught in tributaries to the lake. Moreover, grizzly bears are returning to the lake's spawning streams to feed on cutthroat trout in the spring, but the other impacted species have yet to show any notable recovery, though scientists expected this type of lag time in their response.[26]

The Park Service's Native Fish Conservation Plan is built on the adaptive management concept, recognizing the numerous uncertainties confronting the agency when lake trout were initially discovered in Yellowstone Lake. Recommended by an independent panel of scientists with extensive experience dealing with aquatic species and invasive problems, the adaptive management strategy entails taking initial science-based management actions, monitoring the outcomes, and then adjusting further actions based upon the results. This science-driven strategy has thus far been unaffected by politics, as outside conservation groups, anglers, and the general public have supported it from the outset. The enhanced suppression program is expensive, however, costing roughly $3 million annually and likely requiring perpetual renewal. The Park Service has been able to piece together the funds from several sources, including the park's own budget, Yellowstone Forever (the park's philanthropic arm), and Trout Unlimited—a unique public-private funding partnership that has been essential to the project.[27]

Yet, even as the suppression effort appears to be succeeding, other problems lurk, highlighting the continuing need for adaptive management. No one knows what effect the recent discovery of another nonnative fish—cisco—will have on the lake's ecology; cisco compete with the cutthroat for food, but the cisco is also a major food source for lake trout. Moreover, climate change could affect cutthroat-trout recovery. Cutthroat depend on shallow lake areas and streams, which are affected by both drought conditions and a warmer, diminished water supply. Lake trout, in contrast, inhabit deep lake waters and are therefore unlikely to be similarly affected. When fisheries biologists add disease concerns to the mix of cutthroat threats, the future becomes even more problematic for this imperiled native fish, whose long-term survival will likely require lake-trout netting to continue indefinitely. Although park managers never expected the suppression program to fully remove lake trout from Yellowstone Lake, the quarter-century effort appears to have restored a measure of ecological integrity to these park waters and the surrounding landscape, at least for now.[28]

Grand Teton National Park is seeking to rid the park of nonnative mountain goats that are perceived as a disease threat to the native bighorn sheep population. Not believed to have historically inhabited the

Teton Range, a few mountain goats probably appeared in the park during the 1970s, likely moving there from Idaho's nearby Snake River Range, where the state transplanted them in the late 1960s. As the twenty-first century unfolded, mountain-goat numbers in the park grew, reaching roughly one hundred goats by 2018 while also expanding their range and moving closer to the park's native bighorn sheep. When sampled, several mountain goats tested positive for "bacteria associated with bighorn sheep pneumonia," and they are "known to host several additional respiratory pathogens . . . that collectively pose a high risk of disease to bighorn sheep in the Teton Range." The goats occupy the same type of high-elevation mountain habitat as bighorn sheep, making it likely that the two species would eventually overlap in their habitat use.[29]

Scientists feared this intermingling scenario. Because mountain goats are curious animals, they are likely to touch noses with nearby bighorn sheep, creating a ready path for disease transmission. The park's bighorn sheep—which do not migrate annually probably due to habitat loss from increased human presence and construction activity on adjacent lands—have experienced a notable population decline in recent years. A helicopter survey in 2008–10 counted roughly ninety sheep, but that number dropped to around fifty sheep in the 2015–17 survey. Besides the mountain-goat problem, scientists have documented that the park's bighorn sheep were displaced from prime winter habitat by backcountry skiers, causing them to burn precious calories when they are otherwise in a weather-stressed condition. In response, the Park Service has closed critical backcountry areas to skiers to protect the small big horn population from undue stress during winter months, provoking inevitable criticism from some hard-core recreation users.[30]

In 2018 Grand Teton officials released a proposed mountain-goat management plan and accompanying environmental assessment. To protect the native bighorn sheep population, the plan called for removing 90 percent of the mountain goats from the park, mostly by shooting them from helicopters during wintertime when visitors were absent from the mountainous backcountry. In the final plan the shooting was to be done by trained marksmen under federal contract, but not, as some proposed, by civilian hunters—"qualified volunteers"—who viewed the goat-removal opportunity as a hunting experience. Wyoming officials objected, however, asserting that "aerial gunning flies in the face of all Wyoming values with how we approach wildlife management."[31] When Grand Teton officials ignored Wyoming and proceeded to shoot thirty-eight goats from helicopters, Wyoming's governor angrily contacted Secretary of the Interior David Bernhardt, who instructed the park's acting superintendent to "stand down." After further consultation with the state, the park reversed course and resumed the removal using "qualified volunteers" vetted by the Park Service. They killed another 43 goats, thus eliminating more than half of the goat population during the program's first year. In 2022, without objection from the state, the Park Service reverted to aerial gunners again, who removed another 58 goats, bringing the total killed to more than 100 since the program began.[32]

Notably, none of the dispatched mountain goats tested positive for the types of respiratory illnesses dangerous to the Teton bighorn herds. The Park Service, nevertheless, regarded the removal as a precautionary measure designed to protect a native species by eliminating a nonnative species consistent with the agency's resource-management policies. While largely based upon science and risk-management considerations, the park's ultimate goat-removal strategy was plainly shaped by political considerations. The episode illustrates the significant political power the GYE states can exert in the region, even in the case of the national parks.

Recreation Quandaries: Snowmobiles, Kayaks, and Bikes

The Yellowstone snowmobile controversy unfolded over a nearly forty-year period, presenting the Park Service with the vexing question of appropriate recreational activity in the national park setting. Under the Organic Act, the Park Service is charged with both nature conservation and visitor enjoyment responsibilities, subject to meeting the law's non-impairment mandate. The agency's *Management Policies*, citing the statutory language, are clear: "Congress . . . has provided that when there is a conflict between conserving resources and values and providing for enjoyment of them, conservation is to be predominant." This interpretation of the Organic Act squares with how the courts have interpreted the statute. In a 1980 case that upheld the Park Service's prohibition on hunting in national parks, the court concluded: "In the Organic Act, Congress speaks of but a single purpose, namely conservation." Nonetheless, the powerful political and economic interests invested in Yellowstone snowmobiling ensured a prolonged battle over efforts to curb this mechanical intrusion into the park's winter stillness. It was a battle that transpired in the halls of Congress, the federal courts, and state legislatures, and spanned three different presidential administrations, severely testing the Park Service's commitment to preserving the park environment in an unimpaired condition.[33]

The Yellowstone snowmobile controversy traces to the late 1940s postwar era. Seeing a marked increase in tourism activity, the town of Cody, Wyoming, unsuccessfully implored the Park Service to plow the park's snow-covered roads to establish a new winter visitation season. In 1954 two West Yellowstone entrepreneurs started ferrying visitors into the park on primitive snow coaches. Snowmobiles first entered the park in 1963, quickly growing in number to more than 5,000 visits in three short years. To rebuff renewed pressure from Cody politicians to plow the park roads, Yellowstone officials agreed to allow snowmobiles into the park on groomed (but not plowed) park roads and to open a winter lodge at Old Faithful. Yellowstone superintendent Jack Anderson then started promoting winter snowmobiling in the park, ultimately garnering a special recognition award from the International Snowmobile Industry Association for his efforts. By the early 1970s winter visitation was an established fact as more than 25,000 visitors were annually entering the park. Many of them came on snowmobiles rented from new businesses established in West Yellowstone, which branded itself as the "snowmobile capital of the world."[34]

Within a few years complaints about snowmobile use in Yellowstone began to mount, citing air pollution, excessive noise, and wildlife-harassment incidents. But when a 1972 presidential executive order intended to control off-road vehicle use on public lands compelled park officials to designate roads as either open or closed, the park simply designated all roads as open to snowmobiles without any environmental review. Yellowstone's decision stood in stark contrast to its sister parks, most of whom—including Glacier, Yosemite, Sequoia, and Lassen—closed their roads to snowmobiles after completing an environmental assessment of the matter. Another surge in snowmobile use followed, with winter visitation growing from 70,000 in 1983 to more than 143,000 in 1992. Under new leadership the park prepared its first winter-use plan in 1990, but it imposed few limits on snowmobile activity. Tellingly, its long-term visitation and pollution predictions proved outdated in three years.

In the mid-1990s, as the Park Service began a new winter-visitation management plan intended to finally impose numerical access limits, severe winter weather recast the controversy. Park bison, unable to secure food under the deep frozen snow during the 1996–97 winter, exited the park in record numbers, following the groomed, hard-packed roads. When the bison entered Montana, state officials unceremoniously shot them out of fear they would spread the brucellosis bacteria to nearby

Snowmobiles Passing Bison near Roaring Mountain, Yellowstone National Park.
Credit: Jim Peaco, Yellowstone National Park Photo Collection.

cattle herds. More than 1,000 hungry bison were slaughtered that winter. Incensed, the Fund for Animals and others sued, arguing that the agency's winter-use plan was plainly inadequate. To resolve the matter, the Park Service agreed to prepare an Environmental Impact Statement re-examining its snowmobile and road-grooming policy.[35]

What followed can only be characterized as a prolonged donnybrook, one featuring a series of dramatic policy reversals as well as competing congressional bills and court cases. Upon examining the environmental impacts of its snowmobile policy, the Park Service concluded that it violated the agency's basic management obligations: "Continued [snowmobile] use hinders the enjoyment of resources and values for which the parks were created, most notably natural soundscapes, clean and clear air, and undisturbed wildlife in a natural setting." Embarrassing news photos featured park rangers wearing respirators at the West Yellowstone entrance station to combat the exhaust fumes emanating from a long line of snowmobiles awaiting entrance into the park. In the park the high-pitched whine of two-cycle snowmobile engines penetrated deep into the backcountry, while reports of wildlife-harassment incidents kept growing.

The Park Service responded by announcing in late 2000 that it would phase out snowmobile use over a three-year period, but allow snow-coach access so visitors could continue to enjoy the park in wintertime. The phase-out period and snow-coach option, which gave a licensing preference to local businesses, was designed to ameliorate the economic impact on West Yellowstone and other nearby communities. As the conservation-minded Clinton administration wound down, it appeared that Yellowstone was headed toward a cleaner and quieter winter landscape.[36]

That did not happen, however. Once the Bush administration assumed office, the Park Service announced that it would revisit the closure decision, explaining that new snowmobile technology might reduce the pollution and noise levels coming from the machines. In early 2003, following further environmental analysis, the Park Service reversed its earlier snowmobile ban and adopted a new policy that increased the number of snowmobiles permitted into the park, established new emissions requirements, and imposed a new guide requirement for most trips. Environmental groups promptly sued the agency in a Washington, DC, federal court, which blocked the new plan. According to the court, the Park Service had not adequately explained its policy reversal given the agency's "clear conservation mandate and the previous conclusion that snowmobile use amounted to unlawful impairment." More troubling, the judge found that the snowmobile policy shift "was completely politically driven and result-oriented," while ignoring public input that overwhelmingly preferred the no-snowmobile policy. Rejecting a request to delay its ruling, the court blamed the Park Service for any economic or emotional harm felt by local communities or potential visitors.[37]

Faced with reinstatement of the Clinton era no-snowmobile policy, the International Snowmobile Industry Association and others turned to the Wyoming federal district court for relief. There they convinced a Wyoming federal judge that the Clinton era policy was illegal and also politically driven; it did not adequately examine the environmental and safety effects of snow coaches and provided limited public comment opportunities. Finding that the Clinton policy would have a substantial, harmful local economic effect that outweighed any environmental harm associated with snowmobiles, the court ordered the Park Service to draft a temporary winter-use plan. The ensuing 2004 plan, disregarding overwhelming scientific evidence and public sentiment, authorized nearly 800 daily snowmobile entries into the park, while imposing new pollution-control equipment and commercial-guide requirements. No one thought it would be the final word in the matter.[38]

Soon enough, both the federal courts and Congress were again engaged with Yellowstone snowmobile policy. In 2008, when the Park Service issued another revised winter-use plan that allowed 540 daily snowmobile entries into the park, the DC federal court ruled that the plan violated the Organic Act, emphasizing that "the fundamental purpose of the national park system is to conserve park resources and values" and that "conservation is to be predominant." Citing the agency's own data, the court concluded that "the Winter Use Plan will increase air pollution, exceed the use levels recommended by NPS biologists to protect wildlife, and cause major adverse impacts to the natural soundscape." The Wyoming federal court responded by reinstating the 2004 temporary plan, citing the potential economic impacts if snowmobiling were prohibited in the park. Meantime, Congress was barraged by a series of bills, some seeking to eliminate snowmobiles from the parks and others seeking to ensure access to them. The politically charged issue was now in the hands of the Obama administration.[39]

Somehow, the Park Service's next effort at a winter-use plan managed to strike a workable balance, quelling the litigation while also deterring Congress from action. In 2013 Yellowstone adopted a fourth plan that introduced a flexible "daily transportation event" quota system, which regulated access split between guided snowmobile parties and snow coaches, imposed a best available technology requirement to control emissions and noise, and retained the guide requirement. By then snowmobile technology had improved considerably with the introduction of more efficient four-stroke engines that reduced pollution and noise, though the newer machines could also reach speeds approaching 100 miles per hour. Feeling pressure from Cody politicians, the Park Service agreed to keep Sylvan Pass open at Yellowstone's eastern entrance to support a local snowmobile rental business, notwithstanding the annual $325,000 cost to control the area's extreme avalanche danger with howitzer shells and the small number of entries made from this point.

Apparently weary from the battle, no one sued, the politicians retreated, and the new system has succeeded in reducing pollution, noise, and wildlife incidents. Snowmobiling in the park has tailed off considerably since the Park Service began imposing constraints in the early 2000s, precipitating a significant decline in snowmobile activity in West Yellowstone, which has shifted its attention to the summer season. Meanwhile, snow coaches are displacing snowmobiles as the preferred means to experience Yellowstone's winter wonderland.[40]

Several lessons emerge from the Yellowstone snowmobile controversy. First, despite the Park Service's clear mandate to prioritize conservation over recreation in the national parks, that prioritization cannot be taken for granted. Local as well as national political and economic interests associated with snowmobiling were able to exert extreme pressure on the agency to keep the park open to large numbers of snowmobiles despite compelling scientific data revealing unacceptable impacts on park resources and values. Determined opposition by conservation organizations through litigation and political advocacy proved critical to the final outcome. Second, once a recreational activity is established in a national park, it is difficult to dislodge it, given the inevitable economic and other interests that develop around it. In Yellowstone and elsewhere, this suggests the need to promptly and carefully scrutinize the environmental and other effects associated with proposed new recreational uses before sanctioning them. Third, given that Yellowstone is surrounded by national forest lands where the whine of high-pitched snowmobile engines is not as troublesome, the controversy begged for a regional, collaborative solution among the area's federal land managers. Such an approach could have focused on dispersing snowmobiling across the GYE landscape, thus preserving the park's unique wintertime experience for those seeking a more natural, quieter, and reflective recreational experience. In this era of industrial-scale recreation, not everyone will be able to play in Yellowstone or other national parks on their own terms.

The Yellowstone snowmobile controversy set off a related national controversy with ominous implications for the national parks. Following the 2004 election, amid the Yellowstone imbroglio, Deputy Assistant Secretary of Interior Paul Hoffman drafted proposed revisions to the Park Service's *Management Policies* altering its long-standing interpretation of the Organic Act that prioritized conservation over recreation. Before joining the Interior Department as a political appointee, Hoffman had directed the Cody, Wyoming, Chamber of Commerce, where he led opposition to Yellowstone's proposed ban on snowmobiling. The revisions would have allowed more motorized recreation across the National Park system. When Hoffman's handiwork came to light, it aroused a vigorous public outcry, asserting that the Interior Department was reversing a half century of consistent national park policy for blatant political purposes to appease the recreational vehicle industry. Following a high-profile Senate hearing and strong editorial criticism in the nation's newspapers, the Park Service quietly made a few modest revisions to its *Management Policies* but preserved its integrity and priorities. The entire episode reinforces the constant tension park managers confront between resource conservation and recreational-use proponents, a tension fueled by the powerful economic and political interests behind a rapidly growing outdoor recreation industry.[41]

Both Yellowstone and Grand Teton have faced other recreation-related controversies that have tested the Park Service's commitment to its conservation-first mission. Since the 1950s, when Yellowstone banned boats on its rivers, proposals to open the park's rivers to canoes and kayaks have periodically surfaced but been consistently rejected. In a 1988 decision Yellowstone concluded that paddlers on the park's rivers could damage nearby geothermal resources, displace wildlife, injure streamside vegetation, and pose safety concerns. Since then, whitewater-kayak enthusiasts have advanced other proposals to open park rivers, and even enlisted Wyoming's congressional representative to introduce legislation that would have required the agency to allow hand-propelled crafts on designated rivers. The bill failed in the 114th Congress, much to the relief of opponents who expressed concerns about the impact adrenaline-fueled whitewater kayaking could have on the National Park experience. A related effort to open Yellowstone's and Grand Teton's backcountry lakes to pack rafts has also been consistently rejected by the Park Service for aesthetic and environmental reasons. Nonetheless, paddlers already have access to several park lakes, where they can enjoy a backcountry experience, and whitewater kayaking is available nearby on several national forest rivers where environmental and other impacts are not as problematic.[42]

During the past twenty years, Grand Teton has constructed an extensive multiuse bicycle trail system that parallels the main park roads, providing visitors with an alternate recreational experience while enjoying the park. When Grand Teton officials decided to upgrade the Moose-Wilson Road, a semi-paved road that provides a narrow automobile route connecting the town of Wilson with park headquarters at Moose, bicycle proponents sought to establish a bike trail parallel to the road. They envisioned the opportunity to pedal a loop route running through the southern portions of the park and connecting to the town of Jackson, but others decried the changes it would bring to this primitive setting.[43] Citing the remote area's natural values, including its wildlife habitat importance, as well as the environmental and cultural impacts associated with a larger development footprint, the Park Service rejected the bike-trail option. Despite considerable pressure from wealthy, politically connected Jackson Hole residents, park officials managed to keep their conservation responsibilities foremost.[44]

Escalating Visitation Challenges

Yellowstone has long served as a magnet for visitors, whose presence has presented park officials with an evolving array of challenges over the years. Congress originally conceived Yellowstone as "a pleasuring ground" for the general public, though the park's early visitors were mostly wealthy individuals who arrived via the Northern Pacific Railroad and toured the park in horse-drawn wagons. By the 1920s, with the advent of the automobile, more people began arriving, and their numbers have only grown since then. From its earliest days, the Park Service, under the leadership of marketing executive Stephen Mather, sought to attract visitors through alliances with the railroads, automobile associations, and various business interests. To accommodate its visitors, the Park Service built hotels, roads, campgrounds, stores, and other facilities that endure yet today. And it invited concessionaires into the parks to run the hotels, stores, and other services, giving these businesses a financial stake in attracting visitors and their money. As time passed, the surrounding gateway communities likewise developed a pronounced dependence on park visitors for their economic well-being. With the crowds mounting across the system, park officials have faced considerable pressure to accommodate everyone, even as more people are putting additional strains on park resources and the experience available.[45]

Visitation at the two GYE parks has reached record levels in recent years, following decades of relatively steady increases. Yellowstone first exceeded two million annual visitors in 1965, reached three million in 1992, and then registered more than four million visitors in 2016, the National Park system's centennial year. Since then, the park has generally exceeded four million visitors, except for 2020 when the COVID-19 pandemic was rampant and 2022 when the number dropped to 3.2 million visitors due to flood-related road closures. Grand Teton National Park shows a similar pattern, with annual visitation first reaching two million in 1963, then exceeding three million in 2015, and continuing well above three million except for the 2022 flood year. In addition, both parks have seen a notable uptick in visitation during the usually quiet spring and fall shoulder seasons. These escalating visitor numbers are extracting a toll on the parks, including crowding and environmental damage at popular locations, deteriorating roads, facilities, and waste systems, and increasing visitor–wildlife incidents. The sheer number of people and automobiles at popular sites like Old Faithful or Jenny Lake on a summer day, as well as the bear and bison auto jams that frequently clog Yellowstone's roads, bespeak a visitor experience far removed from wild nature, but one that most visitors seem to regard as satisfying. According to a 2019 Yellowstone survey more than 75 percent of the park's visitors were there for the first time, and more than 90 percent of them were satisfied with their national park experience.[46]

Whether the GYE parks can sustain this level of visitation and accompanying impacts remains to be seen. Although the Park Service has long wholeheartedly embraced the Organic Act's "public enjoyment" mandate by seeking to attract visitors, the agency has the clear legal authority to address the growing visitation challenges at Yellowstone and Grand Teton. The Organic Act and the agency's *Management Policies* prioritize resource protection over visitation. And the courts have consistently sustained Park Service–imposed limitations on backcountry use as well as various recreational activities to protect at-risk park resources.[47]

Nonetheless, the GYE parks are unlikely to invoke this authority to limit visitor numbers, given their vital role as "cash cows" for the park's concessionaires and nearby gateway communities. In 2019 visitors to Yellowstone and Grand Teton spent $1.05 billion locally, a level of spending that supported 16,225 local jobs and put Wyoming among the top five states benefiting from the national park presence.[48] Any effort to curtail visitation will meet almost certain political opposition, as occurred when Yellowstone sought to close Sylvan Pass to snowmobiling, spurring Cody politicians and businesses to derail the plan. Similarly, in the early 1980s, when Yellowstone proposed closing the Fishing Bridge campground to protect critical grizzly bear habitat, Wyoming politicians forced park officials to revise their plans.[49]

Although discussions about alternate transportation systems to reduce the number of cars in the parks have occurred, the parks' vast size and wary neighbors make this a difficult, expensive, and politically fraught option. Indeed, the economic realities of the park-gateway community connection were on display in the aftermath of the destructive summer 2022 floods that closed Yellowstone roads for weeks, creating an economic nightmare for the park's northern gateway communities that saw visitor numbers and spending drop by more than 30 percent for months. Given the powerful economic connection between the GYE parks and neighboring gateway communities, there is little reason to expect the Park Service to contemplate imposing limitations on visitor or automobile numbers, as has occurred at other popular national parks.[50]

Visitation problems, however, extend beyond the matter of legal authority or local community relationships. Faced with years of shrinking budgets, both GYE parks are short of the necessary personnel to handle hordes of visitors or to educate them about the natural setting and proper behavior in the wild. Both parks also report serious deferred maintenance problems with extensive repairs needed to park roads, buildings, waste facilities, and the like. In Yellowstone, for example, the 2018 deferred maintenance costs were estimated at $851 million, with $620 million to cover the park's 392 miles of paved roads and $154 million to cover repairs to the park's 1,483 buildings and 483 housing units. Grand Teton's infrastructure maintenance costs came in at $248 million with nearly half of that devoted to the park's roads.[51]

Congress initially responded to this problem in 1996 by enacting an experimental fee demonstration program that enabled the Park Service (along with other federal land management agencies) to charge and retain entrance fees, and then made the fees permanent in 2004. While helpful, these new fees fell short of alleviating the growing maintenance backlog. In 2020 Congress again responded with enactment of the Great American Outdoors Act, which provided the national parks with five years of funding to address maintenance and infrastructure problems. The act, however, does not address future maintenance needs, limits the amount that can be spent on transportation-related problems, and contains no funds to address personnel needs, such as additional rangers to educate visitors about the GYE, wild nature, and the challenges confronting the parks.[52]

Modern technology has presented both Yellowstone and Grand Teton with new visitor-related concerns, including the use of drones and the installation of cell-phone towers. After a drone crashed in a Yellowstone thermal pool and similar problems arose in other parks, the Park Service adopted a universal "no drones" policy unless approved by individual park superintendents, citing resource damage and visitor conflicts as the underlying rationale. However, the agency and the two GYE parks have been much more receptive to cell-tower installations and related communication equipment, even when obtrusive towers mar the natural scene and experience. Generally justified as necessary to meet changing visitor demographics and expectations as well as public safety and education needs, National Park wireless installations are mostly driven by Verizon, AT&T, and other telecom companies who perceive a market for their services. In several instances Yellowstone officials have approved new cell towers after excluding them from environmental review by invoking NEPA's categorical exception on the often-mistaken ground that they did not significantly impact the landscape. Although usually targeted toward developed areas in the parks, the emitted signals often stretch into the backcountry, altering the wilderness experience available there. While Yellowstone is relocating some obtrusive cell towers, Grand Teton is busy installing towers with signals that spill over into areas designated for wilderness management.

Although many view internet coverage as a near-essential modern convenience, others draw comparisons to the Park Service's historic penchant toward overdevelopment, exemplified by its early approval of roads, buildings, and other services to attract tourists despite impacts to the natural setting and experience. The question is whether in this electronically connected era "windshield tourism" has morphed into "screen-time tourism." Although the cell-tower debate may have limited ecological impacts on the GYE, these National Park decisions set a precedent for backcountry expectations regarding the wilderness experience and the sense of wildness that distinguishes the GYE from other places. Indeed, it is precisely these values that drive ongoing efforts to preserve the GYE intact.[53]

Grand Teton Inholding Controversies

The GYE jurisdictional fragmentation problem is glaringly evident in Grand Teton National Park, where state and private lands—known as inholdings—lie within the park's boundaries. This intermixed land ownership pattern has posed troublesome management challenges for park officials and occasionally strained relations with the state of Wyoming. The problem dates back at least to 1890 when Wyoming became a state. Upon granting Wyoming statehood, Congress transferred two designated 640-acre sections of federal land to the new state to support its public schools—tracts known as state school sections. As the Wyoming territory opened, new settlers were entitled, under the 1862 Homestead Act, to acquire ownership of 160-acre parcels by improving the land. In 1929, when Congress established Grand Teton National Park (and later expanded it), the legislation acknowledged and grandfathered these preexisting property rights into the park. This checkerboard-type ownership pattern has given rise to two questions: Who has jurisdiction over these inholding lands, and what responsibilities does the state owe to the park?

In 2014, when Wyoming asserted jurisdiction over roughly 2,300 acres of state and private inholding properties within the park, the Park Service, somewhat surprisingly, acquiesced to the state's claim, disregarding a long-standing federal-state understanding. According to high-level Park Service officials, government lawyers concluded that Wyoming's jurisdictional claim stood on firm legal ground. This gave the state responsibility for wildlife management on these inholdings, which opened the door for possible wolf and other hunting activity inside the park. The Park Service, of course, prohibits hunting on National Park lands, contrary to the state's policy, although Congress authorized limited elk hunting in the 1950 park expansion legislation. Alarmed at the prospect of additional hunting inside the park and perceiving related management problems, conservation groups contested the matter in federal court. They lost, however, when the court read the Grand Teton enabling legislation to reserve jurisdiction over these inholdings to the state. The resulting arrangement adds more jurisdictional complexity to an area already awash in it. The court's ruling not only complicates wildlife management for park officials, it also endangers park visitors and wildlife during hunting season.[54]

In 2023 the Wyoming Board of Land Commissioners announced it planned to auction the last remaining state school section situated within Grand Teton National Park, a tract known as the "Kelly Parcel." Under Wyoming law, the Land Board must manage these state school sections to support public schools by deriving revenue from these lands. The Kelly Parcel sits several miles north of Jackson and offers stunning views of the Teton Range, giving the land—originally appraised at $62 million— incredible development value in Teton County's red-hot real estate market. The parcel also rests astride an important wildlife corridor and critical winter habitat. Given its strategic location, the Park Service has long sought to acquire the parcel to consolidate its land base, eliminate jurisdictional problems, and protect conservation values.[55]

The Land Board's auction announcement set off alarm bells. Although the Park Service had managed to acquire another school section inholding a few years earlier, which was critical in establishing the Path of the Pronghorn wildlife-migration corridor, the Kelly Parcel was particularly attractive to investors due to its location and the dearth of private developable land within the county. Few believed the appraisal value would be the final selling price at auction. Fearing well-heeled individuals or developers would outbid the Park Service for the parcel with disastrous consequences, the Park Service, Jackson residents, local officials, and conservation groups mobilized to oppose the proposed auction. Tacitly acknowledging the widespread public concern over the Kelly Parcel's fate and potential impacts on the treasured park, the board postponed the matter for another year. The Wyoming legislature then intervened and approved a sale to the Park Service for $100 million. Finally, aided by $37 million in private donations, the Park Service acquired the Kelly Parcel in late 2024.[56]

Taken together these incidents highlight the jurisdictional complexity that complicates land and resource management in the GYE, even in the region's core national parks. Federal-state tensions over wildlife and other issues lurk in the background as both the Park Service and the state are well aware of their sovereign rights, interests, and legal responsibilities. Nor are they indifferent to the park's outsized role in the local economy. The solution, given the conservation values at risk, should give priority to the park's ecological integrity and the need for consistent management standards within the park's boundaries. Accomplishing these objectives requires meaningful intergovernmental collaboration, an essential component of ecosystem management, and occasionally public dollars as well as private philanthropic assistance.

* * *

The past fifty years have witnessed a transformation in the two core GYE national parks—one that resonates throughout the entire ecosystem and beyond. Since the National Park Service reoriented its resource-management policies to prioritize ecological values and to minimize managerial interventions into the natural order, human pressures on these natural values have multiplied in the form of escalating visitation and recreational uses as well as outside development pressures pushing inward. At both Yellowstone and Grand Teton, the Park Service confronts a familiar conundrum, one built into its basic legal mandate: preservation versus use. Within the parks, the presence of more people inevitably means greater impacts on wildlife and their habitat, additional environmental damage, and a diminished park experience due to crowding. Outside the parks, the natural regulation policy ineluctably means more conflicts with neighbors over bears, wolves, bison, and other animals, as well as over wildfire-management policy.

To some degree, the neighboring national forests and abutting wilderness areas can—and do—buffer these impacts and conflicts, providing important additional wildlife habitat and absorbing some visitor and recreational pressures. The GYE counties and communities could assist in these efforts by helping to educate visitors about appropriate behavior in the wild and by controlling new development in fire-prone areas adjacent to the GYE parks and forests. Inside Grand Teton, the state could help too by showing some restraint in exercising its jurisdictional prerogatives. The key, of course, is greater collaboration among the responsible entities to reduce conflict while ensuring the region's ecological integrity, which begins by preserving the GYE parks in an unimpaired condition.

CHAPTER THREE

Restoring Endangered Species
Grizzly Bears and Wolves

Few animals evoke the sense of wildness or raw emotion as a grizzly bear or wolf. These charismatic apex predators occupy a central role in the GYE and the related shift toward ecosystem management. Both animals represent a conservation and ecological restoration success story. After trending toward extinction, grizzly bear numbers have been resuscitated, while the once-exterminated wolf has been restored to the ecosystem. Each is also a source of wonder, manifested in the clusters of visitors, cars, and spotting scopes regularly gathered alongside Yellowstone's roads eager to glimpse a wolf pack on the prowl. A similar scene prevails inside Grand Teton where Bear 399 and her cubs—perhaps the world's most widely photographed grizzly—have enchanted visitors for years.[1] But the GYE grizzly bears and wolves have also provoked controversy, most notably among ranchers and hunters within the three GYE states, who have waged prolonged political and legal battles over their management.

Key to the recovery of both species has been the science-driven Endangered Species Act (ESA), which is expressly designed to "provide a means whereby ecosystems upon which endangered and threatened depend may be conserved."[2] Equally important has been the unflagging commitment among many to preserve these animals from oblivion, thus maintaining the GYE as a natural, functioning ecosystem. Subject to federal protection since 1975 Yellowstone's grizzly population entered a steep decline after the park's garbage dumps closed and only regained its footing following congressional intervention during the mid-1980s that triggered a rejuvenated federal-state recovery effort. Although granted endangered species protection in 1974, wolves were then absent from the GYE, having been eradicated during the 1920s due to an aggressive government-funded predator removal program. When Congress amended the ESA in 1982, creating a new avenue to reintroduce extirpated species, Yellowstone officials and their conservation group allies embarked on a decade-long campaign to restore wolves to the ecosystem. Although population numbers for both species have met initial recovery targets, only the GYE wolves have been removed from federal protection. Despite this progress, deep fissures persist over state wildlife-management policies for both grizzlies and wolves. Indeed, these two animals are enshrouded in legal and political complexities, lucidly illustrating the need for a fully coordinated ecosystem-management approach to both species.

Grizzly Bears: A Defining Symbol

A solitary, wide-ranging omnivore dependent on large wild expanses, the grizzly bear has played a key role in defining the GYE and promoting ecosystem management within the region. Long a prime if elusive attraction for Yellowstone park visitors, the grizzly bear represents unalloyed wildness, evoking images of frontier America while also putting backcountry users on high alert as they enter bear territory. After teetering on the brink of extinction, the Yellowstone bear population has rebounded, thanks to an intensive, federally led conservation effort designed to protect both individual bears and their essential habitat. In the process the grizzly bear recovery effort has reshaped resource-management priorities on the GYE national forests and altered long-standing ranching practices. Once grizzlies were briefly removed from federal protection in 2017, proposed state grizzly bear hunting rules set off alarm bells among the animal's advocates, who succeeded in convincing a court to reinstate federal protection. Meantime, serious efforts are afoot to connect the isolated GYE grizzly population with its northern cousins, an accomplishment that would safeguard genetic diversity and significantly reduce the long-term extinction threat.

The widely publicized Lewis and Clark Expedition introduced Americans to the grizzly bear, which roamed widely across the nation's newly acquired western territory. After hearing about a fearsome bear from local Native Americans, the expedition finally encountered a grizzly on northern Montana's high plains along the Missouri River. In his journals Captain Clark described the bear: "[A] very large and terrible looking animal, which we found very hard to kill. . . . [It] is the largest of the carnivorous kind I ever saw. . . . I think his weight may be stated at 500 pounds." Meriwether Lewis recounted a similar experience: "a most tremendous looking animal, and extremely hard to kill," further relating that it took "five balls through his lungs and five others in various parts" to eventually dispatch the bear. From these and other early encounters, the legend and plight of the grizzly grew. As white settlers advanced westward, they came to view the bear as a threat to themselves and their livestock. Soon a purposeful eradication campaign—one marked by state and local bounties and implemented with the repeating rifle, steel traps, and poison baits—was underway to rid the landscape of the fearsome bear. It succeeded, and by the early 1900s a grizzly population thought to have numbered 50,000 bears was reduced to a few hundred individuals relegated to isolated mountain strongholds, sufficiently removed from the region's new inhabitants to survive as a remnant population.[3]

The remote Yellowstone country offered a secure redoubt for the bear, and the new park with its prohibition on hunting afforded additional sanctuary. Before the Park Service assumed the helm, however, the park's grizzly and black bears were the frequent target of poachers, as well as the park's military caretakers who did not hesitate to eliminate troublesome bears. Once the Park Service came on the scene, it generally sought to protect both bears for the visiting public's viewing pleasure, both along the park's roadways and at the various hotel garbage dumps where the bears had learned food was available daily. In fact, the Park Service even erected bleachers to enable enthralled visitors to view the bears from safety. Once they became habituated to people, the bears began frequenting campgrounds and roadsides, brazenly begging visitors for food and creating the notorious "bear jams" along park roads. For the bears, it was anything but a natural existence in the wild.[4]

During the following decades, however, the Park Service began closing the garbage dumps and discouraging roadside feeding in an effort to return the bears to a more natural existence. Following the 1963 Leopold Report, the agency escalated its efforts. By then the Yellowstone grizzly bears were the most studied bear population in the world, thanks to the Craighead brothers who—under a research agreement with the Park Service—were closely monitoring the bears with sophisticated telemetry devices to better understand their habits and habitat needs. Able to track the bears after fitting them with electronic signaling collars, the Craigheads established that most Yellowstone bears roamed beyond the park boundary in search of food and mates.

The bears, according to the Craigheads, occupied an area they dubbed the "Greater Yellowstone Ecosystem," hence the origin of the GYE term. But when outside the park, bears were in considerable jeopardy since the adjacent states each allowed hunting the grizzly as a trophy animal. When beyond the park, bears were also regularly targeted by ranchers to protect their cattle and sheep from depredation. To the Craigheads, the Park Service's decision to summarily close the remaining dumps imposed a death sentence on many bears, who having become habituated to people would end up being shot as a perceived threat to human safety. Once this disagreement between the Craigheads and the Park Service became more public and personal, the Park Service cancelled the brothers' research agreement, citing problems with their research methods. Consequently, the bears were left to fend on their own, while the Park Service was left without the benefit of timely scientific data concerning population trends under this new-management regime.[5]

In the ensuing years grizzly bear numbers dropped sharply, placing the long-term viability of the isolated Yellowstone population at risk. Nearly ninety bears were killed in the two years following final closure of the garbage dumps, a disturbing trend given the Craigheads put the park bear population at roughly 230 animals.[6] It was appropriate that Congress, when debating adoption of the Endangered Species Act during the early 1970s, highlighted the plight of the Yellowstone grizzly bear population to explain the need to extend stringent federal protection to dwindling species edging toward extinction.

In 1975, two years after the act was adopted, the US Fish and Wildlife Service (FWS) granted the grizzly bear federal protection by listing it as a threatened species, noting that only the Yellowstone and Glacier region bear populations persisted in meaningful numbers within the continental United States. But the FWS, citing social tolerance concerns within the region's communities, decided not to designate critical habitat for the bear, even though the new act recognized that habitat protection was essential to recovery. By then, the FWS and the Park Service had joined the Forest Service and the three states bordering Yellowstone to create an Interagency Grizzly Bear Committee (IGBC) consisting of upper-level managers to oversee the bear and its needs—an early recognition of the greater ecosystem concept and of the need for coordination among the governmental entities responsible for the GYE's wildlife.[7]

Yellowstone's bears remained in jeopardy, however, due to lethal encounters with people and an alarming loss of secure habitat, even in the park. As part of its Mission 66 campaign to expand and upgrade visitor facilities, the Park Service constructed Grant Village on the largely untouched western shore of Yellowstone Lake, promising to mitigate the project by removing aging facilities at nearby Fishing Bridge, a prime feeding area for grizzly bears further east along the lake. But the town of Cody, Wyoming, protested the closure, fearing it would discourage tourists from using the park's eastern entrance and divert them away from the town. Faced with pressure from prominent Wyoming politicians, park officials relented and merely closed a tent campground in the area, leaving the rest of the developed area unchanged. A lawsuit alleging that retention of the Fishing Bridge facilities would jeopardize the bears and their habitat in violation of the Endangered Species Act failed, however. By then, the Park Service understood that the Grant Village project would displace bears from nearby streams that harbored cutthroat trout, a key food source for hungry bears during the trout's spring spawning season. The entire episode represented a double whammy for the bear and its recovery despite federal legal protection.[8]

By the mid-1980s, the Yellowstone grizzly bear's plight was attracting congressional attention as bear numbers continued to fall and bear deaths mounted. Responding to a litany of complaints about federal resource-management practices throughout the Yellowstone region, two US House of Representatives subcommittees—one on Public Lands and the other on National Parks and Recreation—convened a joint hearing in October 1985 to assesses the situation in what they described as the "Greater Yellowstone Ecosystem," This congressional word choice helped legitimize the GYE terminology, which had recently been adopted by the newly created Greater Yellowstone Coalition and its allies in their efforts to reform regional conservation practices.

The upshot of the proceeding was a groundbreaking report identifying the grizzly bear as "the most important indicator of the Ecosystem's health" and lamenting the ongoing loss of bears in "black hole" areas. The report chastised the responsible federal agencies for a "fragmented approach to coordination," concluding that "in virtually all agency decision-making, the whole is subordinated to its fragments." The accompanying Congressional Research Service study explained that the Forest Service's timber, energy, and livestock grazing programs, along with new road construction, was endangering the grizzly population and other wildlife by too often bringing humans into close contact with animals. The committee concluded Yellowstone's bear problem required a more comprehensive, interagency management approach, namely better coordination among the responsible federal and state agencies. Put simply, the hearing gave further legitimacy to the GYE concept, effectively endorsed the ecosystem-management concept, and put political pressure on the agencies to revise their approach to grizzly bear management and related conservation concerns.[9]

In response, the federal agencies took several actions designed to aid bear recovery and improve resource-management practices across the GYE. The IGBC and a subsequently established Yellowstone Ecosystem Grizzly Bear Subcommittee (YES) directly addressed the mounting threats to bears related to timber harvesting, energy development, livestock grazing, and other activities occurring on the national forests surrounding Yellowstone. As recommended by the 1986 congressional report, the IGBC modified its Grizzly Bear Management Guidelines by imposing more rigorous resource-management requirements governing extractive activities and new road construction on federal lands. And the IGBC revised its related "Management Situation" zoning scheme, giving the bear priority in prime habitat areas by either excluding incompatible activities or requiring that they be mitigated.

Further, the Park Service and Forest Service announced a new Vision process initiative designed to improve federal resource-management activities and interagency coordination throughout the GYE. Although the multiyear Vision effort attracted considerable attention across the region and produced an enlightened initial draft document, it ultimately floundered, as we shall see, in the face of stiff local political opposition. Nonetheless, the otherwise disappointing final report committed the agencies to reducing road densities—an important step toward securing grizzly habitat, though not one the region's national forests readily followed.[10]

In 1993 the FWS issued a revised Grizzly Bear Recovery Plan, which described the challenges facing the remaining grizzly bear populations and provided a roadmap for removing the bear from the endangered species list. Under the ESA a recovery plan must contain "objective, measurable criteria" to enable the listed species to be removed from federal protection. To meet this requirement, the FWS established three monitoring criteria to measure grizzly population status in the Yellowstone and other ecosystems, namely the annual number of female bears with cubs over a six-year period, the distribution of these bears over that time period, and the annual number of human-caused mortalities. Because bears reproduce at a very slow rate—on average one or two cubs at three-year intervals—these population criteria provided a rough assessment of how the bears were faring.

Conservation groups disagreed with the plan, however, and went to court, arguing that the criteria failed to ensure that adequate habitat

Grizzly bear sow in Yellowstone National Park. Credit: Jim Peaco, 2016, Yellowstone National Park Photo Collection.

was protected, which was critical to the bear's survival. A federal judge agreed and ruled that the recovery plan must include quantifiable standards to measure effective grizzly bear habitat. But the judge stopped short of requiring the FWS to designate critical habitat for the grizzly bear or linkage zones connecting the separate Yellowstone and Glacier-area grizzly bear populations in order to promote genetic diversity within the isolated Yellowstone population. The decision confirmed that grizzly bear recovery in the Yellowstone ecosystem would require the agencies to protect the wide-ranging bear's habitat needs—an obligation that would be borne principally by the Forest Service and the logging, energy, and ranching interests that used the region's national forest lands.[11]

By then, it was evident that grizzly bear protection and recovery had become the central legal and management concern on the GYE federal lands. Under the ESA, often dubbed the "pit bull" of environmental laws, the FWS must be consulted by federal land-management agencies proposing any action that might jeopardize a protected species or adversely affect its critical habitat. In the GYE this consultation process effectively gave the FWS veto power over timber sales, road construction, oil and gas leases, and livestock grazing permits, though the agency has rarely invoked this power, choosing instead to suggest alternative approaches designed to mitigate problems. In addition, the ESA forbids anyone—whether an individual, corporation, or agency—from "taking" (or killing) a protected species, a prohibition that also prohibits modification of its habitat if a death is likely to result.[12]

In a concerted effort to safeguard the grizzly bear and its habitat, conservation groups turned to the courts. Invoking the ESA and the National Environmental Policy Act, they challenged a litany of timber sales and energy projects across the Northern Rockies, citing a growing list of scientific studies that concluded these disturbances along with the accompanying roads were anathema to bears.[13] In the GYE their efforts largely succeeded in reducing logging on the Targhee and other area national forests, protecting critical grizzly habitat and putting Yellowstone bear numbers on an upward trajectory.[14] Given the grizzly bear's role as an "umbrella species," these efforts also benefited an array of other species and generally improved ecological conditions across the region. But a related effort to reintroduce grizzly bears into Idaho's Selway-Bitterroot wilderness country, which would have facilitated connectivity between the region's dispersed bear populations, failed in the face of staunch local opposition.[15]

By the early 2000s the Yellowstone grizzly population was estimated at 500 bears, a number that prompted the FWS to consider removing these bears from the endangered species list, despite their continued status as an island population. The consequences of delisting the bear would be enormous, both for the bear and the ecosystem more generally. No longer would the Forest Service be obligated to consult with the FWS before approving new timber sales, energy leases, or roads. Similarly, the ESA's "no-take" prohibition would no longer restrain wary ranchers or sheepherders from shooting bears prowling near their livestock—a widely accepted practice in the days preceding the bear's federal listing. Moreover, the bear would become subject to state management, opening the door for sport hunting and potentially reducing the hard-fought habitat protections it enjoyed. Nonetheless, a decision to delist the bear would be quite popular in several quarters and serve as proof that the much-maligned ESA worked as a species recovery law.

Key to delisting the Yellowstone grizzly population was the so-called Conservation Strategy, developed by the FWS and the other responsible agencies to establish recovery criteria and management standards. It delineated a 9,210-square-mile Primary Conservation Area (largely on the region's federal lands) where bears were given priority; set guidelines for managing the bear population and its habitat; established mortality standards; and laid out protocols for monitoring bear numbers upon delisting. Utilizing these standards, the Forest Service amended the region's individual forest plans to better safeguard bear habitat within the Primary Conservation Area. The plan amendments obligated forest managers to maintain secure habitat at 1998 levels, reduce the number of developed sites, constrain motorized access, limit commercial livestock grazing allotments, and phase out existing sheep allotments. In ESA parlance, these standards and guidelines were designed to serve as "adequate regulatory mechanisms" to help sustain delisted bear numbers. At the same time, the states of Idaho, Montana, and Wyoming each developed bear-management plans to guide how they would manage bears outside the designated Primary Conservation Areas.[16]

With these plans in place, the FWS concluded in early 2007 that the once-teetering Yellowstone grizzly bear population was recovered and removed it from the federal endangered species list. The agency's decision rested on its estimate that bear numbers exceeded 600 animals and that the habitat protections now in place would continue to protect them. Citing the Conservation Strategy document and the state bear-management plans, the FWS concluded: "We are confident that these mechanisms provide an adequate regulatory framework within which the Yellowstone grizzly bear population will continue to experience population stability and be appropriately distributed throughout significant portions of the range for the foreseeable future."[17] Its confidence was not shared, however, by the region's conservation groups, who immediately sued in an effort to block the bear's delisting and retain it as a legal tool vital to protecting the GYE through the bear-driven ecosystem-management approach.

In a groundbreaking ruling the Ninth Circuit Court of Appeals affirmed a Montana federal district court decision that found the FWS had illegally sought to eliminate federal protection for the bear by ignoring the effect climate change was already having on the bear's long-term survival. The problem was evident across the GYE. Whitebark pine trees were dying at an alarming rate, due to an ongoing mountain-pine beetle and blister-rust infestation. Yellowstone's grizzlies, according to bear biologists, relied heavily on whitebark pine nuts as a critical late-season food source before hibernating for the winter. But these nuts were increasingly unavailable due to the ongoing infestation, which was linked to warming winter temperatures that no longer consistently dipped low enough to kill the culprit bugs. The biologists agreed that bears unable to find sufficient nourishment from pine nuts faced an increased mortality risk.

To delist a species under the ESA, the FWS must meet the same standards that govern the initial listing decision to ensure the animal is not trending toward extinction.[18] Citing the statute's "policy of institutionalized caution," the court overturned the FWS's delisting decision. It ruled that the agency did not adequately explain how the bear population was recovered in the face of a climate-driven decline in one of its primary food sources, which would bring the bears into increased—and potentially deadly—contact with people. Nonetheless, the court heralded the Yellowstone grizzly bear recovery effort as "a tribute to the comprehensive multi-jurisdictional cooperative effort between federal and state agencies, as well as private interest groups," which constituted a "substantial wildlife conservation planning achievement." On balance, the decision represented a significant victory for the Yellowstone grizzly bear, its advocates, and science-based management.[19]

Though the GYE grizzly bears were spared for the moment, no one expected the court ruling to be the last word on delisting. Yellowstone bear numbers were plainly on the upswing, the FWS wanted to demonstrate that the ESA worked to recover species, and the GYE states were intent on reclaiming control over the bear. To address the court's concerns, the FWS commissioned several scientific studies to assess the impact climate change and the declining whitebark pine nut food source would have on the GYE bears. The results vindicated the agency's earlier delisting decision: the bears would find substitute food sources that did not bring them into contact with people, and many bears did not even depend on the whitebark pine nut as a food source. Armed with these new studies, another FWS delisting proposal seemed imminent, even though the conservation community and others continued to fear the prospect of state bear management and trophy hunting.[20]

As the delisting discussion advanced, the GYE federal agencies and conservation groups expressed serious reservations about the state grizzly bear management plans. Having worked so hard to recover the Yellowstone grizzly population, they were leery of state bear-hunting plans, including the prospect of hunting immediately adjacent to park boundaries, in the linkage zones potentially connecting the Yellowstone bears with their northern cousins, and in the John D. Rockefeller Jr. Memorial Parkway situated between Yellowstone and Grand Teton. They also worried that the states could jettison the current, conservative bear counting method in favor of a less conservative method—an action that could result in an excessive number of bears killed legally.

To address these concerns, the federal agencies pressed the states to provide them a meaningful opportunity to participate in setting their final bear-management and -hunting plans. Although initially united in their opposition to delisting without further concessions from the states, this united federal agency front dissolved after the 2016 election, according to Dan Wenk, Yellowstone superintendent during the Obama administration. With Donald Trump on his way to the White House, the federal agencies concluded that the existing delisting proposal was "the best we can do" and acquiesced to it, with Wenk casting a lone dissenting vote in the IGBC's Yellowstone Subcommittee meeting. Wenk summed up the outcome in these terms: "It was amazing how the election changed the delisting issue, but the world didn't change for the bear."[21]

In June 2017 the FWS announced that it was again removing the Yellowstone grizzly population from the endangered species list. According to the FWS the bear population, estimated at 600–750 animals, met demographic recovery targets, had adequate secure habitat, and verged on overflowing the available habitat. More bears, in its view, would inevitably come into conflict with humans and among themselves; the Yellowstone ecosystem was now full. Under this second delisting proposal, a new Yellowstone Grizzly Bear Coordinating Committee—consisting of the same federal, state, and tribal members currently involved with the recovery effort—would oversee future bear-management, research activities, and financial needs, but the three GYE states would have the final word on bear management.

Although the ESA empowered the FWS to monitor the bear population, mortality, and habitat trends for five years with the authority to "relist" the bear if these trends turned downward, few expected the agency to take such drastic action given the politics involved. By then the states had announced their bear-hunting quotas, with Wyoming allowing twenty-two grizzlies to be shot in its upcoming trophy hunting season. Montana decided not to allow any grizzly hunting during the 2018 season, while Idaho limited hunting to just one bear. Although Wyoming imposed a relatively small no hunting zone adjacent to Grand Teton National Park, it did not do the same for Yellowstone, where many more bears prowled the landscape.[22]

Still largely opposed to delisting, conservation groups, joined by Native American tribes, again turned to federal court to block the states from assuming responsibility for the GYE grizzly bears. And once again, they succeeded, keeping the bears under federal control and also stopping an imminent bear hunt. Viewing the grizzly bear population as a whole, the courts faulted the FWS for not assessing the impact of delisting the Yellowstone bears on the other remnant Lower 48 grizzly bear populations, once more citing the ESA's statutory policy of "institutionalized caution." The courts also concluded that the Yellowstone bear population was genetically threatened by its isolation absent a firm commitment to

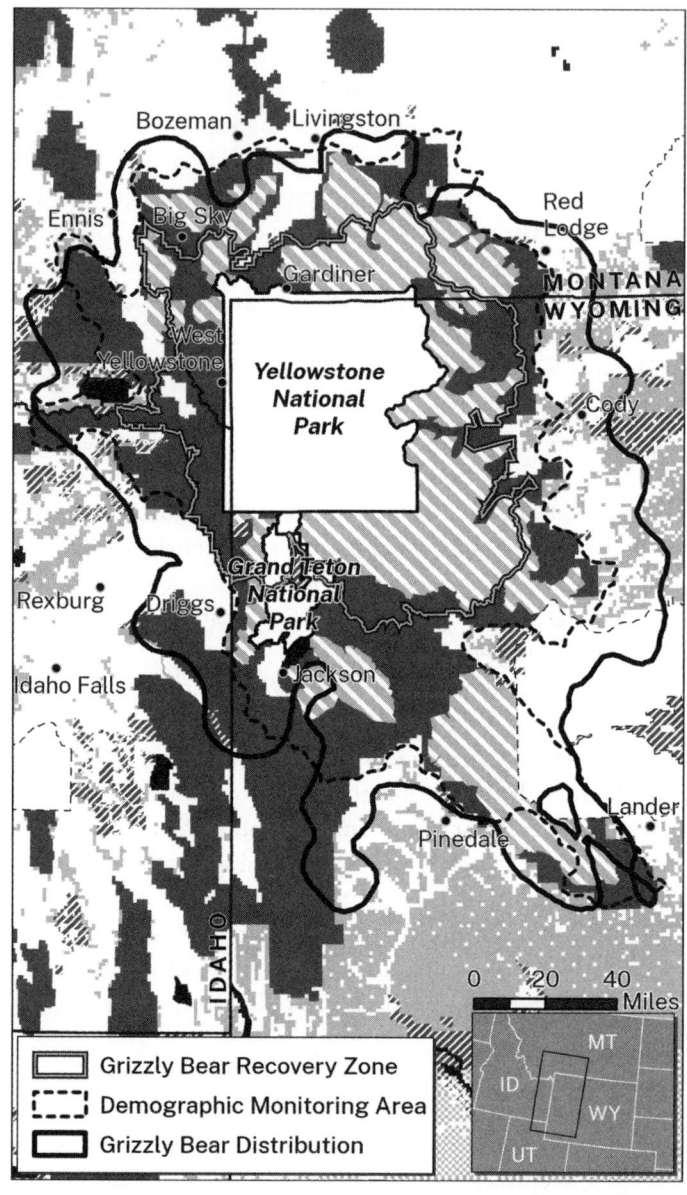

Yellowstone grizzly bear range. The Yellowstone grizzly bear population, now numbering roughly 1,000 bears, is presently protected under the Endangered Species Act. The three GYE states are seeking to remove—"delist"—the bear from federal oversight, which would return management responsibility for the bear to the states. The map shows the expanded scope of grizzly bear range compared to the original Grizzly Bear Recovery Zone and the designated demographic monitoring area. Map by the University of Utah DIGIT Lab.

translocate bears from elsewhere or that ensured the Yellowstone bears could naturally connect with their more northern cousins. Further, the courts ruled that the FWS's failure to require the states to recalibrate acceptable mortality limits in the event the states adopted a new bear counting method posed a real threat to the Yellowstone grizzly population. According to the district court, the FWS dropped the recalibration requirement "not on the basis of the best available science, as demanded by the ESA, but rather as a concession to the states in order to reach a [delisting] deal." Simply put, the ESA's focus on science—informed by conservation biology principles governing island populations—served to safeguard the Yellowstone grizzly bears by emphasizing the need to manage the region as an ecosystem that existed within an even larger, biologically connected landscape. Science rather than politics triumphed, at least in the courts.[23]

The states, predictably, objected to the judicial decision, asserting that the bears were not only recovered but had outgrown the ecosystem, testing the limits of social tolerance within local communities. Long committed to the North American Model of Wildlife Management, the states reconfirmed their intention to manage the bears under that model. They intended to license sport (or trophy) hunting based upon agreed-upon mortality thresholds designed to sustain a viable bear population and thus avoid relisting. Hunting, according to state wildlife managers, was an effective method to condition bears to avoid people, and the revenue generated from license fees would offset their increased management costs. In addition, the states agreed to translocate two bears into the Yellowstone population if the bears did not naturally reconnect by 2025. Animal rights groups and other conservation organizations held a different view, objecting on moral and scientific grounds to trophy bear hunting.[24]

Meantime, conflict intensified between ranchers and bear advocates in Wyoming's Upper Green River country where grizzly bear numbers were increasing along with livestock depredation incidents. A lawsuit seeking to block the Forest Service's proposal to reopen grazing allotments previously closed to protect wildlife failed to stop the agency, even as the appellate court found ESA violations.[25] In an expression of its overall frustration, the Wyoming legislature adopted an obviously invalid law authorizing grizzly bear hunting despite the animal's federally protected status.[26] And the Wyoming congressional delegation filed legislation to overturn the court ruling and to preclude any further judicial review of the Yellowstone grizzly bear delisting decision.[27] Taken together, these legislative actions represent another instance where local politics has collided with the science-driven ESA and the grizzly bear recovery effort.

In early 2021 the FWS completed an ESA-mandated five-year status review for the continental US grizzly bear population, finding that the bear still required federal protection as a threatened species. The independent scientists enlisted for the review concluded that none of the five grizzly bear populations in the continental United States were yet recovered, though the GYE bear population was considered resilient based upon a notable increase in population size and range. The FWS remained concerned, however, about the "uncertainty associated with the stressors of human-bear conflicts, human population growth, and potential reductions in connectivity [that] . . . represent a possible reduction in overall viability of the grizzly bear in the lower-48 States in the foreseeable future." Citing "enough future uncertainty associated with conservation efforts," the FWS found that the bear "remained likely to become in danger of extinction within the foreseeable future throughout all of its range."[28]

Two years later, in response to delisting petitions from Wyoming and Montana, the FWS reversed course and announced it would again review the listing status of both the Yellowstone and Glacier region grizzly bears. Finding that the petitions contained "substantial information that the population size and trends have improved and that threats have been reduced," the agency concluded that delisting "may be warranted," while expressing concern about recently enacted state laws affecting the bears.[29] In 2022, anticipating possible delisting, the Montana legislature had amended the state's hunting laws to allow anyone to shoot a grizzly bear "threatening to kill a person or livestock." Bear scientists and others immediately criticized the new statutory language, which did not define the term "threatening," for its vagueness that seemed to invite bear shooting.[30] A subsequent Montana grizzly bear management plan addressing the state's hunting plans, which would not commence until five years after delisting, did little to assuage the scientists' concerns over the bear's future under state management.[31]

Meanwhile, the IGBC has reassessed its bear-counting method and underlying assumptions, concluding that the actual number of Yellowstone bears likely exceeds 1,000—a figure contested by several bear advocates. If the number is correct, Wyoming appears poised to authorize the hunting of thirty-nine bears annually upon delisting. On the ground, the Yellowstone- and Glacier-area grizzly populations are getting closer to connecting with one another, reportedly having moved within thirty-five miles of one another. Further frustrated by the FWS's delayed response to its delisting petition, Wyoming has turned to the federal courts, seeking a judicial ruling that the Yellowstone grizzly population has recovered.

Without evident objection from the Park Service or FWS, Montana has transplanted two bears from the Northern Continental Divide population into the GYE, hoping to remove genetic diversity concerns from the delisting debate. For now, despite the ongoing political and legal pressures from the states, the Yellowstone grizzly bear remains under federal jurisdiction, though the courts will likely have the final word on the bear's legal fate absent congressional intervention.[32]

Simply put, the grizzly bear recovery effort represents a key step in the movement toward ecosystem management in the GYE. It underscores the critically important role played by the science-based ESA in that effort; its firm, substantive mandates protecting the GYE grizzly population have prompted a major interagency coordination effort involving federal, state, and tribal officials that is framed and implemented in ecosystem terms. The ESA and related litigation have notably reshaped resource-management priorities across the region, reducing timber cutting, energy leasing, livestock grazing, and road construction in the national forests. Ongoing efforts to connect the Yellowstone bear population with its Glacier-area cousins have expanded GYE conservation efforts to a broader, landscape scale and brought private landowners into the effort. This landscape-level conservation effort is not only critical to the Yellowstone grizzlies' long-term survival, but has laid the groundwork for strategies necessary to secure historic wildlife migration routes and to combat the worst effects of climate change on the GYE's wildlife. Although politics has plainly affected grizzly bear policy—most notably the federal-state "deal" involving the 2017 delisting proposal—the courts have properly focused on science and the law when reviewing the delisting decisions. While the three GYE congressional delegations have sought political intervention, Congress has thus far resisted that request, confirming that grizzly recovery is a national and not merely local political concern. In sum, the grizzly bear remains of transcendent importance in the GYE, a widely recognized symbol of the region's wilderness character and a manifestation of its ecological condition.[33]

Wolves: Restoring the Ecosystem

The haunting howl of a wolf at dusk leaves a memorable impression on any Yellowstone visitor. The sight of wolves, perhaps afield on a hunt or pups at play outside a den, makes an equally deep impression—one of wildness and nature unbounded. Indeed, people come from throughout the world for an opportunity to glimpse or hear a Yellowstone wolf. But it is only during the past thirty years that wolves have again inhabited the park and the surrounding terrain, where they remain often unwelcome. The centerpiece of a highly controversial federal ecological restoration initiative, the Yellowstone wolves have repopulated the GYE, bringing with them significant ecological, economic, political, and legal ramifications. An endangered species success story in several quarters, the wolves remain a despised intruder among area ranchers, hunters, and others within the three GYE states. Regardless, their presence has transformed the GYE into a setting where all the species present when Euro-American settlers first arrived are afield and natural processes are operating at a grand scale.

Originally omnipresent throughout the American West, the Lewis and Clark Expedition frequently encountered wolves during their journey of discovery. In their journals the expedition leaders regularly reported seeing "vast assemblages of wolves," "a great number of wolves about this evening," and "a Den of young wolves and a number of Grown Wolves in every direction."[34] Other early westward travelers recounted similar scenes. John James Audubon, describing his 1843 travels up the Missouri River, wrote: "when the Buffaloes fall or cannot ascend and then died, the Wolves are seen in considerable numbers feeding upon them."[35] Francis Parkman, in his writings on the Oregon Trail, recounted what occurred after he killed a nearby buffalo: "that night the wolves made a fearful howling close at hand, and in the morning the carcass was completely hollowed by these voracious feeders."[36]

As settlers arrived with their livestock, the wolf was soon deemed a menace to their livelihood and worse. Having observed its predatory instincts and efficiency, Teddy Roosevelt memorably demonized the wolf as "the beast of waste and destruction." In short order, the western states mounted an all-out extermination campaign against wolves, offering both public and private bounties that ranged from $1 to $15 per hide, and sometimes ran into the hundreds of dollars. But unable to eliminate wolves from the landscape, the states joined with the livestock industry to enlist the federal government in the effort, sealing the wolf's fate.[37]

With federal funding and technical assistance, the wolf eradication campaign swung into high gear, extending even into the new national parks. In fact, the Park Service eagerly embraced the extermination effort in order to protect its "good" animals—elk, deer, moose, and pronghorn—from depredation so visitors could readily view them. The agency's efforts were well received by its neighbors, who did not appreciate having a wolf haven nearby. By 1927 wolves were eliminated from

Yellowstone and a few years later were gone from the rest of Wyoming as well as Montana and Idaho. Grounded in hyperbole and myth, the wolf eradication campaign made the western landscape safe for cattle, sheep, and big game animals, never mind its ecological effects, which were not part of the discussion at this time. The campaign was, in short, regarded as a stunning success.[38]

Soon, though, scientists began to reassess the wolf and its role on the landscape as well as the accepted wisdom underlying the eradication policy. Paradoxically, once the wolf was gone from Yellowstone, the Park Service announced that it would no longer engage in predator control on its lands. In the mid-1940s renowned biologist Aldo Leopold contrite over earlier instinctively shooting a wolf, proposed "restock[ing] Yellowstone" with wolves. In the years following, studies by several respected scientists, including Sigurd Olson and Adolph Murie, offered new insights into predator-prey relationships and the ecological role played by wolves. During the late 1960s the Leopold Report—overseen by Aldo Leopold's son, A. Starker Leopold—prompted the Park Service to reassess its wildlife-management policies and to refocus on "maintaining and when necessary reestablishing indigenous plant and animal life." A related report on federal predator-control policies, also prepared by Starker Leopold, recommended that all native animals be valued and that control efforts be targeted only at troublesome individuals rather than deployed indiscriminately. Gradually, the wolf was being restored to biological if not political respectability.[39]

With passage in 1973 of the Endangered Species Act, the federal government reversed course and soon endorsed wolf recovery as official policy. By then the only wolves remaining in the continental United States were found in northern Minnesota; none remained in the western states. Once the wolf was listed under the ESA as a federal endangered species, the US Fish and Wildlife Service was obligated to prepare a Wolf Recovery Plan, which immediately met with political opposition. Following 1982 amendments to the ESA, the FWS could take advantage of a new "experimental population" reintroduction provision known as section 10(j), designed by Congress to facilitate controversial species restorations. The new provision authorized the FWS to designate a species slated for reintroduction as either "essential" or "nonessential" and to reduce the level of legal protection it enjoyed. Invoking section 10(j), the FWS proposed in 1987 to reintroduce a "nonessential experimental population" of wolves into Yellowstone by translocating wolves from Canada to the park and central Idaho. By then, the so-called Magic Pack had strayed across the US-Canadian border and taken up residence in Glacier National Park, representing the first wolves to inhabit the Rocky Mountain region in more than half a century. Other wolves soon followed, enjoying full legal protection as endangered species.[40]

The FWS's reintroduction proposal drew immediate, deeply felt opposition as well as a similar level of support. Despite the new, more flexible legal status available under the recovery plan, the GYE states and others adamantly opposed bringing the wolf back into the region. State officials generally objected to more federal oversight of wildlife and potential restrictions on economic activity. Ranchers feared for their livestock as well as more regulatory restrictions; many hunters and outfitters opposed the reintroduction, believing wolves would reduce the area's big game herds; and the extractive industries despaired over access limitations and other tighter federal restraints. Deep cultural, moral, and spiritual beliefs also motivated opponents, who saw the return of the wolf as undermining their heritage and the world as they understood it.

Wolf reintroduction proponents saw the matter differently, extolling the wolf as a key missing ecological cog—or "keystone species"—whose absence left the region biologically impoverished. Wolves, in their view, were an essential element of wilderness, and their presence would stir new economic activity by those eager to glimpse these legendary animals. Many also felt a moral imperative to atone for the injustice wolves had suffered during the extermination period. These passionate, conflicting views embraced an array of economic, political, and personal values, which were also reflected in the split between national and local perspectives, with local residents more likely to oppose the reintroduction proposal.[41]

In the face of such bitter contention, the region's congressional delegations fended off wolf reintroduction for several years through budgetary riders, study committees, and other maneuvers. But eventually wolf advocates prevailed, and Congress attached a rider to an appropriations bill directing the Secretary of the Interior to prepare an Environmental Impact Statement (EIS) on wolf reintroduction to Yellowstone and central Idaho. In 1994 the FWS released its long-awaited draft EIS, which proposed employing section 10(j) to reintroduce wolves into Yellowstone National Park and the central Idaho wilderness areas. The proposal called for translocating wolves from Canada and designating them a "nonessential experimental population," thus eliminating potential land-use restrictions and enabling ranchers to shoot wolves found actually depredating on livestock. The Defenders of Wildlife (a staunch reintroduction proponent) sweetened the federal proposal by establishing a Wolf Compensation Fund to help ranchers cover livestock losses. The draft proposal generated a record 160,000 mostly supportive comments, the most ever received on a federal conservation initiative. Under the proposal, recovery would be achieved when three separate wolf populations with ten breeding pairs of wolves had inhabited the recovery area for three consecutive years, and the three states had in place wolf-management plans deemed federally acceptable.[42]

Once the FWS gave final approval to the wolf-reintroduction plan, eight Canadian wolves were soon on their way to Yellowstone, oblivious to the Wyoming Farm Bureau's last-ditch lawsuit that failed to block their return to the park. The wolves arrived at the park's northern gateway on January 12, 1995, accompanied by Secretary of the Interior Bruce Babbitt, attesting to the national significance of the event. Several hundred enthusiastic onlookers along with a cluster of news reporters greeted the wolf caravan. Babbitt summed up the occasion: "We are showing our children that restoration is possible, that we can restore a community to its natural state." As planned, the wolves—soon joined by six others—spent ten weeks in holding pens to acclimate them to their new environment before finally being released into the park—a "soft release" process. The courts were not finished, however. A Wyoming federal judge subsequently ruled that the section 10(j) reintroduction violated the statutory "experimental population" requirements since some wolves already inhabited the park. On appeal, his decision was soon overturned, putting the final legal stamp of approval on this remarkable species recovery saga.[43]

Within a few years, the reintroduction proved a stunning success, consistent with the various projections that proceeded it. Because of their brisk reproductive rate—litter sizes annually average 4–6 cubs per alpha female—the GYE wolf population grew steadily, numbering eighteen packs and more than 177 wolves by 2001. Their presence provided scientists with an extraordinary opportunity to study the animal and its ecological impact. Several early studies concluded that the wolf's presence helped restore the ecosystem to health, reestablishing vital predator-prey relationships with cascading ecological effects. Willows, aspen, and other vegetation that elk had heavily browsed were reappearing along with absent beavers and songbirds.[44] Livestock losses proved relatively light, while big game herd counts remained constant. Visitation to Yellowstone notably increased once wolves were on the ground, supporting the original projections that wolf-related tourism would annually add $7–$10 million to local economies.[45]

Mike Phillips, Jim Evanoff, Director of USFWS Molly Beattie, YNP Superintendent Mike Finley, and Secretary of the Interior Bruce Babbitt carrying first crate with wolf in it to the Crystal Bench pen, January 12, 1995. Credit: Jim Peaco, Yellowstone National Park Photo Collection.

With wolf numbers mounting steadily due to the species' rapid reproductive rate, the GYE states were soon exhorting the FWS to remove the wolves from the endangered species list. Although the original Northern Rockies Wolf Recovery Plan set a goal of ten breeding pairs of wolves in each of three zones (Yellowstone, central Idaho, and northern Montana), the FWS had modified the plan in 1994 to require a total population of 300 wolves for three consecutive years with evidence of genetic interchange between the subpopulations. By 2000 the Northern Rockies wolf population exceeded 300 wolves and thirty breeding pairs spread across the region, but there was yet no evidence of genetic interchange between the separate populations. Nonetheless, faced with intensifying pressure from the GYE states, the FWS proceeded in 2008 to designate the Northern Rocky Mountain wolves a "distinct population segment" and removed them from federal protection, even though Wyoming's wolf-management plan treated wolves as predatory animals subject to being shot on sight throughout much of the state.[46]

Wolf advocates lost no time taking the FWS's delisting decision to court, where they secured a preliminary injunction blocking the proposed delisting. According to Montana federal district court judge Donald Molloy, the FWS did not demonstrate genetic connectivity between the Greater Yellowstone wolf population and the more northern wolf populations as required in the recovery plan. Moreover, the court found that Wyoming's revised wolf-management plan, which designated nearly 90 percent of the state as a shoot-to-kill predator zone and relied largely on the two national parks to meet future wolf population goals, did not satisfy the state's obligation to maintain fifteen breeding pairs of wolves. Noting little change in Wyoming's management-plan revisions, the court observed: "Armed with the same information, the [FWS] flip flopped without explanation" when it approved the revised plan. Just as the powerful, science-based ESA originally facilitated the federal wolf-reintroduction initiative despite local opposition, it now stood as a roadblock to terminating federal management for similar antiwolf political reasons.[47]

The FWS, intent on shedding federal responsibility for at least some of the wolves, responded to the court ruling by peeling off the Montana and Idaho wolves and delisting them as a "distinct population segment." Another lawsuit ensued, again before Judge Molloy, who was growing ever more familiar with the legal, biological, and political dimensions of wolf recovery under the ESA. The FWS's decision to divide the Northern Rockies wolf population into state-based subpopulations presented him with a new legal issue: whether the ESA allowed the agency to divide the growing regional wolf population into smaller segments for purposes of delisting. Noting that the FWS had consistently treated the Northern Rockies wolves as a single "distinct population segment," Judge Molloy's answer was a resounding "no." While acknowledging that the FWS's division of the region's wolves into subpopulations was a "pragmatic" response to Wyoming's inadequate wolf-management plan, the judge concluded it was "at its heart a political solution that does not comply with the ESA." Once again the science-based ESA carried the day, while highlighting the landscape-scale dimensions of the GYE wolf-recovery effort and the need for interstate coordination.[48]

Stymied by the courts, the GYE states turned to Congress for assistance to unyoke themselves and the wolves from federal management. Not only were state officials beset by complaints from the politically powerful ranching community over wolf-related livestock losses, but the equally influential hunting community bemoaned the drop in elk numbers in several locations. When a settlement proposal engineered by members of the conservation community was rejected in court for failure to bring everyone on board, Senator Jon Tester of Montana and Congressman Mike Simpson of Idaho added section 1713 as a rider to the Defense Continuing Appropriations Act for 2011, directing the FWS to reissue its wolf delisting rule for Montana and Idaho while adding that "such reissuance shall not be subject to judicial review."[49] The rider effectively overrode the ESA and sealed the wolves' fate, at least in Montana and Idaho. A legal challenge asserting a separation of powers argument—that Congress was unconstitutionally intruding on the judiciary in pending litigation—failed when the courts concluded that Congress was within its right to amend its own law. Judge Molloy held a quite different view of Congress's action: "Section 1713 sacrifices the spirit of the ESA to appease a vocal political faction." Consequently, the wolves were now delisted in Montana and Idaho, and wolf hunting soon resumed in both states.[50]

Still saddled with federally protected wolves, Wyoming filed suit in Wyoming federal court seeking approval of its hybrid trophy-predator zone wolf-management approach. The court obliged, ruling that the FWS could not require that the entire state be designated a trophy game area, which essentially endorsed Wyoming's plan to treat the wolf as a predator across most of the state.[51] Once Wyoming modified its wolf-management plan to specify the state's northwestern quadrant—15 percent of the state—a trophy hunting area and committed to maintaining a wolf population of 100 animals and ten breeding pairs, the FWS again proposed delisting the Wyoming wolves. More litigation by wolf proponents ensued, this time in Washington, DC, federal courts, but ultimately failed. Soon thereafter, the FWS delisted the Wyoming wolf population, vesting management responsibility with the state.[52]

State control over the Northern Rockies wolves has not abated controversy, however. Outside of Yellowstone and Grand Teton national parks, as contemplated during the delisting imbroglio, the states are managing the wolves under the North American Wildlife Management model, which makes them fair game for hunters and only obliges the states to maintain minimum population levels. Wolf hunting immediately outside the parks has particularly agitated wolf supporters and biologists. Several radio-collared wolves were soon shot outside Yellowstone, disrupting ongoing scientific research, reducing wolf-viewing opportunities for park visitors, and prompting negative media attention and public anger. Although Montana originally responded to these concerns by reducing its wolf quota from fifteen to three animals north of the park, the other two states did not do the same. Park scientists have suggested establishing wolf sanctuary areas adjacent to the parks where hunting would be

Aerial of the Leopold Wolf Pack, Yellowstone National Park. Credit: Yellowstone National Park Photo Collection.

restricted, arguing that the proposal would not significantly affect hunting opportunities or control over livestock depredation. The states have disregarded the proposal, even though better coordination between the two parks and the states would help to quell public concern and to assist research efforts, which could improve wolf-management practices.[53]

That does not seem likely, however, as antiwolf sentiment remains strong within the GYE states, which are intent on driving the wolf population down. Since the delisting, wolf numbers have increased in Montana and Idaho, which each host more than 1,000 wolves, while Wyoming's wolf population outside the national parks hovers around 350 wolves. Livestock depredation continues to be perceived as a problem, despite state compensation programs to cover these losses. Whether wolves are significantly impacting big game numbers is a hotly contested matter; in 2020 Montana's elk population exceeded the state's target population goals, but hunters have complained about localized herd declines. Regardless, the powerful ranching and hunting communities have successfully importuned Montana and Idaho to loosen wolf-management restrictions.[54]

In 2021 both states adopted legislation authorizing the killing of wolves up to the ESA-dictated minimum population floor of 150 wolves and fifteen breeding pairs. Montana also extended its wolf-trapping season, allowed the use of snares and spotlights, and will reimburse wolf hunters and trappers for their costs, creating a quasi-bounty system according to critics.[55] Reversing its earlier policy, the Montana Fish and Wildlife Commission dramatically increased wolf-hunting quotas adjacent to Yellowstone. As a result hunters shot twenty-five park wolves—20 percent of the park's wolf population—during the early weeks of Montana's 2021–22 hunting season, plainly irritating Yellowstone's superintendent who implored Montana's governor to intervene.[56] Idaho no longer limits the number of wolf tags per individual, permits year-round hunting and trapping on private lands, authorizes the usage of private contractors, and sanctions aerial shooting as well as the use of ATVs and snowmobiles to chase down wolves.[57] Public opposition to these new wolf-hunting laws was strong in both states, including former wildlife managers who argued they violated widely accepted wildlife-management practices and sportsmanship norms.[58]

With these aggressive hunting and trapping laws, the states opened a new chapter in the GYE wolf saga. Convinced that the states could no longer be trusted to manage this controversial animal, conservation groups petitioned the FWS to restore the Northern Rocky Mountain wolves to the endangered species list, citing the threat liberalized hunting posed to the wolf population and the absence of regulatory mechanisms to safeguard the wolves. The FWS responded by rejecting the relisting request, finding that the expanding western US population of wolves was resilient and thus not endangered despite the new Montana and Idaho wolf-hunting laws.[59] To no one's surprise, wolf advocates turned to federal court, contending that the FWS's relisting denial ignored the impact the liberalized state wolf-hunting laws were having on the wolf population. In separate federal lawsuits, conservation groups also challenged the wolf trapping and snaring provisions in the Montana and Idaho laws, arguing that they violated the ESA by imperiling—or illegally "taking"—federally protected grizzly bears. The courts agreed and limited wolf trapping to areas and periods when grizzlies were likely denning for the winter.[60] Meantime, Montana has released a draft wolf-management plan that contemplates a statewide target population of 450 wolves, but endorses no specific seasonal or geographic mortality quotas.[61] Were the courts to order the FWS to relist the Northern Rockies wolves, the restoration saga will have come full circle, with the tug of war between the science-driven ESA and local political sentiments beginning again.

In early 2024 the wolf saga took another turn in Wyoming, one that has injected humane treatment concerns into the state's wolf-management policies. A local news outlet triggered a public storm of protest after reporting an incidence of evident animal cruelty. The incident involved a Sublette County resident who had recently run down a lone wolf in the state's predator-management zone with his snowmobile, captured the severely injured animal, and then took it to a local bar where he put it on public display before eventually shooting it outside the bar. Although Wyoming officials condemned the man's actions, predators like wolves are not covered by the state's animal cruelty laws, so the culprit was merely fined $250 for live possession of wildlife. The incident and the state's tepid response sparked national as well as international outrage, including calls for a tourism boycott and a congressional bill to outlaw the use of snowmobiles to run down wolves. Whether Wyoming will revise its predator management or animal cruelty laws in response appears unlikely after a legislative panel decided not to move forward with such legislation. Nonetheless, the incident squarely calls into question the adequacy of these laws, while also raising a potential ESA issue involving the "adequate regulatory mechanism" listing criteria. It also calls into question the state's sensitivity to the issue of animal cruelty—a matter of obvious concern to many people involved in the ongoing GYE wolf-management debate.[62]

The GYE wolf-reintroduction and subsequent delisting controversy offers important lessons in the realms of ecology, law, and politics. As an ecological matter, wolf restoration has been a true success. Not only is a robust wolf population now roaming the GYE, its presence has restored historic predator-prey dynamics and fostered related ecological changes across the landscape. Visitors to Yellowstone and Grand Teton eagerly seek the opportunity to observe the wolves, an experience that both thrills and enriches them, while also generating important local tourism revenues. The ESA proved the driving force behind the federal reintroduction initiative and then insulated it from initial, politically motivated efforts to return wolf management to the states. The courts were also instrumental throughout the controversy, refusing to separate the wolves into subpopulations and requiring proof of genetic interchange for delisting purposes. The judicial rulings interpreting the science-based ESA can only be read as endorsing the ecosystem-wide dimensions of the restoration effort. But in the end, powerful local political forces triumphed, securing congressional approval to delist the wolves in Montana and Idaho. And local politics, still awash in antiwolf sentiment, is again driving the states to revise their wolf-control laws in ways that endanger the restored wolf population as well as the states' own hard-won management role.

Other lessons have also emerged from the GYE wolf-reintroduction initiative. The ESA has proven a powerful tool for promoting landscape-scale conservation that extends beyond the GYE. The courts and the FWS (at least initially) both interpreted the ESA to link the three separate Northern Rockies wolf populations together for recovery purposes, including the exchange of genetic material between the populations. As interpreted, the ESA legitimizes the landscape conservation concept and embeds it in law. Wolf management at the landscape level will require meaningful coordination between the states and with the GYE national parks. But that is not occurring. The Wyoming plan, predicated on separate trophy and predator zones, stands in sharp contrast to the Montana and Idaho wolf-management plans. Only Montana has shown any willingness to coordinate with the national parks—where wolves are managed protectively with visitors in mind—by originally adjusting its hunting quotas adjacent to the park boundary. But that concession did not continue in the wake of the state's revved-up wolf-hunting legislation. In fact, the wolf-sanctuary idea of restricting hunting on lands immediately proximate to the parks has fallen on deaf ears.

This lack of coordination puts the wolf-restoration accomplishment in the Northern Rockies in apparent jeopardy, given Montana's and Idaho's liberalized wolf-hunting laws, along with Wyoming's predator-based management law that has kept its wolf numbers at a minimum level and brought animal cruelty concerns into the wolf debate. A severely reduced wolf population largely confined to the two GYE parks is not sustainable over the long term. This unassailable biological fact poses a looming ESA legal violation while also reaffirming the need for a coordinated ecosystem-management approach to the region's wolves.

* * *

Grizzly bear and wolf restoration efforts are entwined in the GYE as we ponder the future for these two extraordinary yet highly controversial animals. The impending effort to again delist the GYE grizzly bears from federal protection cannot ignore what has transpired politically at the state level in the aftermath of the wolf delisting. Because trophy hunting remains a major sticking point in the grizzly bear delisting debate, the Montana and Idaho accelerated wolf hunting laws along with the Wyoming predator laws counsel caution whether the states can be trusted to oversee the bears and constrain hunting. Conversely, the current, joint federal-state grizzly bear management approach provides a positive model of coordinated management that could greatly improve management of the delisted GYE wolf population. There is no apparent reason why the states cannot work with the Park Service to sustain a viable park wolf population while also addressing legitimate local concerns. The same holds true for interstate coordination; failure to meet wolf or grizzly bear population commitments in one state could lead to relisting of the species across the three GYE states. If the GYE states are to be responsible for these two high-profile species that are so closely identified with the GYE's special wild character, then they must demonstrate more of a commitment to an enlightened and coordinated management approach.

In sum, the GYE grizzly bear recovery and wolf restoration stand as remarkable, though still-tenuous, ecological management achievements. The science is clear that sustaining these charismatic apex predators over the long term requires a minimum viable population with regular reproduction, secure habitat with adequate food sources, and opportunities for genetic exchange with nearby populations. The law, namely the ESA, is also clear that the GYE grizzlies and wolves merit federal protection until fully recovered and their long-term survival ensured. The politics governing the GYE grizzlies and wolves are not so clear, however, as reflected in the steadfast local pressure for state management notwithstanding strong national support for protective federal management under the ESA.[63] At bottom, the controversy over the GYE grizzlies and wolves is grounded in profound individual social, moral, and political beliefs that transcend biology and law.[64] The challenge, then, is to acknowledge and respond to these beliefs in order to achieve a level of social tolerance that will enable these wild animals to persist in the GYE. A coherent, coordinated, and humane ecosystem-based management approach would go far toward meeting this challenge as well as overarching biological and legal requirements.

CHAPTER FOUR

Maintaining Migratory Wildlife
Bison, Elk, and Pronghorn

The GYE landscape abounds in hoofed wildlife—bison, elk, pronghorn, moose, and other ungulates—regularly referred to as "charismatic megafauna." Once ubiquitous across most of the American West, these large, wide-ranging mammals stand as an enduring symbol of the region's frontier heritage. Since its inception Yellowstone National Park has served as an important refuge for bison and elk, providing them with a historic sanctuary as well as legal protection against hunting. In fact, Yellowstone's wild bison, facing extinction during the waning days of the nineteenth century, represent an impressive early wildlife-restoration achievement, one that has preserved the nation's only genetically pure bison herds. Elk likewise found refuge in the park with distinct elk herds growing in number over the years and providing animals to restock depleted elk numbers elsewhere. Both species regularly enthrall park visitors during summer months, which masks a harsher reality when they migrate outside the park to escape notoriously severe winters by moving to lower-elevation lands in quest of nourishment.

Indeed, the migratory instincts engrained in the GYE's bison, elk, and pronghorn offer tangible evidence of the need for a coordinated, ecosystem-based management approach to conserve these eminent wildlife populations. The animals blithely tread across the boundary lines we have imposed on the landscape, unaware that by crossing the national park boundary they are now subject to state authority and fair game for hunters. The fact that the region's bison and elk herds are infected with brucellosis—a bacterial disease transmissible to domestic livestock with serious economic consequences—further complicates their migratory presence outside the national park setting. But bison are treated much differently than elk when they venture outside Yellowstone. In Montana the park's bison are constrained, often tested for disease, and even slaughtered, treated more like cattle than free-roaming wild animals. Elk, on the other hand, are free to roam at will and to frequent state feedgrounds in Wyoming, the beneficiary of a powerful, politically connected hunting constituency. Lately, the region's Native American tribes have reconnected with their buffalo heritage and offered Yellowstone's bison a new sanctuary on their reservations, while Wyoming's feedgrounds are under attack. In short, the political and legal challenges confronting the GYE's migratory bison, elk, and pronghorn remain formidable in the region's jurisdictionally fragmented landscape.

Bison: In Political and Legal Limbo

Before Euro-Americans set foot in what is now the western United States, bison herds teemed across the Great Plains and well beyond. The bison—known by early French trappers as "buffalo" and by the Lakota as "tatanka"—provided the Plains tribes with food, shelter, clothing, weaponry, and other daily essentials, bestowing the animal with a mythological status in native culture. During the Lewis and Clark Expedition, Captain Merriwether Lewis, writing in his journal on June 3, 1805, observed that "the country in every derection around us was one vast plain in which innumerable herds of Buffalow were seen attended by their shepperds the wolves." Scientists estimate that the bison population then likely exceeded 30 million animals, with travelers reporting it often took days for the massive herds to pass into the distance.[1]

The bison was doomed, however, as settlement pushed relentlessly westward. Recounting his 1832 journey up the Missouri River, artist George Catlin described the dispiriting changes he encountered: "Nature has nowhere presented more beautiful and lovely scenes, that those on the vast prairies of the West, and of *man* and *beast*, no nobler specimens than those who inhabit them—the *Indian* and the *buffalo*—joint and original tenants of the soil, and fugitives together from the approach of civilized man; they have fled to the great plains of the West, and there, under an equal doom, they have taken up their *last abode*, where their race will expire, and their bones will bleach together." Catlin responded to this unfolding tragedy by proposing "A *nation's Park*, containing man and beat, in all the wild and freshness of their nature's beauty," widely regarded by historians as the first reference to the national park idea.²

As western settlement proceeded, abetted by completion of the transcontinental railroad system, the once-abundant bison herds were soon brought to the brink of extinction. Buffalo Bill Cody and other market hunters after valuable buffalo robes, joined by rail passengers taking sport at shooting bison from the moving trains, slaughtered the seemingly plentiful bison without apparent thought. In 1873 Colonel Richard Dodge reported from his base in Fort Dodge, Kansas: "the air was foul with a sickening stench, and the vast plain, which only a short twelve months before teemed with animal life, was a dead, solitary, putrid desert." Efforts in Congress to stop the slaughter failed when the military blocked proposed legislation, arguing that eliminating the buffalo would aid their efforts to keep the Indians on federal reservations. Congress did, however, enact legislation in 1872 creating Yellowstone National Park, which soon became the bison's last refuge.³

In the early 1890s, as the western bison herds dwindled to fewer than 1,000 animals, 200 or so bison sought refuge in Yellowstone, where they still faced rampant poaching by wayward local residents. As the new park's bison numbers dipped to just twenty-three animals, Yellowstone's military caretakers not only took aggressive action against the poachers, but also supplemented the herd with bison imported from Texas and Montana. They constructed the Buffalo Ranch in Lamar Valley to care for and breed bison behind secure fences, while putting the animals on display for visitors to enjoy. Though saved from extinction, the park's bison were treated much the same way ranchers managed their cattle, which by then, ironically, had replaced the bison across the western landscape. The frontier was deemed closed, with the buffalo and the Indian both removed from the landscape and relegated to federal reservations.⁴

The newly created Park Service arrived in Yellowstone in 1916 following passage by Congress of the National Parks Organic Act. Expressly charged with conserving wildlife, the Park Service continued to corral Yellowstone's bison and to control the herd size to meet the park's perceived range-carrying capacity. As bison numbers grew, reaching more than 1,000 animals by 1930, park officials continued culling the population; by 1967, it is estimated, more than 9,000 bison had been slaughtered and removed from the park population. These practices, however, changed following the 1963 Leopold Report and the fundamental shifts it brought to the agency's wildlife-management policies with its call to maintain and restore natural biological conditions. Having already closed the Buffalo

Bison corralled in Yellowstone National Park during early twentieth century. Credit: Albert Noyes, photographer, Yellowstone National Park Photo Collection.

Ranch in 1952, Yellowstone ceased culling the herd and adopted a hands-off approach to its wildlife, allowing the park's bison to roam freely with minimal human intervention. Faced with fending for themselves during Yellowstone's harsh winters, the bison soon discovered that more suitable seasonal habitat was readily available beyond the park boundary on nearby lower-elevation national forest and privately owned ranchlands.[5]

Though merely following their natural instincts, the bison carried with them the *brucella abortus* bacteria that posed a threat to domestic livestock present on these same adjacent lands. Animals exposed to *brucella* bacterium—primarily through the ingestion of infected reproductive material—can develop brucellosis, a troublesome disease that can cause spontaneous abortion in the host animal. For ranchers, this can be costly. Besides losing a valuable yearly calf, the rancher's herd can be quarantined and potentially slaughtered to eliminate the disease threat. A brucellosis outbreak also puts a state's ability to sell its livestock on the interstate market at risk of embargo. By recent estimates, roughly 60 percent of the GYE's more than 5,000 bison test seropositive for the brucellosis organism, but more accurate testing reveals that far fewer bison are actually infected with brucellosis, and even fewer female bison carry it in their reproductive tracts—making the risk of disease transmission relatively low.[6]

Ironically, Yellowstone's bison are believed to have originally contracted brucellosis from nearby domestic cattle, but the park's bison have become the villain in what has become an acrimonious, decades-long wildlife disease controversy. Although Yellowstone's bison have never been found to have transmitted brucellosis to a local cattle herd, the Park Service has been forced to institute an array of intensive bison-management policies reminiscent of the Buffalo Ranch days in an effort to prevent intermingling between park bison and local cattle. And the irony does not stop there. The GYE elk herds are also infected with brucellosis, and elk have been deemed responsible for at least twenty-seven brucellosis outbreaks in area cattle herds, yet the elk have been largely ignored as a disease vector.[7]

As it has unfolded, the Yellowstone bison-brucellosis controversy has been driven more by the region's jurisdictional fragmentation and political considerations than by prevailing scientific realities. By the mid-1980s the end seemed within reach for a prolonged, multibillion-dollar federal-state brucellosis eradication campaign designed to rid the nation's cattle herds of the disease. Led by the US Department of Agriculture's Animal and Plant Health Inspection Services (APHIS), which was vested with the authority to restrict statewide the interstate shipment of cattle following an outbreak, campaign officials turned their attention to the Yellowstone region, where brucellosis persisted in the area's bison and elk. Fearing potential export restrictions, Montana adopted legislation in 1991 that reclassified bison as "a species in need of management" and transferred management responsibility for bison that posed a contagious disease threat from the state's wildlife agency to its Livestock Board. No longer did Montana regard bison exiting Yellowstone to be wildlife; rather, the state was legally treating them as livestock. At the same time Montana adopted a zero tolerance policy, subjecting Yellowstone's bison to being shot once they crossed into the state, whether on national forest or private lands. The state then implemented this policy by sanctioning a bison hunt outside the park, but it soon turned into an abysmal slaughter scene, triggering a loud public backlash that forced Montana officials to abruptly cancel the hunt.[8]

Federal and state officials, each bound by seemingly irreconcilable legal mandates, struggled during the 1990s to find common ground in an effort to resolve this complex transboundary resource-management problem. Amidst a litany of lawsuits, they managed to adopt a series of interim bison-management plans that pleased hardly anyone. In one lawsuit the state of Montana sued the Park Service and APHIS, alleging that Yellowstone officials were obligated to aggressively manage the park's bison to eliminate the risk of disease transmission, and that APHIS was illegally threatening to downgrade the state's brucellosis status. Although the two federal agencies operate under quite different statutory mandates, the case was settled when the parties committed to implement intensive bison-management policies pending completion of a long-term bison-management plan. The Park Service agreed to begin capturing, testing, and slaughtering its bison to prevent them from leaving the park—a significant concession to Montana and APHIS that ran counter to the agency's hands-off wildlife-management policy. Predictably, conservation groups responded by suing the Park Service, asserting that the agency was violating its Organic Act conservation responsibility by agreeing to kill the very animals it was charged with protecting. The courts disagreed, however, ruling that the Organic Act did not prohibit the Park Service from removing individual bison to control herd size and to prevent the animals from leaving the park.[9]

In 2000 the Park Service, APHIS, the Forest Service, and Montana's wildlife and livestock agencies finally managed to reach agreement on an Interagency Bison Management Plan (IBMP), which would govern bison management in the northern GYE. Although based on a risk-management strategy, the plan set a long-term goal of eradicating brucellosis from GYE wildlife populations. To do so, the IBMP established a herd target size for Yellowstone's bison of 3,000 animals, limited when and where bison would be tolerated outside the park, continued the controversial test and slaughter program inside the park, endorsed an experimental bison-vaccination program, and included adaptive management protocols. To reduce the risk of disease transmission, the plan instituted a spatial and temporal zoning system designed to keep bison separate from cattle. Some bison would be allowed to exit the park but would have to be back inside the park before cattle were released onto the range for the spring grazing season. The plan approved the use of hazing to help drive the bison back into the park, prompting some to subsequently complain about inhumane hazing practices. And the plan permitted some bison to be quarantined, tested for brucellosis over an extended period, and then transported elsewhere upon testing negative for brucellosis.[10]

From the outset the IBMP appeared contradictory, scientifically flawed, and insensitive to Yellowstone's vaunted bison population. The long-term goal of eradicating brucellosis from GYE wildlife populations is difficult to square with the plan's bison control, risk-management strategy. In fact, an eradication goal seems ultimately unattainable given the prevalence of the disease in the park's elk population, a reality ignored in the plan. The plan's dubious bison-vaccination experiment has since been rejected by the Park Service following an environmental assessment that concluded the vaccine was unproven and would be impossible to administer in the park's rugged terrain. Pointedly, the plan did not contain any measures designed to manage the region's livestock more intensively, even though cattle are more easily managed than wild bison. Readily available options included delaying the date when cows were turned out on the range to eliminate the likelihood of contact with bison as they seasonally moved back into the park or altering national forest grazing allotments to eliminate intermingling opportunities. Instead, the IBMP mandated intensive management of Yellowstone's bison during their instinctual seasonal migration outside the park, affording them only limited tolerance and condemning many to slaughter. Moreover, the plan completely ignored Native Americans and their historical connection to the bison.[11]

Nonetheless, the IBMP is noteworthy as a federal-state interagency effort to address the transboundary nature of the bison issue, essentially reflecting a collaborative ecosystem-based management approach, albeit a flawed one. In 2009, responding to tribal concerns, the original Interagency Bison Management Group expanded its membership by adding three Native American tribal representatives, further recognizing the expansive scale of the bison issue. Following adoption of the IBMP, APHIS helped to relieve some of the political pressure around brucellosis transmission concerns, when it revised its rules governing a state's brucellosis-free status to allow states to subdivide themselves and focus quarantine protocols on the area affected by a disease outbreak. Each of the three GYE states have since established designated surveillance areas that serve to confine the impact of a brucellosis outbreak to the immediate area. Montana's entire livestock industry is thus no longer at risk in the event of an outbreak in the GYE, which has reduced the political pressure to eradicate (rather than control) brucellosis in the region's wildlife populations.[12]

Since the IBMP's adoption, much has changed for Yellowstone's bison that has gradually helped to improve their situation on the ground. There have been no recorded cases of brucellosis transmission from bison to cattle in the area, even as Yellowstone's bison population has grown on

Bison near Mud Volcano in Yellowstone National Park, 2012.
Credit: Diane Renkin, Yellowstone National Park Photo Gallery.

average to 5,000 animals—far exceeding the IBMP's 3,000 target number. Although the Montana legislature reinstated public bison hunting in 2003 and several Native American tribes have asserted their treaty hunting rights on public lands outside the park, efforts to control the population through renewed hunting have plainly proven insufficient. Although Montana law imposes fair chase requirements on hunters, bison hunting outside Yellowstone still often resembles a firing line or shooting gallery scene, prompting continued public objections to hunting as an acceptable means to control population numbers. Besides, faced with a phalanx of hunters at the park boundary, many bison retreat back into the park where they are safe. One apparent but still contentious option would expand bison hunting opportunities more broadly across the landscape by allowing bison more room to roam.[13]

As time has passed, bison have been afforded access to additional habitat outside the park, reducing utilization of the plan's capture, test, and slaughter protocols. Conservation groups—working with the Park Service, the state of Montana, and others—have secured easements and leaseholds on key ranchlands, opening thousands of acres of habitat for bison north of the park. Bison are therefore now able to graze as far north as Yankee Jim Canyon in the Gardiner Basin, roughly 13 miles beyond the park border. West of the park, in response to a Citizen Working Group proposal, Montana's governor opened 300,000 acres to bison year-round following the retirement of key national forest grazing allotments, permitting the animals to roam onto Horse Butte in the Lake Hebgen region. By reducing the risk of contact between bison and cattle, the park's bison have gained some freedom to follow their natural instincts, no longer strictly confined by ecologically artificial boundary lines.[14]

Moreover, bison translocation has surfaced as a nonlethal means to help control the Yellowstone bison population, though Montana originally objected to moving possibly diseased bison from the park to elsewhere in the state. But in 2010 the state relented and allowed eighty-seven bison certified as brucellosis-free to be transferred from federal quarantine facilities to billionaire Ted Turner's nearby ranch, which was also a haven for wildlife. The transfer represented the first time the state had allowed any bison to be transported outside of the park, setting the stage for a broader bison translocation program that has provoked additional controversy.[15]

Enter the state's Native American tribes, deeply interested in reconnecting with their cultural and spiritual heritage by restoring bison to their reservation lands. In 2012, in response to a tribal request, sixty-one disease-free bison were transferred from Yellowstone's quarantine facility to the Fort Peck Indian Reservation, where the tribes maintained a certified quarantine facility. The following year, another thirty-four disease-free bison were transported to the Fort Belknap Indian Reservation. Unhappy with these transfers and distrustful of tribal intentions, ranching interests sued to block them. But the Montana Supreme Court responded by ruling that the state had adequate statutory authority to transfer disease-free Yellowstone bison to these Indian reservations, seemingly paving the way for additional transfers as a means to relieve the GYE's bison population pressures. A year later, in 2014, twenty-eight tribal nations from across the United States and Canada signed "The Buffalo: A Treaty of Cooperation, Renewal, and Restoration," pledging to jointly work toward restoring bison on their lands and in their cultures.[16]

The road forward has not been entirely smooth, however. After consulting with tribal organizations to gauge their interest in receiving Yellowstone bison, the Park Service completed an environmental assessment in 2016 recommending that the park's bison be allowed to be transported outside the area without first being tested and quarantined.

Instead, the bison would be translocated directly to the Fort Peck reservation's quarantine facility for disease testing and surveillance before being made available to help restore bison on tribal, federal, and private lands in Montana and elsewhere.[17] In 2017 the National Academy of Sciences reviewed the GYE brucellosis problem and released a report endorsing this approach.[18] The Montana legislature balked at this option, however, and killed a bill that would have allowed untested bison to be transferred within the state. Because the park's quarantine facilities had limited capacity, only a modest number of bison—roughly eighty animals annually—were eligible for transfer, which relegated the remaining captured bison to possible slaughter.

Despite these notable developments under the IBMP, the most controversial aspects of the plan remain unchanged, and park bison continue being killed merely for following their instincts. Efforts to revise the IBMP have thus far proved unavailing, though adjustments have been made by mutual agreement as circumstances changed on the ground. Consequently, the plan's target population number of 3,000 bison remains unchanged, along with the plan's capture and slaughter policies and related hazing practices that have injured and even killed some bison. In extreme winter years, between 600 and 1,200 bison have been sent to slaughter, while more than 10,000 Yellowstone bison have been sent to slaughter or killed by hunters since the IBMP's inception. Groups opposed to these practices have sought judicial relief in several lawsuits, but without notable success except in one instance when the federal agencies agreed to prepare a supplemental environmental analysis to the IBMP after the Blackfeet tribe announced it would begin hunting Yellowstone bison.[19]

Climate change and warmer winter temperatures could change things, however. During the unusually mild 2020–21 winter, when fewer bison than normal left the park, no bison were captured for test and slaughter purposes, and the following winter less than fifty bison were killed outside the park. But two years later, the unusually harsh 2022–23 winter reminded everyone that weather remains variable, the result being roughly 1,200 bison perished outside the park. Yet, if warming trends continue, climate changes may accomplish what bison advocates have been unable to do: stop the capture, test, and slaughter practices directed toward park's migratory bison.[20]

With efforts to revise the IBMP at a standstill, the Park Service has expanded its "Bison Transfer Conservation Program" in an effort to further reduce resort to the slaughter option. With financial assistance from conservation groups, Yellowstone has increased its quarantine facilities to accommodate nearly 250 bison, almost tripling the number of animals it can handle on-site.[21] In early 2022, the Custer Gallatin National Forest finalized its revised forest-management plan, which identified bison as a keystone species and "allows for expanded tolerance of bison on the national forest," while committing to improve bison habitat. Also in 2022 the Park Service announced it was preparing a new "Bison Management Plan for Yellowstone National Park" and an accompanying Environmental Impact Statement. While reaffirming its commitment to working with the IBMP group, the agency stated that scientists believe the park contains sufficient forage to support 5,000–8,000 bison, signaling that it no longer supported the original IBMP goal of 3,000 bison. Further, the initial management alternatives identified in the announcement endorsed the use of additional quarantine facilities and the need to redirect bison hunting well beyond the immediate park boundary area.[22]

There was good reason to suspect, however, that Montana would not be receptive to the park's impending plan. In 2020 Montana had completed its own Bison Management Plan under the state's Environmental Policy Act, which essentially approved the blossoming bison-transfer and restoration efforts statewide.[23] But the plan was challenged in court by the United Property Owners of Montana (UPOM), a group of ranchers and other landowners opposed to emergent bison restoration efforts across the state, including the high-profile American Prairie reserve initiative that was actively acquiring ranchlands and bison in an effort to restore a frontier landscape in central Montana's Missouri Breaks country.[24]

After the Montana governorship changed hands in the 2020 election, the state's new Republican administration swiftly settled the lawsuit. The state, agreeing with UPOM's dubious allegations of procedural violations, proceeded to vacate its Bison Management Plan and committed to not preparing another EIS on bison restoration for ten years.[25] At the same time, the Montana legislature redefined and narrowed the term "wild bison" in the state's statutes, thus expanding the state Department of Livestock's jurisdiction over bison, and it granted county commissions across the state the power to review and reject any state proposal to transplant wild bison into the county.[26] While the legislation excludes sovereign tribal entities from its coverage, it nonetheless reflects a further hardening of the state's position toward bison relocation efforts.

When released in mid-2024, the Yellowstone Bison Management Plan and attendant EIS represented a clear departure from the 2000 IBMP, reflecting new information and developments during the intervening years. The purpose of the new plan, according to the Park Service, "is to preserve an ecologically sustainable population of wild, migratory bison while continuing to work with partners to address brucellosis transmission, human safety, and property damage and fulfill tribal trust responsibilities." Acknowledging its jurisdiction ends at the boundary line, the plan indicates the Park Service intends to continue coordinating with its IBMP partners. But citing the absence of any brucellosis transmission events, the plan increases the acceptable bison population number to 3,500–6,000 animals, and provides the park will manage bison similarly to how Montana manages elk—the source of all GYE brucellosis transmission events—in nearby Paradise Valley. Noting the Park Service's responsibility to tribal partners, the plan will expand the bison-transfer program and increase bison hunting opportunities outside the park as larger numbers of bison migrate there. It also endorses adaptive management enabling officials to adjust the plan in response to changing conditions, and it retains the test and slaughter option though at a much lower level. Thus, the new plan rests upon the notable changes that have occurred since the 2000 IBMP was adopted, including the fact that the brucellosis transmission risk is quite low, the tribes have a profound interest (which was ignored twenty years ago) in the matter, and it supports the park's basic noninterventionist management policy.[27]

The reaction from Montana, as expected, has been strident and unreceptive to the Park Service's plan. An earlier letter from the governor and other state officials commenting on the draft plan expressed their disapproval of the three alternatives, asserting that the Park Service had disregarded the state's concerns and failed to collaborate during preparation of the draft plan. Specific concerns included increasing the targeted bison population level above 3,000 animals as agreed to in the IBMP, disregarding contrary science regarding the brucellosis transmission risk, not committing to a bison-vaccination program, ignoring the impact increased numbers of bison could have on elk, and the absence of concrete standards or timelines for various management actions. The state's letter also suggested it would consider eliminating or reducing the bison tolerance zones outside the park, effectively taking bison management back to pre-2000 days when any bison exiting the park was at risk of being shot. In response to the final plan, Montana's governor chastised the Park Service for failing to collaborate during the planning process and has sued to block implementation of the plan. Bison advocates, on the other hand, generally supported the plan while pressing to afford bison more space outside the park, eliminate the test and slaughter practice, and further improve the tribal role.[28]

The state of Montana, however, may not have the final word on the fate of bison outside Yellowstone. In 2014 the Buffalo Field Campaign and others filed a petition with the US Fish and Wildlife Service seeking to protect Yellowstone's unique, genetically pure bison under the Endangered Species Act, asserting that the park's bison population constituted an imperiled species that merited federal protection. Although the FWS has twice rejected the petition, federal courts have twice ordered the agency to reconsider its decision, concluding in each instance that the agency irrationally rejected a scientific study documenting important genetic differences within the park's bison herds and calling into question the viability of the smaller herd. Were the ESA listing petition to be granted, it would relieve the state of any authority over the park's bison when they venture beyond the boundary.[29]

To date, the state of Montana has taken only modest steps to address the related elk brucellosis problem, even though elk have been identified as the transmission culprit in more than twenty-five cases of cattle brucellosis.[30] What explains this seemingly illogical differential treatment between the region's bison and elk? Certainly not science: According to the National Academies of Sciences report, "additional aggressive control measures in bison seem unwarranted" until the elk brucellosis problem is addressed.[31] Which suggests that it is politics, as driven by Montana's powerful ranching and hunting interests.

Indeed, Montana's hunters, guides, and outfitters have a major stake, economically and recreationally, in the state's elk herds, giving elk a powerful constituency not enjoyed by the bison. This political reality is reflected in the actions of the Montana legislature, which initially shifted responsibility for Yellowstone's bison from the state's wildlife agency to its livestock agency, and since then has tightly controlled efforts to translocate the park's excess bison. The obvious solution, as noted by the National Academies' report, is meaningful, region-wide coordination among the agencies designed to comprehensively address the bison, elk, and cattle dimensions of the disease problem. In short, the debatable bison problem cries out for an ecosystem-based management approach.

Until then, the nation's largest, genetically pure, free-ranging bison herd remains at risk, still treated more like livestock than treasured wild animals, symbols of our frontier heritage and revered by Native American tribes. Though Yellowstone's bison are infected with brucellosis, the risk of disease transmission from bison to cattle has proven negligible, given the absence of any such cases. With cattle now largely absent from federal and private lands outside Yellowstone, the park's bison have enjoyed greater freedom to roam during the winter months. The Park

Service's recent expansion of its quarantine facilities provides another safety valve for migratory bison. Yet park bison still face the prospect of the unseemly test and slaughter practices that have been the principal means for addressing the disease transmission risk. The Park Service's finalized Yellowstone Bison Management Plan should improve this situation, especially if Montana can be brought on board. Moreover, as tribal interest in bison restoration has grown, the option of translocating disease-free bison from Yellowstone to Native American reservations or tribal quarantine facilities has emerged as a viable alternative to slaughtering the park's wayward bison. The bison translocation program reaffirms the park's long-standing role in wildlife restoration, acknowledges the historic Native American connection with the park, and also advances Indigenous cultural and historic ties to this fabled animal.

Elk: Natural Regulation, Feedgrounds, and Disease Management

Elk have long been a major presence in the GYE, while elk management has posed a number of challenges. Overall, the GYE hosts somewhere between 30,000 and 40,000 elk, making it the most abundant mammal in the ecosystem. Roughly 10,000–20,000 elk grouped in six or seven herds spend summer months in Yellowstone National Park, many of which then migrate outside the park as winter arrives. Within the park the Northern Range and Jackson elk herds constitute the largest herds and have stirred the most controversy. As the park's elk population mounted during the early twentieth century, park managers felt compelled to regularly cull excess animals from the herds to sustain range conditions.

By the late 1960s, however, the agency had reversed course and stopped actively managing the park's elk, allowing nature to regulate the population. Elk numbers soon mounted, alarming some observers that the animals were devastating their habitat due to mismanagement, yet the Park Service stood by its revised hands-off policy. But once wolves arrived in the park, the elk noticeably altered their behavior. Although protected from hunting while in the park, the elk are fair game for hunters under state law when they leave the park, following ancestral migratory instincts. Today, ecosystem conservation efforts to protect these migratory pathways present some of the GYE's thorniest management challenges.

By all accounts, elk were abundant across the western landscape when the region was initially explored and then settled. The Lewis and Clark journals contain repeated references to large elk herds during the expedition's westward journey along the Missouri River: "This scenery already rich pleasing and beautiful was still farther heightened by immense herds of Buffaloe deer Elk and Antelopes which we saw in every direction feeding on the hills and plains." Other early explorers offered similar observations. In 1837 trapper Osborne Russell reported the Yellowstone Lake area was "swarming with elk" during his trip there. But as settlement proceeded, market hunters and early residents decimated many of the region's large elk herds, just as occurred with the bison. Once Congress established Yellowstone National Park, it seemed the area's wildlife would be secure in the new park, since the legislation directed the Secretary of the Interior to "provide against the wanton destruction of the fish and game found within said park, and against their capture or destruction for the purposes of merchandise or profit."[32]

Nevertheless, the park's elk were still subject to hunting for sport and subsistence, as well as by market hunters, who reportedly killed thousands of elk during the park's earliest years. In 1883, to stop the widespread killing, the Secretary of Interior issued regulations "to prohibit absolutely the killing, wounding, or capture at any time of any bison, buffalo, moose, elk" within the park. Though that slowed elk losses in the park, local residents continued to poach elk, bison, and other wildlife, despite efforts by the park's early military caretakers to stop them. Following national publicity highlighting the capture of a recalcitrant local poacher by a winter army patrol, Congress was persuaded in 1894 to adopt the Lacey Act, designed "to protect the birds and animals in Yellowstone National Park." By the turn of the century, elk and other wildlife, except predators, were relatively secure inside the park.[33]

Whether elk were initially abundant in Yellowstone has stirred passionate debate in response to the Park Service's evolving management approaches. Some scientists have argued that few elk were present during the park's early days, while others find convincing evidence that large numbers of elk made the park home. As settlement proceeded outside the park, traditional elk migratory routes were severed, limiting their migration opportunities and confining more elk to the park even during wintertime. By the early 1900s settlement in the Jackson Hole area had so badly plugged annual migration routes that elk were literally starving in plain sight, prompting the community to initiate a supplemental winter-feeding program and to support establishment of the National Elk Refuge. Once wolves were gone from the park and other predators subdued, elk numbers were no longer constrained by depredation, and the population continued to grow. By the early 1920s, the Northern Range elk population had reached an estimated 15,000 animals, prompting some to suggest they were overrunning the park and damaging range conditions.³⁴

The perceived overabundance problem focused particularly on the Northern Range, which Yellowstone officials concluded was being badly overgrazed and could not support so many elk. Hunting the elk as they migrated seasonally from the park helped to reduce the population, but park officials believed it did not adequately control the burgeoning elk numbers. Drawing upon the still-nascent discipline of range management, the Park Service embarked upon an aggressive population-control program, annually directing rangers to shoot "excess" elk to maintain the range's preconceived carrying capacity. Between 1934 and 1967, more than 67,000 elk were dispatched from the Northern Range herd; 13,735 elk were shot by rangers, while hunters killed 43,400, and the remainder either perished from winter conditions or were transshipped to restock elk in other states. In 1962 alone, more than 4,600 elk were slaughtered by Park Service marksmen. The prevailing policy was to manage aggressively to achieve what park officials believed necessary to maintain healthy conditions on the Northern Range, consistent with the prevailing "balance of nature" theory that then held sway among ecologists.³⁵

By the early 1960s the annual cull had become increasingly controversial while its scientific underpinnings were being seriously questioned. Television images of rangers shooting—many would say slaughtering—the park's elk evoked a loud public outcry, enough to move Secretary of Interior Stewart Udall to enlist a group of scientists led by A. Starker Leopold to review the policy. The ensuing Leopold Report, as noted, prompted a complete overhaul of National Park wildlife-management policy, including the reversal of Yellowstone's culling practices. Although the Leopold Report did not reject the notion of active management or of culling excess animals to sustain ecological conditions, the Park Service interpreted it to largely let nature take its course without human intervention. Thenceforth, Yellowstone's elk were allowed to proliferate limited only by weather, forage conditions, and hunting outside the park.³⁶

As elk numbers multiplied, critics were soon convinced that this new natural regulation policy was unworkable within Yellowstone's limited confines. Noting that the Northern Range elk herd grew from less than 5,000 elk in 1969 to roughly 19,000 animals in 1994, the Park Service's critics asserted that range conditions were being depleted to the point that the park's ecology was unalterably changed. They also contended that the absence of wolves and other predators exacerbated the problem. Of course, local hunters were the beneficiaries of the enlarged elk popu-

Bull elk in Yellowstone's Lamar Valley during winter, 1998. Credit: Jim Peaco, Yellowstone National Park Photo Collection.

lation; they annually lined the park's northern boundary during hunting season to shoot elk as they migrated from the park to forage at lower elevations. A 1997 congressionally funded report on the condition of the Northern Range largely rejected these criticisms, however, and affirmed the natural regulation policy. According to the report, the range was not ecologically damaged, and elk population fluctuations were consistent with "density dependence" theory, namely that the population naturally fluctuated in response to the number of elk present at any time.[37]

As the twenty-first century got underway, the debate over the Park Service's management policies and Northern Range conditions took another sharp turn following reintroduction of wolves to the park. Once wolves were returned to the park in 1995, they promptly set about predating on the abundant elk, just as scientists had predicted. The elk population declined sharply, dropping from roughly 12,000 in the late 1990s to around 5,000 by 2020. Elk behavior also changed as the park's elk dispersed from open spaces and riparian areas to higher-elevation, vegetated areas in an effort to protect themselves from the wolves. The result, according to several scientists, has been an ecological transformation—or "trophic cascade" in the language of ecology—on the Northern Range. Aspen and willow have regenerated, which has prompted the return of beavers as well as songbirds attracted to these new habitat conditions.

But other scientists are reluctant to link these ecological changes entirely to the wolves, arguing that hunter harvest levels, other predators, and the weather are also key factors in the continuing evolution of range conditions.[38]

Beyond these ecological consequences, the elk–wolf connection prompted related legal and political changes. As the Northern Range elk numbers declined, local hunters, who had long enjoyed a surfeit of elk to shoot, lamented the growing absence of elk to hunt. Montana officials were soon compelled to reduce annual elk-harvest quotas adjacent to the park. In response, Montana's hunters, guides, and outfitters— joined by the state's ranching community that had long opposed the wolf reintroduction—turned to the state's congressional delegation, urging them to legislatively remove the wolves from federal endangered species protection. In 2011, Congress obliged, turning wolf management over to the state, which promptly implemented a wolf-hunting season. Although the state initially limited wolf hunting adjacent to the park, it changed course in 2021, revising its wolf-hunting laws and adopting more robust wolf-harvest quotas. Exactly how this will affect the park's wolf population over the long term and, in turn, its elk population and related ecological conditions continues to unfold.[39]

Meanwhile, another elk-management debate is underway, one connected to wildlife diseases. The GYE elk, like the park's bison, carry brucellosis, and elk have been identified as the sole culprit in multiple brucellosis outbreaks within local cattle herds. Thus far, the state of Montana has dealt with elk brucellosis by creating a disease surveillance area in southwestern Montana in an effort to contain the disease, which it has gradually expanded as more distant elk have tested positive for brucellosis. At the same time, the state continues to impose severe limitations on bison leaving the park, even in the aftermath of the 2017 National Academies of Sciences report, which concluded that further restrictions on bison were futile until authorities effectively addressed the elk brucellosis problem. To complicate matters, chronic wasting disease (CWD) has made its way across Wyoming into the GYE, and has arrived in both Yellowstone and Grand Teton national parks. This untreatable and always fatal disease is believed readily transmitted among congregating animals, which can decimate previously healthy elk and deer herds. Ironically, according to some scientists, a vigorous wolf population could help to disperse GYE elk and thus reduce disease transmission opportunities, but the GYE states are intent on reducing wolf numbers. In any event, these twin elk-disease problems reinforce the need for a coordinated ecosystem-based management effort among federal and state officials as well as between the three GYE states.[40]

Elk-management issues are also prevalent in the southern reaches of the GYE, involving the Park Service, National Elk Refuge managers, the Forest Service, and the state of Wyoming. Somewhere between 40,000 to 50,000 elk populate western Wyoming, including the Jackson elk herd that consists of about 11,000 animals. Historically, these elk herds migrated seasonally from the region's high mountain country to more hospitable habitat at lower elevations during the winter months, but these migratory pathways were increasingly cut off once settlers began arriving during the late nineteenth century. In the case of the Jackson elk herd, settlement in the valley blocked its traditional southward migration routes, diverting the animals into the community where many perished from starvation or to nearby ranchlands where the stressed animals consumed valuable winter-stored hay. Faced with this dire situation, Jackson residents began feeding the starving elk and then implored Congress to establish a refuge to sustain them through the winter months. In 1912 Congress obliged by establishing the National Elk Refuge (NER), which is managed by the US Fish and Wildlife Service (FWS). Since then, the FWS has fed the elk during the winter months, a practice that today attracts thousands of tourists who thrill to see the congregated elk up close. Though the NER and its supplemental feeding program may have saved the Jackson elk herd, it also legitimized a most unnatural practice—artificial feeding of wildlife—in the midst of the largely wild GYE area.[41]

As time passed, other elk herds in the southern GYE presented similar problems for area residents, prompting the state of Wyoming to also adopt a winter-time elk-feeding program on state-run feedgrounds. As western Wyoming filled with ranches and towns, the elk found themselves displaced from their winter habitat and increasingly unable to navigate historic migration routes, tempting them to turn to ranchers' stored haystacks for nourishment meant for cattle. After Wyoming, in 1939, approved a wildlife damage law obligating the state to pay ranchers for these elk-related losses, the Wyoming Game and Fish Department embraced the NER model and began establishing winter feedgrounds. With twenty-two state feedgrounds spread across western Wyoming on a mix of federal, state, and private lands, the Department now feeds about 14,000 elk annually. The state-run feedgrounds have not only eliminated most elk–cattle conflicts, but also support a large elk population, which pleases the state's politically powerful hunting community and generates important economic activity in the region. Nonetheless, the feedgrounds represent another unnatural addition to the landscape as well as a troublesome disease reservoir.

In fact, the National Elk Refuge and the state feedgrounds—where roughly 21,000 elk congregate during the winter months—have become prime disease incubators. The crowded feeding conditions enable brucellosis and other diseases to persist and spread among the elk and bison that feed there. The risk of transmission between animals is actually exacerbated on the smaller state feedgrounds, where the animals are in closer contact with one another. At the NER the FWS is seeking to curtail and eventually eliminate wintertime feeding but, facing pressure from Wyoming, has been reluctant to commit to a specific timetable for this change. Wyoming, on the other hand, asserts that its feedgrounds help to prevent the spread of brucellosis to still-wild elk and to keep elk away from ranch cattle and haystacks. And the feedgrounds help ensure elk are plentiful during the fall hunting season. Yet, several brucellosis outbreaks in local cattle herds have been traced to the state's feedgrounds, and scientists fear the feedgrounds will accelerate the spread of CWD, which has steadily made its way westward across the state.[42]

Beginning in the early 1990s, these federal and state-run feedgrounds have prompted rancorous litigation focused on prevailing management practices. First, following a brucellosis outbreak in a local cattle herd, Wyoming federal and state courts rejected the affected rancher's lawsuits claiming his cattle contracted brucellosis from mismanaged wildlife, though the federal judge chastised the NER and National Park Service officials for not being sufficiently aggressive in controlling the disease. Second, a conservation group successfully sued the FWS to block an annual bison hunt on the NER, convincing the court that federal officials had disregarded the effect of the refuge's winter-feeding program on bison. Third, troubled by the prevalence of brucellosis among wildlife on the NER, Wyoming sued refuge officials for prohibiting the state from vaccinating elk on the NER against brucellosis, arguing that the state had authority over wildlife found within its borders. Although the courts rejected Wyoming's vaccination argument, the appellate court admonished the federal agencies for "threaten[ing] the wellbeing of a neighboring sovereign's livestock or game industry."[43]

Amid this flurry of litigation, the federal agencies began drafting a bison and elk management plan for the NER and the nearby national parks. The overarching issue, according to conservation groups, was whether the refuge was prepared to phase out its supplemental feeding program. The federal agencies' answer in the 2007 plan was a disappointing "no." While acknowledging the disease and ecological risks associated with the concentrated feeding of some 7,500 elk as well as the need to transition away from supplemental feeding, the plan deferred that decision indefinitely. Instead, noting local economic concerns, the plan focused on collaboration with state wildlife managers and called for additional hunting opportunities along with habitat enhancements to reduce the number of elk feeding on the refuge during winter months.[44]

Dissatisfied by the absence of a firm phase-out commitment, conservation groups took the FWS to court. They argued that the laws governing refuge management required the FWS to "conserve" wildlife and to "ensure that the biological integrity, diversity, environmental health of the [system] are maintained"—obligations that the agency was ignoring by continuing the unnatural practice of feeding elk and bison. In a 2011 decision, however, the court rejected these arguments, at least insofar as requiring the FWS to establish a date certain to phase out its supplemental feeding program. But the court did not stop there, concluding its opinion with a pointed message to refuge officials. Noting that "the agencies are committed to ending supplemental feeding," the court then, quite tellingly, remarked on "the Secretary's duty to end a practice that is concededly at odds with the long-term health of the elk and bison in the Refuge."[45]

Since then, the refuge has continued its supplemental winter-feeding program, while elk and bison hunts remain annual rituals in the area. Vaccination is no longer an option, as Wyoming discontinued its refuge vaccination program when it proved ineffective at reducing the incidence of brucellosis in the elk. Meantime, refuge officials have begun spreading wintertime feed more broadly across the landscape to reduce elk density on the feedlines and stopping supplemental feeding earlier in the year to encourage elk to disperse before the birthing season begins. These revised management strategies are calculated to minimize the risk of disease transmission by reducing the opportunity for elk to congregate and forcing them to look elsewhere for food. Recent studies suggest, however, that supplemental feeding is not an effective management tool for wild elk, while other studies conclude that neither supplemental feeding nor vaccination or test and slaughter strategies are cost effective. In 2018 brucellosis was again detected in local cattle for the first time since 2004, further calling the strategy's effectiveness into question.[46]

Citing changed conditions and new scientific information, the FWS announced in mid-2023 that it would prepare a new bison and elk management plan for the NER that would address supplemental feeding, disease management, and habitat conditions. Pending at the time was another lawsuit by conservation groups, who had returned to court in a further effort to establish more natural conditions on the NER. Noting that twenty years had passed since the court sent refuge officials a clear message that supplemental feeding practices were inconsistent with their wildlife-management obligations under existing law, the conservation groups requested a court order setting a firm phase-out date. Asserting that the refuge has illegally given Wyoming a veto power over its elk-feeding plans, they further argued that the arrival of CWD requires a clear phase-out plan. Regardless of this litigation, climate change could assume a prominent role in the feedground controversy. Warmer winter conditions should shorten the time supplemental feeding is required to maintain the animals, reducing the time elk and bison congregate during birthing season when the risk of disease transmission is greatest. Yet, despite the refuge's revised strategy designed to reduce supplemental feeding and discourage elk congregation, winter elk numbers on the refuge are actually increasing.[47]

Even if supplemental feeding is stopped on the NER, Wyoming continues to operate twenty-two feedgrounds scattered across its portions of the GYE on federal, state, and private lands. Roughly 14,000 elk congregate on these state feedgrounds during the winter months, creating a heightened risk of brucellosis transmission among themselves. Though recognizing this reality, the state has steadfastly continued its supplemental feeding program for decades to assuage local ranchers, outfitters, and hunters. In 2007, in an effort to remove the state feedgrounds from federal lands, conservation groups mounted a legal challenge to decisions by the Forest Service and BLM granting the state special-use feedground permits. According to the lawsuit, the agencies did not carefully analyze the environmental consequences involved in maintaining these unnatural feedgrounds. The federal court disagreed, however, and sustained the federal permitting decision, leaving feedground concerns to be resolved at the state level.[48]

A decade later, with CWD knocking at the GYE doorstep, the Wyoming feedground issue took another turn on the region's federal lands. In 2018, when the Bridger-Teton National Forest renewed Wyoming's Alkali Creek elk feedground permit on forestland, conservation groups again sued claiming that the Forest Service had not met its environmental analysis obligations, specifically by failing to consider the alternative of phasing out supplemental feeding to prevent the spread of this new deadly disease. A Wyoming federal judge, observing that chronic wasting disease portended "the irreversible and irretrievable loss of wildlife and habitat," agreed. She faulted the agency for not considering the cumulative impacts linked to Wyoming's entire feedground system or the alternative of phasing out the feedgrounds as the NER was contemplating. In response, the Forest Service temporarily extended the Alkali Creek feedground permit through 2024, but the state subsequently decided to phase out its use of the site. Somewhat surprisingly, the Forest Service is contemplating keeping two other state feedgrounds open for another twenty years based on an initial NEPA environmental review. But by sending the Forest Service back to the drawing board on the Alkali Flat permit, the judge's message was clear. Given the magnitude of the risk CWD presents on feedgrounds, the western Wyoming supplemental feeding program should be reviewed in its entirety through a cooperative federal-state effort with serious consideration given to phasing out the feedgrounds.[49]

Since first detected in southeastern Wyoming during the mid-1980s, CWD has steadily advanced westward across the state, and has finally arrived in the GYE. In 2020 an elk in Grand Teton National Park tested positive for the disease, which has also been detected in elk situated near state feedgrounds. In late 2023 Yellowstone announced its first case traced to a buck deer that died in the park's southeastern corner. Always fatal, CWD is a degenerative neurological disorder linked to prions that, scientists believe, are transmitted between animals and through contaminated soil, plants, and animal feed. The disease causes brain lesions, weight loss, abnormal behavior, and eventually death in elk, deer, and moose. Not only is there no effective treatment, but the culprit prions remain in the environment indefinitely, contaminating soil and plants and making it impossible to control transmission. Wyoming's own biologists believe it is only a matter of time before CWD infects elk on the state's feedgrounds.[50]

The principal options for addressing CWD are limited to reducing elk herd sizes and scattering the elk across the landscape to reduce intermingling—which essentially means eliminating the feedgrounds. Paradoxically, a decision to close the Wyoming feedgrounds to control CWD could facilitate the spread of brucellosis to domestic livestock, since the feedgrounds function to separate elk and cattle. Closure of the feedgrounds would also anger the state's hunting and guiding communities, since estimates are that the elk population would decline by more than 50 percent without artificial feeding. And it would impact the region's tourism industry that draws visitors to the easy elk-viewing opportunities available at the feedgrounds. Some scientists believe that wolves and scavengers could help by dispersing the elk and cleaning up diseased carcasses, but wolves are deeply despised in the state. Yet, taking no action imperils the state's world-class elk herds and will hasten the disease's spread throughout the region, and potentially even threaten human health.[51]

There is little evidence the state of Wyoming is prepared to reconsider its feedground policy. Politicians—rather than scientists and wildlife managers—are now driving the matter in the state. In 2020 the Wyoming Game and Fish Department issued a new CWD management plan that focuses on controlling the disease through monitoring elk and deer herds for the infection, dispersing feed at the feedgrounds to deter crowding, reducing the length of the feeding season, acquiring additional elk winter habitat, coordinating with federal counterparts, and more research. Following the plan's release, the Wyoming legislature—alarmed over possible feedground closures—stripped the Department of its closure authority and vested that authority with the governor, while also requiring consultation with the state's Department of Livestock. And should the Forest Service decide to close a state feedground, the legislation directs the Department to seek a replacement feedground on nearby state or private lands. These same political forces, as we shall see, are also evident when it comes to wildlife migration corridors, where they have limited the Department's role in designating such corridors.[52]

Not surprisingly, the state continues to signal it is not yet prepared to reconsider its elk feedground program. Released in early 2024 on the heels of a widely circulated draft plan, the Wyoming Game and Fish Department's Elk Feedgrounds Management Plan provides long-term guidance based on two goals: "promote elk health by limiting disease transmission while providing supplemental feed" and "reduce reliance of elk on supplemental feed while adhering to the Sideboards." Recognizing different conditions prevail at each feedground, the plan calls for individual, collaboratively developed feedground management action plans that use "the best available science, expertise, and local knowledge." The plan identifies potential management strategies that include altering feedground feeding practices, acquiring additional elk habitat, working with private landowners and federal officials, additional research, and more funding. It also elaborates "Sideboards" seemingly designed to moderate any significant change in elk management by requiring action plans to maintain elk-population objectives and prioritize hunting as the primary elk-management tool while also seeking to minimize property damage and disease transmission.

Although the draft management plan expressly acknowledged that the "practice of controlling elk distribution in western Wyoming utilizing supplemental feeding is likely not sustainable," the final plan retreats from this view and simply acknowledges that it may be possible to phase out individual feedgrounds consistent with the plan's strategies and sideboards.[53] The plan notes, moreover, that any decision to phase out an individual feedground would require considerable planning as well as the governor's approval. That future may be growing nearer, however. A US Geological Survey report prepared by eight experts, including three Wyoming Game and Fish Department employees, predicts a nearly 50 percent population decline during the next twenty years in several winter-fed elk herds as well as notable population declines if the state delays just three years before closing the feedgrounds.[54]

The overarching question is whether the political will exists to phase out the NER and state feedgrounds. The state of Montana, confronting the threat of CWD in its elk herds, has implored Wyoming to close its feedgrounds. The NER, facing renewed litigation over its supplemental feeding program, is engaged in revising the refuge's Bison and Elk Management Plan and is at least considering the phase-out option. With adoption of an Elk Feedgrounds Management Plan, Wyoming wildlife officials finally took a focused look at the feedground management options available to address the state's wildlife disease problems, but backed away from any meaningful closure commitment. In 2022 Congress passed the Chronic Wasting Disease Research and Management Act, which provides $70 million in funding to support further research and cooperation at the federal and state levels. Of course, any solution will have to balance the relative value of the GYE's high-profile elk herds against the impending risks associated with both brucellosis and CWD, including the benefits and costs involved in maintaining or ultimately closing the feedgrounds. Given the problem's ecosystem-wide implications, the solution will require even more coordinated action at the federal, state, and tribal levels, as well as among the three GYE states.[55]

Protecting Migratory Corridors

Centuries-old migratory instincts remain intact within much of the GYE's ungulate populations, giving rise to growing migration corridor conservation efforts and simultaneously expanding our conception of the ecosystem. Although biologists, ranchers, and other observers have long appreciated the annual migration ritual among the region's elk, deer, and pronghorn, recent research has uncovered substantial new information about the range, routes, and challenges confronting the region's migratory wildlife. Moving seasonally during the fall from lush high-elevation mountain meadows to lower-elevation lands for forage to sustain themselves during the harsh winter months, the area's ungulates then repeat the journey in the late spring as mountain grasses again "green up." In the process, the animals blithely cross jurisdictional boundaries moving from the region's national park and forest lands to BLM and private ranchlands, encountering numerous barriers—roads, subdivisions, energy fields, fences, and even feedgrounds in Wyoming—that can disrupt their learned migration behavior.

The GYE is one of the few places remaining in the country where such a grand migration spectacle is annually on full display despite the challenges confronting the animals. Scientists, using new GPS technological research techniques, have recorded numerous migration routes in the GYE, prompting greater public appreciation for the annual rite as well as increasing efforts to protect the routes. In the southern GYE scientists have documented a mule deer herd's 300-mile round-trip migration route from the Hoback area outside Jackson southward to the Red Desert, during which the deer navigate roads, fences, and subdivisions as well as oil and gas projects. Further north, pronghorn summering in Grand Teton National Park annually undertake a roughly 100-mile journey that takes them to the Upper Green River sagebrush country during the winter season. Several elk herds spill out of Yellowstone during the fall, heading to more hospitable lands in the surrounding states.[56]

These annual migrations constitute a crucial ecological process shaping the GYE, while the pulsing herds draw tourists and hunters alike. According to one widely respected biologist, the "flow of elk in and out of Yellowstone sustains the entire ecosystem."[57] Scientists have documented nine migration routes that different elk herds utilize to exit the park before winter arrives, taking the animals into each of the three surrounding states. These elk herds help sustain grizzly bears, wolves, and other predators, as well as scavengers, with elk calves being particularly targeted during the spring season when the newborn are defenseless against large predators. Big game hunting is a time-honored pursuit in the GYE states, and the transitory elk herds represent prime quarry for both trophy and meat hunters. Tourists and locals alike marvel at these stalwart animals, both during their arduous journeys and when gathered at feeding grounds, like the National Elk Refuge. For the GYE states and local communities elk and other migratory wildlife represent an impor-

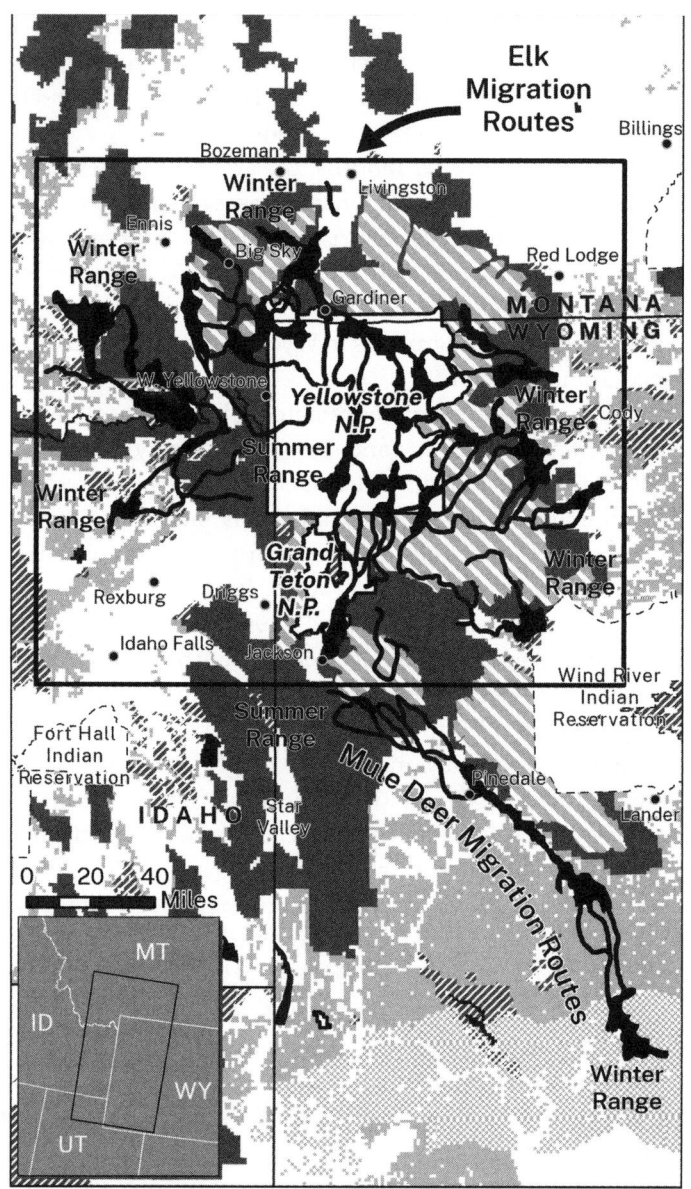

Elk and mule deer migration routes in the Greater Yellowstone Ecosystem. Seasonal migration is a regular phenomenon for the GYE's elk, mule deer, and other ungulate populations. The map shows the extensive scope of these seasonal migrations, which cross multiple legal boundary lines and effectively expand the GYE concept to a broader landscape scale for wildlife-management purposes. Map by the University of Utah DIGIT Lab, courtesy of Arthur Middleton (University of California, Berkeley), Hall Sawyer (WEST, Inc.), and University of Oregon Infographics Lab).

tant economic asset, sustaining guides, outfitters, hotels, restaurants, and other local businesses.

As a result, efforts are underway within the GYE states to conserve vital migration corridors before more habitat is lost due to ongoing development activities. Estimates are that 75 percent of elk and other ungulate migration routes in the GYE have been lost since the 1970s. In Wyoming the state's feedgrounds have disrupted elk-migration patterns, along with sprawling energy projects, new subdivisions, roads, and fences. Amid swirling political tensions, the state of Wyoming has taken affirmative steps toward designating and protecting wildlife corridors at the state level. In 2018, seeking to build public support for corridor conservation efforts, the Wyoming Migration Initiative—spearheaded at the University of Wyoming by biologists, writers, and others committed to wildlife conservation—produced a *Wild Migrations Atlas* documenting and illustrating the myriad migrations occurring annually across the state with a particular focus on the GYE. The book—specifically designed to capture the public's attention, build political support, and promote cooperation to protect the state's remaining vital migration corridors—has helped to mobilize public support and prompt political action.[58]

Which is what occurred, though not without some political drama. In 2019 the Wyoming Game and Fish Department proposed to officially designate several wildlife migration corridors, including two in the GYE. But the Wyoming legislature, acutely sensitive to the oil and gas industry's concerns, responded by proposing to curb the department's authority over wildlife corridors. Hoping to avoid a showdown, the governor intervened and convened a statewide citizen advisory group to make recommendations over corridor designation. Upon receiving the group's proposals, the governor signed an executive order vesting himself with the authority to designate mule-deer and antelope migration corridors within the state. The order also officially designated three corridors (including the Red Desert to Hoback mule-deer corridor in the GYE region), constrained development in bottlenecks and high-use areas, and promoted public involvement and interagency collaboration over future corridor designations. In Montana the Department of Fish, Wildlife and Parks has adopted a Terrestrial Wildlife Movement and Habitat Strategy that emphasizes "voluntary incentives and the development of local partnerships to facilitate wildlife movement and migration in a way that works for people who are directly affected." Idaho, instead of developing a statewide wildlife corridor policy, has adopted an Action Plan for a few identified migration corridors, including two in the extended GYE area.[59]

National efforts are also underway to protect migration corridors that complement these efforts. In 2007 the Western Governors' Association (WGA) adopted a farsighted resolution committing the member states to protect wildlife corridors. Subsequently, the WGA issued a policy statement endorsing locally driven wildlife corridor identification and protection initiatives and requesting federal agencies to consult with the states before taking any corridor conservation efforts. In 2012 the Forest Service adopted new planning regulations that instructed forest managers to consider habitat connectivity in their forest plans. In 2017 the Trump administration endorsed wildlife corridor conservation when Secretary of the Interior Ryan Zinke, a lifelong Montana hunter, signed an order directing the federal land management agencies to collaborate with the states to "enhance and improve the quality of big game winter range and migration corridor habitat on Federal lands."

In response, the US Geological Survey prepared a series of maps detailing wildlife migration corridors across the western states. In 2022 the Biden administration committed significant federal funding to support voluntary, landowner-protected migration corridors and related conservation initiatives in Wyoming that could be used as a model elsewhere across the country. A year later it extended the program to Montana and Idaho. Since 2016 Congress has been considering a Wildlife Corridors Conservation Act that seeks to bring together federal, state, and tribal agencies along with private landowners to designate and protect migratory corridors. Moreover, a few states have already adopted corridor conservation legislation, though not the GYE states where landowners and citizens harbor long-standing suspicions toward the federal government and land regulation generally.[60]

Consistent with its historic role in nature conservation, the GYE boasts the nation's first federal wildlife corridor designation, which took shape in the early 2000s through a series of federal, state, and private landowner actions that established the aptly named "Path of the Pronghorn." Several hundred antelope spend summer months in Grand Teton National Park and then embark on a nearly 125-mile migration journey to the Upper Green River Basin south of Pinedale, Wyoming, where they spend the winter months. Along the way, the antelope file through the Gros Ventre river drainage in the Bridger-Teton National Forest, cross BLM and private ranchlands where they regularly confronted unfriendly fencing and other development, and then had to navigate across a major state highway to reach winter habitat. When the BLM approved opening the Pinedale Anticline oil field to leasing during the 1990s, concern mounted over the pronghorn herd's fate given the scale of development anticipated in what was regarded as the second-largest natural gas reserve in the continental United States. Ongoing subdivision and other development on the area's private lands added a further sense of urgency. Conservation groups joined with concerned local citizens and agency officials in an effort to safeguard the imperiled migration pathway.[61]

Ultimately, their efforts succeeded with the establishment of the Path of the Pronghorn migration corridor, patched together across the region's federal and private lands. Grand Teton National Park contributed by acquiring a critical state school trust land section at the corridor's northern terminus. Bridger-Teton National Forest supervisor Kniffy Hamilton took a seminal step by amending the forest plan to formally designate a wildlife corridor on forest lands, requiring that future development or activities must "be designed, timed and/or located to allow continued successful migration of the pronghorn that summer in Jackson Hole and winter in the Green River basin." The BLM helped too by designating Areas of Critical Environmental Concern (ACECs) on the southern end of the migration route, precluding industrial activity on 9,500 acres there. Conservation groups collaborated with government officials to purchase conservation easements on private ranchlands situated in the migration pathway to prevent further development or construction on these lands. They have also worked collaboratively with private landowners to remove and alter problematic fencing that endangered passing pronghorn. In addition, the Wyoming Highway Department expended federal and state funds to construct a highway overpass at the Trapper's Point chokepoint west of Pinedale, which has enabled the antelope to safely cross this busy roadway. But when signing the executive order establishing a state corridor designation process, Wyoming governor Mark Gordon chose not to imprint the Path of the Pronghorn lands with an official state corridor designation.[62]

Though widely heralded as a groundbreaking achievement, the Path of the Pronghorn corridor is threatened by a new natural gas project. In 2018 the BLM approved development of the Normally Pressured Lance gas field, which involves drilling 3,500 new wells to the south and west of the Anticline and Jonah fields, without any specific limitations to protect the Path of the Pronghorn corridor.[63] Once the BLM acknowledged that the project "could reduce the viability of the pronghorn Herd Unit 401 in perpetuity," conservation groups sued to block the project, arguing that the BLM violated the National Environmental Policy Act by not considering an adequate range of alternatives or sufficient mitigation measures

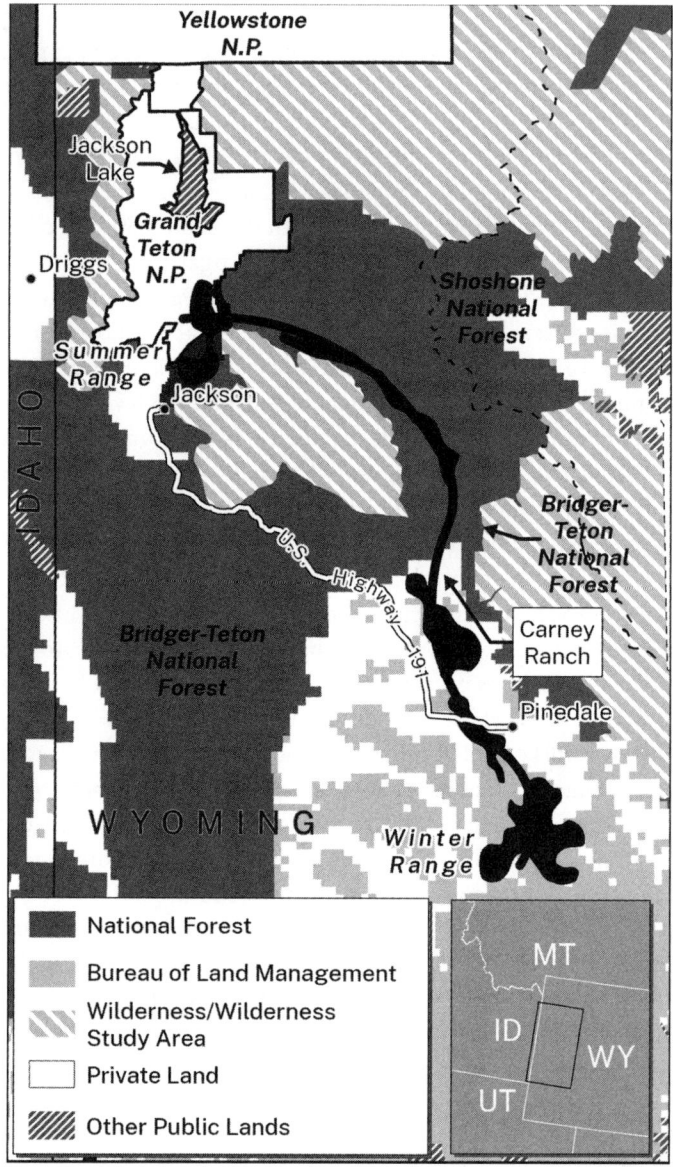

Path of the Pronghorn Migration Corridor. The Path of the Pronghorn is the first federally designated and collaboratively established wildlife migration corridor. It is designed to protect the pathway used by a regional pronghorn herd as it seasonally travels about 250 miles from summer to winter range and back. The Path extends across a jurisdictionally complex landscape from Grand Teton National Park, through the Bridger-Teton National Forest, across private ranchlands, over a busy state highway, and then terminates on Bureau of Land Management lands that are the site of a major natural gas field development. Map by the University of Utah DIGIT Lab.

to protect migrating pronghorn or resident sage grouse. A Wyoming federal court disagreed, however. Noting that the Wyoming Game and Fish Department had not designated any migration corridors in the project area and that the developer had agreed to work with the BLM to address wildlife concerns as the project unfolded, the court ruled that the agency's decision adhered to NEPA's procedural requirements. The court also ruled that the BLM's decision to authorize the project without sufficient research on its impact on sage grouse was permissible, because the agency would evaluate such impacts when reviewing specific drilling requests. On appeal, the Tenth Circuit affirmed the district court's decision, effectively deferring to state wildlife officials and entrusting the BLM and the drilling company to follow through on future, unspecified mitigation measures with no assurance they will adequately protect migrating pronghorn or local sage grouse.

Relatedly, the Wyoming Land Board has approved an oil and gas lease on a state school parcel situated in the Path of the Pronghorn corridor, despite objections that development there would create a bottleneck for the migrating animals. Soon thereafter, concluding that the pronghorn corridor is at "high risk" of being lost due to impending development impacts, the Wyoming Game and Fish Department announced it is considering an official corridor designation recommendation.[64]

Wildlife corridor conservation initiatives are underway elsewhere in the GYE. Using his executive authority, Wyoming's governor designated the Sublette mule deer corridor as an official wildlife migration corridor, providing some protection to this 160-mile route that stretches southward from the Hoback Basin outside Jackson to the Red Desert area near Rock Springs. In the High Divide portion of the GYE, northwest of Yellowstone, federal and state agencies have joined with conservation groups and landowners in a concerted effort to protect a critical corridor connecting the GYE with more northern ecosystems. Working together, the Wyoming Game and Fish Department and Wyoming Highway Department have completed several highway underpasses near Pinedale and Big Piney to safeguard pronghorn and mule-deer migration pathways.[65] But in Idaho, despite support from the state's highway department and wildlife agency as well as conservation groups, a proposed wildlife highway overpass project in Island Park designed to reduce wildlife–vehicle collisions during annual migration periods was derailed by vociferous local opposition driven by antigovernment sentiment.[66] Although neither uniform nor coordinated across the GYE, these various wildlife corridor protection efforts—highlighted by the multiagency Path of the Pronghorn initiative—illustrate a growing recognition of the critical role corridors play in GYE conservation efforts and a mounting willingness to collaborate across jurisdictional boundaries to protect these critical ecological processes at the landscape scale.

* * *

The GYE's elk, bison, and pronghorn constitute key ecological components in this dynamic landscape that even in today's world retains much of its original wildland character. Following deeply engrained migratory instincts, these iconic species still roam across the GYE, oblivious to the legal boundary lines we have imposed on the landscape. In doing so, they often find themselves at serious risk due to their mere presence and the disease threat they pose to cows and other animals. Notwithstanding these concerns, the GYE boasts relatively robust ungulate populations—a testament to the conservation and restoration efforts undertaken during the past century. But these efforts remain incomplete, as evidenced by the plight of Yellowstone's bison upon exiting the park, the unnatural feedgrounds that dot Wyoming's portion of the GYE, and related brucellosis and chronic wasting disease concerns.

The ongoing challenge is largely one of ecological restoration to ensure adequate habitat and migration corridors for GYE ungulates amid a growing human presence and escalating development pressures. The answers do not come easily in the region's jurisdictionally fragmented and conflicted environment. Not only is the Park Service at odds with neighboring states over bison, but the states remain at odds over Wyoming's feedgrounds. Wildlife corridor proposals continue to evoke controversy. Plainly, solutions to these devilish wildlife conservation matters, given the region's ecological and political realities, must be forged at the relevant landscape level. They will require understanding, tolerance, and collaboration across boundary lines between federal, state, and tribal officials as well as among the GYE states themselves.

Recent efforts to identify and safeguard wildlife corridors present a politically viable opportunity to work together to breach boundary lines and collaborate at the landscape scale. Indeed, the first-of-its-kind Path of the Pronghorn migration corridor represents a true wildlife conservation success story and a tangible model—one built upon federal, state, and local collaboration that emerged organically from the ground level and has attracted national attention. In the same vein, the budding relationship between the Park Service and tribal nations over Yellowstone's bison holds promise, too, if Montana can be brought aboard. And the recent federal funding commitment to encourage Wyoming landowners to participate in conservation and corridor protection efforts holds promise. Additional collaborative efforts built upon the lessons gleaned from these initiatives are essential if Greater Yellowstone's bison, elk, and pronghorn populations are to reclaim lost habitat and persist as wild animals.

CHAPTER FIVE

The Multiple-Use Lands
From Clearcuts to Playgrounds

The GYE overlays an extensive federal land complex that includes substantial acreage managed by the Forest Service and the Bureau of Land Management under multiple-use principles. The Forest Service oversees much of the GYE's mountainous acreage, while the BLM manages lower-elevation lands, portions of which are leased for energy development. Although abidingly committed to its multiple-use tradition, the Forest Service's policies have shifted noticeably throughout its 120-year history. As the twentieth century drew to a close, the agency underwent a remarkable transformation, discarding its timber-dominated agenda and charting a new course driven largely by ecological restoration and recreation. Change has come more slowly to the BLM; while energy production still occupies a dominant position on its agenda, the agency has acquired new conservation responsibilities. These policy shifts have not come easily, however, amidst competing economic and political pressures. Pitched battles have been waged in Congress and the courts as well as with different presidential administrations, reflecting deep national and local divisions over each agency's primary responsibilities. How this has occurred and what it means in the GYE is a fascinating chapter in the region's history.

By any measure, the national forests occupy a central position in the GYE critical to the region's ecological integrity. Covering nearly 11 million acres, the five GYE national forests buffer the region's two national parks, contain critical wildlife habitat, and boast extensive wilderness lands. They have supported a robust commercial timber industry now nearly disappeared, attracted periodic interest from the oil and gas industry, hosted mining activity still evident in several drainages, sustained the region's livestock operators, and afforded hunters, anglers, hikers, skiers, and others boundless adventures. In one form or another, the GYE's scattered communities are connected economically and otherwise to the region's national forests, though those connections have changed over time. Similarly, Yellowstone and Grand Teton national parks are deeply entwined with the GYE national forests, as reflected in the ecosystem-based grizzly bear recovery program and other joint wildlife-management efforts.

All of these resource uses have occupied the Forest Service's agenda during the past several decades, which has seen profound changes in management priorities throughout the GYE national forests. Once a dominant use, commercial timber harvesting has slowed to a crawl; oil and gas exploration is now largely gone from the forests, but not on Wyoming's BLM-administered lands; new mining activity is at a standstill on national forest lands despite efforts to reopen old mines; and livestock grazing continues to recede. Meantime, wilderness designation debates persist unresolved, recreational activity is at peak levels, the region's ski areas have become four-season resort settings, and wildlife concerns have grown in public importance. As the Forest Service confronts this changing natural and human landscape, it also must address climate change impacts, an escalating wildfire threat, and mounting recreation conflicts, while engaging with the neighboring states, local communities, and a vigilant environmental community.

Moving Beyond the Timber Wars

In the years preceding World War II the GYE national forests supported a modest timber industry primarily to meet local needs. But once the war ended with the country's private timber lands depleted, the nation turned to the West's national forests for lumber to meet surging demands across the country. The Forest Service obliged and prioritized timber production on its largely untouched lands. Even in regions like the mountainous GYE, where aridity, elevation, and remoteness presented major timber management challenges, agency officials were intent on cutting trees—a policy shift justified in part by new jobs and other economic benefits for the region's largely rural communities.

In the GYE, the Targhee National Forest, located just west of Yellowstone in eastern Idaho, took this policy shift to heart, ultimately precipitating what amounted to a timber war. During the 1960s the Forest Service completed a massive timber sale on the Targhee, then the agency's largest timber sale outside of Alaska. Billed as an effort to control a rampaging pine beetle outbreak as well as an opportunity to bring good-paying jobs to this rural area, the Forest Service sold more than 318 million board feet of timber (MMBF) for harvest over a multiyear time period. The cutting would occur across 100 square miles and involve constructing an extensive network of new roads in the forest. The sale, in turn, prompted the construction of two large lumber mills in the small towns of St. Anthony and Rexburg to process the harvested logs, providing additional local jobs and economic benefits.[1]

Once firmly wedded to timber, Targhee officials continued to promote logging over other forest uses, though many of its timber sales operated at a financial loss to the agency due in large part to road construction costs. In 1985 the Targhee completed its initial National Forest Management Act (NFMA)–mandated forest plan, projecting a harvest level of 860 MMBF per decade, mostly as salvage sales due to the ongoing pine beetle epidemic. The timber cutting that ensued stood in stark contrast to the other GYE national forests. In 1989 the Targhee harvested 91.6 MMBF of timber compared to the 82.6 MMBF total harvested in all of the other GYE national forests.[2]

By then, it was evident that this accelerated level of timber cutting was not sustainable and was degrading other forest resources and values. Idaho Fish and Game officials ordered portions of the forest closed to elk hunting, citing the lack of cover for the animals. The Targhee was unable to meet its commitment to the GYE grizzly bear recovery effort because bears were absent from most of the forest's bear-management units due to the absence of secure habitat. Satellite photos revealed a distinct, miles-long straight edge on Yellowstone's western border with the forest, the result of an immense clear-cut conducted with no regard for the adjacent park or wildlife impacts. Anyone driving the principal north-south highway through the Targhee could not help but notice extensive clearcuts visible from the roadway, usually marked with signage dating the cutting activity in the name of forest restoration.[3]

By the mid-1980s, conservation organizations were aggressively responding to the ecological carnage. The newly created Greater Yellowstone Coalition, along with other groups, challenged the Targhee's 1985 forest plan and also appealed individual timber sales, alleging violations of the Endangered Species Act, National Forest Management Act, and the

National Environmental Policy Act (NEPA). Unrefuted scientific studies showed that clear-cut logging and the accompanying access roads into the backcountry endangered grizzly bears, destroying secure habitat while bringing people into contact with bears and facilitating illegal poaching. Confronted with this scientific evidence in a northern Montana case, the courts had limited new road construction in grizzly bear habitat, establishing a precedent directly relevant to the Targhee situation. Forest Service officials had little choice but to blink.[4]

In 1997 the Targhee released a revised forest plan that dramatically reversed course in deference to the grizzly bear and other wildlife concerns. The new plan curtailed timber harvest levels to 80 MMBF per decade (nearly a 90 percent reduction), closed 408 miles of roads (reducing the forest's road density by 20 percent), and eliminated 233 miles of motorized trails (a 30 percent reduction). By then, the St. Anthony and Rexburg lumber mills had closed, marking the end of unrestrained logging in the forest and concluding the region's most acrimonious timber war.[5]

A similar battle over timber production played out during the 1980s in Wyoming's Bridger-Teton National Forest involving the Louisiana Pacific sawmill in the town of Dubois. Following extensive logging in the forest near the mill, Bridger-Teton officials stopped approving new sales in that portion of the forest, citing the loss of elk habitat and erosion concerns. Asserting it would be forced to close its mill and terminate its workforce, Louisiana Pacific sued, arguing that the Forest Service was legally obliged to make timber available under governing law and its existing timber plan. A Wyoming federal district court disagreed, however, ruling that neither the agency's organic legislation nor its timber-management plan gave logging priority on the forest or guaranteed access to public timber.

Earlier, Louisiana Pacific had gone to court seeking to upgrade the Union Pass Road, a poorly constructed dirt road that ran across the Continental Divide, to enable its logging trucks to reach timber in a largely unharvested area of the forest. The courts, however, sustained the Forest Service's authority over the road, which effectively blocked the company's access to this timber source. Not long after these setbacks, Louisiana Pacific closed its Dubois mill, forcing the community to ponder an alternative future with the departure of its largest employer. As with the Targhee, the Bridger-Teton National Forest was in transition away from commercial logging in deference to other forest resource values.[6]

The movement away from commercial timber harvesting in the GYE took a different path in the Gallatin National Forest. Large portions of the southern Gallatin Range north of Yellowstone National Park were in the hands of Louisiana Pacific and Big Sky Lumber, who had acquired more than 80,000 acres of checkerboard railroad lands originally granted to the Northern Pacific Railroad in 1864. Given the proximity of these mostly undisturbed lands to Yellowstone, conservation groups were intent on protecting them as prime wildlife habitat, particularly for grizzly bears that had expanded their range northward from the park. They feared the companies would either log their lands, bringing clearcuts and new roads into this backcountry area, or subdivide and sell the land to private developers. Already, a few miles northward, the growing Big Sky ski area complex was spreading throughout a formerly undeveloped mountain valley. More development seemed increasingly inevitable in the southern Gallatin Range, further fragmenting the landscape and rendering management of the forest lands more difficult.

To prevent this outcome, conservation organizations joined with the Forest Service and the state of Montana to facilitate two large land exchanges. Under the terms of the 1993 and 1998 congressional authorizing legislation, the Forest Service acquired 83,000 acres of checkerboard inholdings in exchange for 54,030 acres of federal land elsewhere plus additional cash. The legislation also included a restoration provision designed to bring heavily logged watersheds up to existing Forest Service standards. The land that the Forest Service received in the exchanges protected critical wildlife habitat and migration corridors extending beyond Yellowstone's northern boundary, while also stopping the drift southward from Big Sky of subdivision development on the former railroad lands. With the Gallatin Range fully in federal ownership, the Forest Service has managed the area primarily for wildlife and recreational purposes, a fact confirmed in the 2022 Custer Gallatin forest plan.[7]

Coincident with the Gallatin land exchanges, timber harvest levels were dropping across the forest. Although the initial 1980s Custer and Gallatin forest plans projected an annual timber harvest sale quantity at 17.2 MMBF, the actual amount of timber sold and cut had hovered around 10 MMBF annually since the early 1990s, while the forest had also moved away from clearcutting as a harvest technique. The 2022 joint Custer Gallatin forest plan maintains these trends; it sets the annual projected sale quantity of timber at just 10 MMBF, a nearly 50 percent reduction from the prior forest plans. Though several factors have undoubtedly contributed to this notable downward turn in timber production—public sentiment, administrative appeals, litigation, budget concerns, and workforce capacity—the result is an undeniable shift in the forest's priorities toward wildlife and recreation.[8]

Elsewhere in the GYE, the Beaverhead-Deerlodge National Forest covers nearly 3.4 million acres, and it historically maintained a high-volume timber program. When the Forest Service undertook in 2005 to revise its earlier forest plans, the volume of timber actually being sold was dwindling due to budgetary constraints and resource considerations. Concerned about the forest's unhealthy condition owing to bug infestations and wildfire danger, the revised 2009 forest plan adopted "an approach to managing vegetation that focuses more on ecological restoration than timber production." It identified 284,000 acres as suitable for commercial timber production and allowed timber cutting on another 1.6 million acres to reduce hazardous fuels, restore ecosystem conditions, and address aquatic system concerns. Although the forest contained 1.8 million acres of inventoried roadless areas potentially available for wilderness designation, the plan recommended only 322,000 acres for wilderness designation. As the planning process evolved toward the final plan in 2009, local timber companies—already stressed over the drop-off in timber availability—were quite unhappy with the proposed commercial harvest limits, while conservation groups lamented the agency's reluctance to make significant new wilderness recommendations.[9]

Disappointed in the original draft plan, representatives from the timber companies and three environmental organizations came together and initiated discussions over the forest's future direction. Proceeding as the Beaverhead-Deerlodge Partnership, they managed to resolve their differences by negotiating an entwined set of timber and wilderness recommendations. Their agreement, however, met with mixed reactions across the local political spectrum. Because new wilderness designations were involved as well as timber harvesting quotas, the partnership's recommendations required congressional approval.

At the group's urging, Montana US senator Jon Tester introduced the Forest Jobs and Recreation Act in 2009 and continued to pursue the bill in subsequent congressional sessions. Senator Tester's bill proposed several additions to the wilderness system covering roughly 577,000 acres, designated another 310,000 acres for recreational use, and instructed the Forest Service to sell a minimum of 7,000 acres annually for timber production to stabilize the local timber industry and protect forest jobs. However, the provisions guaranteeing an annual timber-harvest quota along with those releasing wilderness study areas and roadless lands for multiple-use management provoked intense opposition from environmental groups that had been excluded from the collaborative negotiations. Moreover, the Forest Service was not on board with the legislation, which basically disregarded its revised forest plan.[10]

After reintroducing the bill in several congressional sessions, Senator Tester ultimately conceded defeat in 2014. In the aftermath, timber-management decisions on the Beaverhead-Deerlodge forest are being dictated by the 2009 forest plan, the forest's wilderness study areas retain legal protection, and its roadless area lands remain administratively protected by the agency's roadless area rule. The episode underscored the fierce passions surrounding timber cutting as well as wilderness protection in the GYE. It revealed deep fissures within the environmental community over forest-management policy and wilderness-designation strategy, as well as over the appropriate role and design of collaborative processes to resolve forest-policy differences. Even the allure of substantial new wilderness designations—a matter that had eluded the Montana conservation community for decades—was not enough to bridge the gap. In addition, the forest-specific nature of the Beaverhead-Deerlodge legislative proposal raised serious concerns about locally negotiated, place-based forest legislation that removed an individual forest from some or all of the overarching national laws governing forest management.[11]

In sum, commercial timber harvesting on the GYE national forests is a shadow of its former self and the timber wars have largely subsided. The current forest plans confirm this fact. In Montana and Idaho, as noted, the revised forest plans have significantly reduced harvest levels. In Wyoming the 2015 Shoshone forest plan identified a mere 127,000 acres in the 2.4-million-acre forest as suitable for timber production, while the Bridger-Teton's 1990 forest plan classified just 279,000 acres as suitable timber lands within the 3.3-million-acre forest. Further proof is manifest in the number of mill closures within the three GYE states. In Montana the number of active lumber mills declined from eighty-seven in 1988 to twenty-five in 2018, while the Townsend mill—a long-standing mainstay in the GYE—closed in 2020. In Wyoming the number of sawmills operating dwindled from fifty in 1976 to just twelve active mills in 2018. Idaho's experience is the same: the number of active sawmills plummeted from ninety in 1985 to twenty-seven in 2011. However, the transition away from commercial-scale timber programs has not quelled controversy over forest management and logging in the region's national forests.[12]

In the new age of climate change, persistent drought, and runaway wildfire events, the Forest Service is enmeshed in what it dubs a "wildfire crisis." Since the early 2000s most of the western states have experienced their worst fire seasons ever; the seasons are lasting much longer and considerably more acreage is being consumed in flames than since white settlers arrived on the scene. The problem is exacerbated by the

Forest Service's previous all-out fire suppression policy that has left the national forests in an unhealthy condition and ready to burn. Congress has sought to help, passing the Healthy Forests Restoration Act in 2003, which prioritized hazardous fuel reduction projects in the wildland-urban interface zone while waiving legal compliance requirements for smaller projects. In 2009 Congress adopted the Collaborative Forest Restoration Program legislation, creating a forest restoration fund and empowering the Forest Service to enter into contracts to reduce the wildfire threat through locally approved forest-thinning projects that were otherwise commercially unviable. But the fire situation has continued to worsen, such that more than 50 percent of the Forest Service's annual budget goes toward wildfire management. In 2022 the Forest Service invoked the term "crisis" to describe the wildfire problem, calling for "decisive action to protect people and communities" with hazardous fuel treatments to "reduce dangerous fuel levels and restore forest health and resilience." The agency also stressed the need for "a paradigm shift in land management across jurisdictional boundaries to reduce risk and restore fire-adapted landscapes."[13]

The GYE, with its expansive forests, is not immune to grave wildfire events. Memories of the explosive 1988 fires that raged across the region without regard to legal boundaries or full-throttle extinguishment efforts have not faded. With temperatures warming, the GYE national forests have experienced extensive tree mortality attributed to an ongoing bark beetle epidemic that has created a serious wildfire danger, particularly for the growing wildland-urban interface zone where new homes are proliferating adjacent to the forest boundary. Several devastating wildfires have raged within the region, including the 2001 Green Knoll fire just outside Jackson, which threatened hundreds of high-end homes, and the 2018 Roosevelt fire near Bondurant, Wyoming, which destroyed more than fifty homes. In Montana the 2020 Bridger Foothills fire destroyed thirty homes near Bozeman. Astonishingly, more than a quarter of the Custer Gallatin National Forest—roughly 810,000 acres—has burned since 1980. The Forest Service, while acknowledging the role of wildland fire in sustaining ecosystems, is therefore taking a more active management approach to forest health, conjuring visions in some quarters of a return to rampant timber cutting.[14]

The revised GYE forest plans are focused on ecosystem restoration and hazardous fuel reduction projects to protect at-risk communities while restoring fire-adapted ecosystems. For example, the 2015 Shoshone National Forest plan calls for managing "forested ecosystems . . .

to maintain healthy, diverse stands that are resilient to endemic insects, wildfire, and changes in climate, while providing for viable populations of all native and desired nonnative vertebrate species." Similarly, the 2022 Custer Gallatin forest plan "promotes landscapes resilient to fire-related disturbances," accomplished by reducing hazardous fuels to protect "watershed health, wildlife habitat, community values at risk, air quality, and [public] safety." The plan enumerates three tools to accomplish these objectives: prescribed fire, noncommercial thinning, and timber sales, and it "prioritizes hazardous fuel treatments in wildland-urban interface areas." However, prescribed burning in tinder-dry forests near populated areas presents serious runaway fire risks, and residents regularly complain about the accompanying smoke. Forest thinning enjoys public support, though it is viewed skeptically—and frequently challenged—when commercial-grade timber or old-growth trees are included in these projects. Nonetheless, given the flammable condition of GYE forests in a warming environment, the agency has little choice as a political and ecological matter, but to take a proactive approach to fire management.[15]

While some Forest Service hazardous fuel reduction projects proceed without objection, others have ignited controversy within the GYE and continue to do so. In Montana conservation groups have aggressively challenged several fuel-reduction projects involving timber removal in or near wildland-urban interface areas. Arguments that the projects failed to comply with NEPA or the Endangered Species Act have generally failed, however, with the courts giving deference to the agency's environmental analysis.[16] In the Custer Gallatin National Forest the South Plateau Area Landscape Treatment Project is designed to address a pine beetle infestation and to reduce hazardous fuels in the West Yellowstone area. The project covers 16,400 acres and involves more than 5,500 acres of clearcuts, 6,500 acres of commercial thinning, and 56 miles of temporary roads, while harvesting 83 MMBF of timber over a fifteen-year period. Given the scale and commercial dimensions of the project, conservation groups have filed a lawsuit challenging it, citing NEPA deficiencies, negative grizzly bear impacts, and the carbon-sink implications of so much tree cutting. In Wyoming's Bridger-Teton National Forest, the Teton to Snake Fuels Management Project, which was designed to reduce wildfire risks west of Jackson, provoked opposition due to cutting in a wilderness study area, but the project has moved forward. In short, the wildfire danger across the GYE has reached the point where the region's growing populace generally supports the Forest Service's more active approach to the problem, at least outside the GYE backcountry.[17]

Over the past several decades timber policy in the GYE national forests has noticeably shifted away from large-scale commercial logging and extensive clearcutting. The Forest Service today is focused on restoring the region's forested ecosystems, acknowledging that past timber management and fire-suppression policies have left the forests in an ecologically damaged condition that has contributed to the rising fire danger. Climate change and recurrent insect infestations have further exacerbated the fire danger, while the growth in new homes proximate to the region's national forests presents new fire-management problems that constrain the use of prescribed burns as a forest-restoration strategy. Nearly everyone acknowledges that fire plays an important ecological role in the region's forests, that prescribed fires can mimic natural ones, and that wildfires are indifferent to political boundaries. Few local citizens, though, are ready for a repeat of the 1988 fires.

This leaves the Forest Service to walk the proverbial tight rope as it pursues a new hazardous fuel reduction policy that promotes ecological integrity and public safety without straying into the commercial timber business of yesteryear or reducing the forest's carbon-sequestration capacity. The agency's task is only getting more difficult as the climate warms, the wildland-urban interface (WUI) zone continues to spread, and wildfires show no sign of abating. Restoring the GYE's forests to ecological health is proving to be as rife with controversy as earlier clear-cut logging practices. At the very least, an effective forest restoration policy must rest on a sound scientific foundation, extend across existing jurisdictional boundaries, and garner widespread public support—all without unduly damaging the region's wildlife and recreational opportunities or exacerbating the climate crisis.

Energy Development in Wyoming

During the past forty years oil and gas development has evolved from a looming threat throughout the GYE to an industrial activity concentrated in Wyoming mostly on the periphery of the ecosystem. By the 1970s Wyoming was on its way to becoming a major energy-producing state, a transition hastened by the initial Middle Eastern oil embargo that roiled the nation's economy and sparked profound international security concerns. The oil industry responded to the shortage by pressing to explore the promising Overthrust Belt running along the eastern flank of the Rocky Mountain range. This put the GYE national forests in the crosshairs for industrialization, imperiling the region's wildlife and natural attributes.

The Forest Service, while acknowledging these environmental impacts, nonetheless approved a bevy of ten-year mineral leases in the GYE national forests, inviting the eager companies to explore for oil and gas. The leasing frenzy proved short-lived, however, as the oil crisis subsided and energy prices slid downward, dampening the industry's enthusiasm.[18]

Meantime, the prospect of oil rigs dotting the landscape adjacent to the GYE national parks provoked a flurry of litigation that established critical environmental standards governing oil and gas development on public lands. In separate cases involving the Bridger-Teton and Gallatin national forests, the courts ruled under NEPA that the Forest Service must prepare an environmental analysis—either an Environmental Impact Statement (EIS) or an environmental assessment—before issuing a lease. In requiring prior NEPA analysis, the courts explained that the lease conveys the right to explore and develop oil deposits, absent a stipulation prohibiting activity on the surface of the leased lands, commonly known as a nonsurface occupancy stipulation or "NSO stip." The environmental analysis must consider not only the impacts of an exploratory well but also those associated with full field development in the event of a discovery. Whenever grizzly bears or other species protected under the federal Endangered Species Act were present, the Forest Service was required to consult with the US Fish and Wildlife Service to determine whether the proposed project might jeopardize the protected species, considering the full cycle of possible development impacts. In short, the prevailing legal standards plainly required a rigorous environmental review of potential development impacts before leasing and exploratory drilling could proceed in the national forests.[19]

Subsequent developments have further clarified the legal landscape governing oil and gas development on both national forest and BLM lands in the GYE and elsewhere. Under the Mineral Leasing Act of 1920, the Secretary of the Interior oversees oil and gas development on all federal lands and has the discretionary authority to decide whether or not to lease specific parcels.[20] In 1987 Congress passed the Federal Onshore Oil and Gas Leasing Reform Act, which granted the Forest Service a veto power over leasing on its own lands.[21] During the late 1990s an intrepid forest supervisor in the Northern Rockies named Gloria Flora issued a decision effectively closing a wildlife-rich portion of the Lewis and Clark National Forest to additional oil and gas leasing, citing the public's interest in the area's natural qualities as the basis for her decision. Her no-leasing decision, though challenged by the oil industry, was ultimately upheld by the courts.[22] In 2005, facing another perceived energy crisis, Congress adopted the Energy Policy Act in an effort to expedite energy development on public lands by accelerating the drilling permit process and reducing environmental review obligations when surface disturbance impacts were limited.[23] Since then, federal policies as well as court decisions have required the agencies to analyze and disclose the greenhouse gas emissions linked to fossil fuel projects on public lands as part of their environmental analysis obligations.[24]

Against this legal backdrop and with the nation facing uncertain international energy markets during the early 2000s, a new controversy involving oil and gas development flared in the southern reaches of the GYE. In the so-called Pinedale Anticline, a 198,000-acre area extending southwestward from the town of Pinedale on sagebrush-covered and mostly undeveloped federal BLM lands, oil companies realized they had a major discovery on hand. With new horizontal drilling and fracking technologies, the area's oil and gas deposits were now more easily accessible and represented one of the largest natural gas fields in the country. But the area also contained important wildlife habitat that provided lower-elevation winter range for pronghorn, mule deer, and elk, as well as critical greater sage grouse habitat. In 2000, under the Clinton administration, the BLM approved the Pinedale Anticline project, authorizing 700 well pads and accompanying roads, processing facilities, and pipelines, with a projected ten- to fifteen-year development horizon. At the same time, the BLM imposed seasonal no-drilling wildlife mitigation measures and monitoring requirements to detect potential environmental damage.[25]

As the Pinedale Anticline project unfolded and the companies realized the full extent of the area's natural gas deposits, they sought to expedite and expand development. The BLM, under the energy-focused George W. Bush administration, agreed and approved 4,399 additional wells (while reducing the number of well pads), eliminated the seasonal drilling restrictions, and required development to be concentrated in a core area. The net effect was to further convert this remote landscape into a year-round industrial zone. To reduce the impact of lifting the seasonal drilling restrictions, the BLM imposed new wildlife mitigation measures, including phased development requirements in core areas and a $36-million monitoring and mitigation fund. Unpersuaded that these new measures would adequately safeguard the region's wildlife, the Theodore Roosevelt Conservation Partnership sued in an effort to block the revised project. It was rebuffed by the courts, however, which ruled the BLM had met its federal environmental analysis obligations and its mitigation measures were sufficient, noting the Wyoming Game and Fish Department endorsed them.[26]

With development in full swing, the small town of Pinedale soon

Aerial view of the Pinedale Anticline natural gas field. Credit: Ecoflight.

turned into a classic boomtown with the attendant alcohol, drug, and other social problems, along with escalating housing costs. As the number of operating wells increased, ozone levels soared in the winter atmosphere, bringing serious air-pollution problems to the town and the nearby Wind River Range wilderness areas. The EPA responded by establishing the Pinedale ozone nonattainment area, an unexpected development for an area long regarded as having near-pristine air quality. The requisite wildlife monitoring has revealed a decline in the area pronghorn, mule deer, and sage grouse populations.[27] Moreover, the state of Wyoming has yet to officially designate or otherwise safeguard the much-celebrated Path of the Pronghorn Wildlife Migration Corridor, which extends into the project area.[28]

In 2018, notwithstanding these concerns, the BLM approved the Normally Pressured Lance (NPL) Natural Gas Development Project that sprawls across another 141,000 acres of mostly public land southwest of the Pinedale Anticline project. The Jonah Energy company proposed drilling 3,500 wells over a ten-year period, with the project expected to extend another thirty years during the production phase. Wildlife problems were evident, because the project impacts Path of the Pronghorn migration routes as well as greater sage grouse habitat. To address the problem the BLM again adopted wildlife mitigation measures, but this did not satisfy conservation groups. They went to court, asserting that the project would block the migration route for Grand Teton National Park's pronghorn herd, a view also expressed by park officials. Further, citing the BLM's own environmental analysis that sage grouse would be displaced by the initial drilling, they contended the agency could not approve the project without assessing its impact on sage grouse throughout the development process.[29]

A Wyoming federal judge disagreed, however, and sustained the BLM's decision and environmental analysis. Explaining that NEPA merely establishes procedural requirements, the court ruled that once the BLM adequately analyzed the environmental impacts of the project, it was free to choose any alternative among those studied, even if its choice adversely affected the pronghorn or greater sage grouse. The court acknowledged and deferred to the Wyoming Game and Fish Department's comment that the project would only disrupt some pronghorn migration routes, noting that the state had yet to designate an official migration corridor in the area. Citing the lack of scientific data on the sage grouse, the court approved the BLM's wait-and-see plan, merely requiring the agency to acquire additional information about sage grouse impacts once drilling plans were more specific. On appeal, the Tenth Circuit affirmed the district court's decision, concluding that the BLM adequately addressed pronghorn migration and sage grouse habitat concerns in its NEPA analysis. Simply put, the courts were extremely deferential to the BLM, the state, and the company, satisfied that the proposed mitigation measures would suffice to reduce but not eliminate harm to pronghorn and sage grouse. As was true with the Pinedale Anticline project, national energy concerns buttressed by Wyoming's economic interests again triumphed over serious wildlife concerns in this corner of the GYE.[30]

Significantly, a quite different energy development story has emerged in the Wyoming national forests. In 2003, with the Pinedale Anticline project underway, Bridger-Teton National Forest supervisor Kniffy Hamilton placed more than 375,000 acres in the forest's northern reaches off-limits to oil and gas leasing. Hamilton's decision mirrored the earlier actions of her counterpart in Montana's Lewis and Clark National Forest, where forest supervisor Gloria Flora halted leasing on forest lands along the Rocky Mountain Front to safeguard the area's prized wildlife and natural values. Hamilton's decision removed more than 10 percent of the

Bridger-Teton forest from potential energy development and effectively created a no-leasing buffer zone around Grand Teton National Park that extended southward and eastward to nearby wilderness areas. While her decision would have met resistance thirty years earlier when this area was initially under consideration for energy development, it was instead greeted with strong public approval, revealing a striking shift in attitudes among local residents as well as Wyoming politicians.[31]

A similar scenario soon played out in two other Wyoming national forest settings. The Wyoming Range, lying in the southern portion of the Bridger-Teton National Forest and just west of the Pinedale Anticline project area, had long been eyed by energy companies already holding leases there. Fearing industrialization of the area similar to what was occurring in the nearby Pinedale Anticline, a once unlikely alliance that included ranchers, environmental groups, guides and outfitters, sportsmen, and local residents emerged. They banded together as "Citizens for the Wyoming Range" and initiated a "Too Precious to Drill" campaign to block further development in these locally esteemed mountains. Wyoming's governor and the state's congressional delegation ultimately joined the campaign, which succeeded in convincing Congress to adopt the Wyoming Range Legacy Act of 2009. The legislation approved a buyout of existing leases and withdrew these national forest lands from further mineral leasing. Aided by several conservation-oriented foundations and wealthy Jackson Hole residents, the Citizens group raised the $8.75 million needed to purchase the outstanding leases, thus removing this threat to the mountains and area wildlife.[32]

But the Forest Service still faced a problem due to its earlier lease offer, which involved thirty-five parcels on 44,720 acres in the Wyoming Range. Would it honor the offer or choose not to lease these lands? The issue reached the highest levels in the US Department of Agriculture, where Undersecretary for Natural Resources and Environment Robert Bonnie resolved the matter. His carefully elaborated no-leasing decision drew upon extensive public comments extolling the area's outstanding "wildlands, recreation opportunities, wildlife, biodiversity and watershed values" as well as its "large expanse of backcountry." He explained that the public placed much greater importance on the region's natural, cultural, wildlife, and recreational values than on any local or statewide economic benefits that might be derived from oil and gas development. The decision confirmed the remarkable magnitude and diversity of opposition to drilling in the area despite the role mineral development occupies in the state's economy. More broadly, it reflected an evolving consensus within the GYE that the region represented a "special place" prized primarily for its natural and recreational values rather than its development potential.[33]

The Shoshone National Forest has historically hosted oil and gas activity. But with half of this steep, rugged forest classified as wilderness and with grizzly bears roaming the area, the Shoshone has not drawn the same level of exploratory interest as the Bridger-Teton. Nonetheless, in 1995, the forest supervisor released a forest-wide oil and gas leasing EIS that opened 950,000 acres to leasing while imposing a "no surface occupancy" stipulation on roughly half of this acreage to protect wildlife habitat and other sensitive resources. Environmental groups, fearing any development would adversely impact federally protected grizzly bears, challenged this expansive leasing decision, but were rebuffed by the courts.[34]

Twenty years later, however, Shoshone officials reached a quite different decision when revising the original forest plan—one that significantly curtailed leasing in the area known as the Absaroka-Beartooth Front (AB Front). The AB Front covers transitional lands extending from the forest's high-elevation wilderness areas to its eastern lower-elevation lands, most of which constitutes critical grizzly bear and wildlife habitat. The Forest Service, noting widespread public opposition to energy development in this area—a sentiment endorsed by Wyoming's governor—included no surface occupancy and other protective stipulations in the revised forest plan to govern any future leasing in this sensitive area. The decision extended additional protection to the eastern reaches of the GYE in northwestern Wyoming, providing transitory wildlife with safe haven and preserving important recreational opportunities.[35]

Significantly, there is little oil and gas activity on the other GYE national forests. In Idaho, although oil and gas leases "covered almost every available acre of the [Caribou] Forest in the early 1980s," the forest reported no lease applications in its revised 2003 forest plan.[36] In 2000 the Targhee National Forest opened only a small portion of the forest to future leasing, but imposed surface occupancy stipulations on most of the acreage.[37] In Montana the Custer Gallatin National Forest drew considerable interest from oil companies during the 1980s, but a 1985 court order suspended most of those leases for inadequate environmental analysis, and there has been no movement since then to reinstate them.[38] Although the 2009 Beaverhead-Deerlodge forest plan concluded its lands did not have "high potential for oil and gas," forest officials approved the Tendoy drilling project in 2021 after finding the proposed single well would not "significantly impact regional or national resources," despite concerns expressed by conservation groups about adverse wildlife and watershed impacts.[39] Besides the evident lack of industry interest in the Montana and Idaho GYE national forests, well-grounded climate change concerns and related environmental analysis requirements militate against future leasing in these forests. Barring another major energy crisis—one that runs deeper than the problems created by COVID-19 and the Ukraine war—the GYE national forests appear largely free from this industrial threat to the region's conservation values.[40]

During the past fifty years energy development in the GYE has steadily moved away from the region's core national parks and national forest lands, preserving these ecologically vital areas from industrialization. How has this occurred, particularly in energy-oriented Wyoming? Perhaps most importantly, a diverse array of local citizens concluded that industrial development was inappropriate in much of the Bridger-Teton and Shoshone national forests; rather, wildlife, aesthetic, and recreational values merited priority for both cultural and economic reasons. They succeeded in convincing the Forest Service as well as Wyoming political leaders to support a no-leasing position and even secured federal legislation to safeguard the Wyoming Range. Litigation pursued by conservation groups clarified the Forest Service's environmental assessment obligations related to leasing, while ensuring the public would be involved in the agency's leasing decision process. Private philanthropy was also important, providing the funds to buy out existing leases in the Wyoming Range.

The picture is different, however, on the GYE's federal BLM-administered lands in Wyoming. The BLM has issued energy leases in much of the Upper Green River country, giving rise to industrial activity that sprawls across the Pinedale Anticline, Jonah Infill, and Normally Pressured Lance project areas. The resulting drilling rigs, roads, pipelines, and processing facilities have notably affected the region's wildlife, impacting pronghorn and mule deer migratory routes as well as critical sage grouse mating areas. Although mitigation measures have reduced the level of impact to some degree and funded habitat enhancement projects on-site and elsewhere, the fact remains that this peripheral portion of the GYE has been industrialized and is no longer intact, though efforts continue to safeguard migration routes and critical sage grouse habitat. The energy imperative, driven by powerful political and market forces, has carried the day on these GYE multiple-use lands, albeit with some effort to address conservation concerns.

Mining: Echoes from the Past

Once a vibrant industry in the Yellowstone region, mining is today more a historical legacy than a fiscal mainstay in the region's economic life, though mines have not disappeared from the scene. Several Montana towns, including Cooke City, Virginia City, and Red Lodge, trace their origins to early gold and silver discoveries, but the mines have long been dormant while the communities have gradually embraced tourism and recreation as their economic lifeline. Given this transition, it is not surprising that efforts to revive mining at several sites across the GYE with the use of modern technologies have spawned intense opposition that has succeeded in forestalling ill-advised projects. In one instance presidential intervention helped to kill the project, and in another congressional intervention has banned future mining on Yellowstone's northern edge. Nonetheless, industrial mining activity continues apace in corners of the GYE, namely phosphate extraction in the Caribou-Targhee National Forest and chromium mining at the Stillwater complex in the Custer Gallatin National Forest.

Mining on the public lands is governed by the 150-year-old General Mining Law of 1872. Under this antiquated statute, mining is granted a priority position on federal multiple-use lands relative to other resources and uses. Though widely criticized as a throwback to a bygone era of rampant exploitation of the West's natural resources, the General Mining Law invites anyone onto the public lands to explore for minerals, such as gold, silver, copper, and the like. The law empowers the miner to stake a so-called unpatented mining claim upon discovering a valuable mineral deposit, which then constitutes a legally protected property right. Although the Forest Service and BLM have authority to regulate mining activity on their lands, they do not have the apparent ability to say "no" to a legitimate mine proposal that meets federal and state environmental requirements. Where mining activity has occurred in the past, the mine site may have been patented, which would transfer ownership of the land from federal to private hands. In this case, state rather than federal law governs any ongoing or new mining activity. Cyanide leaching and other modern technologies have enabled mining companies to profitably reopen former mines—a practice that has triggered several high-profile controversies in the GYE region given the prevalence of old mining sites.[41]

That is what occurred during the early 1990s on a high mountainside above Cooke City, Montana, three miles outside Yellowstone National Park. Until the 1950s, for more than half a century, the sprawling New

World Mining District produced an abundance of gold, silver, and copper in a boom-and-bust cycle that provided good paying jobs and drove the local economy. Once the mines played out, however, the miners jumped ship, leaving a devastated landscape contaminated with toxic wastes and polluted waterways. In 1989 a large Canadian mining company named Noranda appeared on the scene when it announced plans to construct a large underground mine topped by a massive tailings holding pond. The project, when completed, would enable Noranda to extract $800 million worth of minerals using new technologies. Despite the promise of new jobs and extensive tax and royalty revenues, Noranda's proposal evoked an intense negative reaction from within the nearby communities as well as conservation groups and the Park Service. Even the state of Wyoming balked at the proposal, concerned that it would bear the heightened risk of downstream pollution from the mine yet derive few economic benefits. Nonetheless, the Forest Service, whose lands were involved in the mine proposal, indicated it was prepared under the General Mining Law to approve the project.[42]

The proposed New World Mine presented an assortment of environmental, aesthetic, and economic concerns that may have been ignored during an earlier time but were deeply troubling in a community and region no longer hitched to mineral extraction. The proposed tailings pond, which Noranda asserted could be trusted to retain forever toxic mining waste deposits, was to be constructed in a seismically active area at the headwaters of three tributaries to nearby rivers. Should it fail, pristine waters in Yellowstone National Park, the Absaroka-Beartooth Wilderness Area, and Wyoming's Clarks Fork River, the state's first federally designated wild and scenic river, could be severely damaged. In fact, local streams coursing the mine site were presently sterile from contaminated mining waste runoff, but Noranda rejected any suggestion it bore legal responsibility for cleaning them up. The proposed mine was situated in prime grizzly bear habitat of critical importance to the region's slowly recovering bear population. Moreover, the sights and sounds of industrial mining activity would be evident from within Yellowstone, not what most visitors expected from their national park experience. Given these myriad impacts, significant local and national opposition soon surfaced, intent on blocking construction of the mine.[43]

Mine opponents ultimately succeeded through a combination of relentless public advocacy, strategic litigation, and presidential intervention. In 1995 President Bill Clinton drew national attention to the issue when he highlighted the controversy during a Billings, Montana, town meeting, where he observed that the New World Mine posed a serious threat to Yellowstone National Park. He then took action, announcing a 19,000-acre federal land withdrawal to protect the at-risk watersheds from further mining activity. In late 1995, troubled by the mine location, the United Nations' World Heritage Committee listed Yellowstone as an "in danger" World Heritage site, further galvanizing local and national opposition to the project. On another front, conservation groups prevailed in a lawsuit against Noranda under the Clean Water Act, establishing that the company was liable for ongoing acid-mine discharges and thus faced massive cleanup costs that threatened the project's economic viability. National and local news media universally editorialized against the project, citing the danger posed to Yellowstone as well as general environmental concerns. As public opposition mounted, politicians in Montana and Wyoming weighed in and announced their opposition to the mine, essentially signaling that Noranda's mine project was doomed.[44]

The matter was finally settled through negotiations that removed the mine threat. With their Clean Water Act litigation victory, conservation groups succeeded in bringing Noranda to the negotiating table, where they were soon joined by federal officials. The final agreement—announced in Yellowstone by President Clinton—involved a contingent $65 million payment to Noranda that enabled the United States to secure ownership of the mining district lands in exchange for other federal lands and assets. Noranda agreed to place $22 million in escrow to cover cleanup costs for the polluted mine lands, and conservation groups agreed not to press additional pollution cleanup claims against either Noranda or the federal government. Despite the powerful General Mining Law and the allure of local jobs and related economic benefits, the broad coalition of mine opponents succeeded in stopping the New World project, drawing on the threat to Yellowstone along with an unmistakable evolution in local attitudes toward national forest management priorities and the area's natural attributes. As the New World controversy unfolded, it starkly revealed the mining site's ecological connections to the park as well as the surrounding landscape, reinforcing the importance of an ecosystem approach to resource management within the GYE, regardless of existing boundary lines.[45]

Less than twenty years later, troublesome new mine proposals surfaced on other old mining claims located near Yellowstone, prompting another round of controversies. As in the New World Mine matter, conservation groups, local citizens, and the Park Service rallied in an effort to block the projects, convinced any such intensive industrial activity did not belong near the park where it imperiled critical wildlife habitat and vital watersheds. One proposal by Lucky Minerals sought to explore old mining properties in Emigrant Canyon situated above the popular Chico Hot Springs resort roughly thirty miles north of the park. Another sought to revisit the old Jardine mining district above the gateway town of Gardiner and less than a mile from the park's northern boundary. Public opposition was intense. A 400-member Park County Business Council was swiftly assembled to oppose the mines, while a vocal opponent of the mines succeeded in being elected to the Park County Commission, which traditionally had favored such economic development activity. When mine opponents convinced the Forest Service to conduct an environmental assessment of the Emigrant proposal, Lucky Minerals promptly dropped its plans to mine on national forest lands, but pressed forward on nearby private lands. Meanwhile, two Interior secretaries—Sally Jewell in the Obama administration and Ryan Zinke in the ensuing Trump administration—signed formal land withdrawal orders that prohibited mining activity for twenty years on more than 30,000 acres of national forest lands north of the park. Montana senator Jon Tester followed by securing congressional passage of the Yellowstone Gateway Protection Act that permanently withdrew these lands from mining.[46]

Though the congressional legislation blocked future mining activity on national forest lands north of Yellowstone, it did not extend to privately owned lands or valid existing mining claims. When Lucky Minerals received state approval to explore its privately owned claims in Emigrant Canyon, conservation groups went to the Montana state courts. There they secured an injunction prohibiting mining activity pending completion of a legally valid environmental assessment, one that adequately addressed potential wildlife and road construction impacts, which appears to have derailed the project.[47] Although the proposed Crevice Mine in the historic Jardine mining district generated similar opposition, the mine owner appeared intent on exploring the claim on privately owned lands under Montana's small-mine exemption. But to mine on nearby national forest lands, he had to establish a valid preexisting mining claim, and was prohibited by the Yellowstone Gateway Protection Act from expanding his mining activity onto the withdrawn national forest lands. Nonetheless, the possibility of an active mine adjacent to Yellowstone raised alarm bells for the Park Service, Greater Yellowstone Coalition, and other conservation groups. Faced with potential legal and financing challenges, the owner agreed to a buyout arrangement with the coalition to resolve the matter. After raising $6.25 million, the coalition consummated the deal, putting an end to the mining threat. In the process, the coalition borrowed a strategic page from earlier oil-and-gas controversies, choosing to buy out the mining rights rather than engage in uncertain litigation.[48]

In 2017 another worrisome mining proposal surfaced in the Caribou-Targhee National Forest near the small Idaho town of Kilgore, roughly fifty miles west of Yellowstone. Although distant from the park, the exploratory project—estimated to involve recovery of 825,000 ounces of gold—would bring unwelcome industrial activity and new roads to this corner of the ecosystem. The site sits in the Centennial Mountain Range, which affords GYE wildlife a migratory pathway to central Idaho wilderness areas and boasts pristine streams. Seeking to protect the area, conservation groups twice sued the Forest Service in Idaho federal court, alleging potential harm to the area's water quality, its Yellowstone cutthroat trout population, grizzly bears, and related wildlife corridor concerns. The court ultimately rejected these claims, however, upholding the agency's decision and NEPA analysis, finding adequate mitigation measures to protect the area's surface and groundwater. With the court's approval, the mine operator is poised to begin exploratory drilling at 131 sites over a five-year period, though limited to working only six months each year.[49]

A quite different situation prevails elsewhere on the Caribou-Targhee National Forest in southeast Idaho, where an entrenched phosphate mining industry continues to thrive. This remote corner of the GYE harbors immense phosphate deposits that have been mined in the area since the late 1800s. Today, under the Mineral Leasing Act of 1920, major corporations—including Simplot, Monsanto, Bayer, and Agrium—hold federal leases blanketing more than 46,000 acres of national forest and BLM lands currently pockmarked by large open-pit phosphate mines. Phosphate is a valuable ingredient in an array of agricultural and consumer products, including fertilizers, herbicides, toothpaste, and dietary supplements. Not only does southeast Idaho's aptly labeled "Phosphate Patch" provide well-paying jobs and tax revenues in this rural area, but it also supports several processing facilities, including a large plant situated more than eighty miles away in Pocatello, Idaho. In short, communities in both Idaho and Wyoming are economically tied to the phosphate mines, a fact not lost on local politicians as well as the federal land management agencies overseeing mine operations and expansion.[50]

Large-scale phosphate mining in the Caribou-Targhee National Forest has produced a troubling environmental legacy manifested in poisoned streams and toxic groundwater, the result of selenium contamination from mine waste material. In fact, more than a dozen area mines are designated federal Superfund sites under ongoing cleanup orders. Nonetheless, the Forest Service and BLM continue to approve mine expansions, and the courts have proven reluctant to constrain additional mining activity. In 2009, in an effort to limit further selenium pollution, a coalition of conservation groups mounted an unsuccessful legal challenge to the Simplot Company's request to expand its Smoky Canyon Mine, asserting various Clean Water Act violations. Although two judges on the Ninth Circuit Court of Appeals rejected their arguments over the adequacy of proposed remediation plans, the dissenting judge discerned the controversy from a different perspective. In her opinion, Judge Betty Fletcher candidly observed the powerful economic forces at work in the matter, noting the parties intervening in the case included several towns and communities in Idaho and Wyoming, J. R. Simplot Company, the Idaho Farm Bureau Federation, and the United Steelworkers Union. Since then, the BLM has approved additional phosphate mine expansions, including the Bayer Company's Caldwell Canyon Mine expansion plan. However, an Idaho federal court vacated the BLM's Caldwell Canyon decision, finding the agency did not adequately analyze the impacts on sage grouse and the nearby processing plant. Despite this setback, the company is still pursuing the project, and the BLM is beginning a new NEPA analysis. Phosphate mining plainly remains a dominant local industry and political force in this remote area.[51]

A somewhat different picture emerges at the Stillwater mining complex located on Custer Gallatin National Forest lands in the northeastern reaches of the GYE. Situated in the Boulder River watershed at the end of a lengthy roadway penetrating deep into the Absaroka-Beartooth Wilderness Area, the Stillwater Mine began operation in 1986 to extract platinum and palladium. Since then, the mine owners have opened the East Boulder Mine to develop a second ore body. The two mines represent the only sites in the United States where these rare precious metals, which are used in catalytic converters as well as electronic and health care products, are being mined. Once the Stillwater Mine opened, local residents and conservation groups, concerned about potential water-pollution problems linked to the toxic waste holding pond, filed a lawsuit that was eventually dismissed when Montana amended its water-quality standards. Reversing course, the groups—Northern Plains Resource Council, Cottonwood Resource Council, and Stillwater Protective Association—met with the mining company to address their concerns.

A year later, in 2000, the parties announced a "Good Neighbor Agreement" (GNA) that established an ongoing contractual relationship between themselves designed to protect the nearby communities and residents as well as downstream water quality. Unique in the GYE region, the GNA committed the parties to negotiate rather than litigate their differences; it established a respectful, transparent relationship that obligated the company to provide the groups with outside experts to assess the impacts flowing from future mining projects. Since then, the GNA parties have usually managed to resolve their differences, offering a different model for addressing future GYE mining conflicts.[52]

The GNA model, however, is being tested following the Stillwater Mining Company's request to expand its East Boulder Mine. The proposal, which involves expanding the mine's toxic waste tailings pond and the waste rock disposal site, would double the size of the mine's footprint to 729 acres of national forest land. During the NEPA scoping process, conservation groups outside the GNA parties have raised a series of environmental concerns, including surface and groundwater pollution, wildlife and fisheries impacts, incursion into an inventoried roadless area, and reclamation requirements. The GNA signatories have expressed similar concerns. In late 2023 the Forest Service released its final EIS, approving the project with modifications designed to better protect water quality and to improve reclamation requirements.[53]

Industrial-scale mining activity, though still present in the GYE's distant corners, is otherwise mostly absent from the ecosystem, reflecting the profound shift in public values that has occurred throughout the region. Despite the powerful General Mining Law, conservation organizations and local advocacy groups have succeeded in stopping several economically attractive mining projects, even succeeding in enlisting the president and Congress to support their efforts. There is clearly little appetite for new industrial activity in the shadow of Yellowstone National Park—the core of the GYE—where nearby communities have attached their economic and cultural futures to the park, wildlife conservation, recreation, and tourism. Further from the park, in the Phosphate Patch and perhaps the Stillwater and Kilgore projects, the environmental impacts associated with mining are perceived locally as a lesser concern when balanced against the substantial economic benefits available. The conclusion is inescapable: Well-organized local opposition abetted by strategic litigation and national political alliances have succeeded in reversing the region's historical connections to mining, helping to reorient the GYE toward a new era—one that recognizes the region's ecological connections and attaches real value to its natural attributes.

Wilderness: Preserving the GYE's Wild Character

Wild nature is a defining characteristic of the GYE, captured most visibly in the region's designated wilderness areas. More than 3.7 million acres of national forest land within the GYE enjoy official wilderness designation, putting them out of bounds to new roads, industrial activities, vehicles, and even mountain bikes. Another 2.5 million acres—including 338,000 acres of national forest land and 2.1 million acres in the two GYE national parks—have been officially designated wilderness study areas (WSAs) and are thus eligible for wilderness designation. Yet the GYE wilderness debate has dragged on for more than forty years and remains yet unresolved. A hotly contested political issue, the wilderness question can only be settled by congressional legislation, which will require a broad and durable local political consensus. Although Congress passed statewide wilderness bills in 1984 for most of the western states, including Wyoming, neither Montana nor Idaho was included in the tranche of bills. Since then, wilderness proposals involving the GYE's national forest lands have surfaced periodically, generating the hope for a breakthrough that has not materialized; rather, the battle lines have hardened, particularly over recreational access for off-road vehicles (ORVs) and mountain bikes. Meantime, most of the GYE's undisturbed national forest lands derive a layer of protection from the Forest Service's 2000 roadless area rule.

The Wilderness Act of 1964 established a national wilderness system consisting of largely undisturbed federal public lands that today cover nearly 112 million acres. Over the course of eight years in the aftermath of the intense 1950s Echo Canyon dam controversy, wilderness advocates—led by the Wilderness Society—doggedly petitioned Congress to adopt legislation that would limit the rapidly expanding human imprint on the nation's public lands. And that is what Congress finally did when it passed the Wilderness Act, which defined "wilderness" as "an area where the earth and its community of life are untrammeled by man . . . [that] retain[s] its primeval character and influence without permanent improvements or human habitation."[54] The act vested Congress and the president with the final say over wilderness designation, thus ensuring that the process was inherently political. To protect the character of designated wilderness areas, the act prohibited roads, motorized vehicles, commercial facilities, logging, mining, and other industrial activities. The legislative history underlying the act suggests a primary purpose was to provide a different recreational experience for those intrepid enough to seek solitude and adventure in the backcountry, though the act also references "ecological" and "scientific" features in the "wilderness" definition. Over time, scientists and others have recognized that undisturbed wilderness lands are profoundly important for conservation purposes, providing refuge for human-sensitive species like grizzly bears, cougars, and wolves.[55]

In combination, the GYE's national parks and wilderness lands constitute a largely intact landscape as well as a defining regional characteristic. Yellowstone National Park—the two-million-acre core of the GYE—is bounded on three sides by large designated wilderness areas covering roughly 2.4 million acres: the Absaroka-Beartooth, North Absaroka, Washakie, and Teton wilderness areas. Only the park's western flank, abutting Idaho's Caribou-Targhee National Forest and Montana's Custer Gallatin National Forest, is not buffered by wilderness lands, though a segment of Montana's Lee Metcalf Wilderness adjoins the park boundary. Grand Teton National Park is likewise bordered by wilderness areas—the Teton, Gros Ventre, and Jedediah Smith—and by the National Elk Refuge and John D. Rockefeller Parkway. The Wind River mountain range, which extends roughly 100 miles southeastward from the GYE core, contains three large wilderness areas: the Bridger, Fitzpatrick, and Popo Agie. In total, these legally protected public lands represent one of the largest wildland complexes in the Lower 48 states.

Today the GYE wilderness areas afford the region's wildlife valuable sanctuary while also offering the local populace and visitors a cherished recreational setting. The region's wilderness lands provide critical habitat as well as migratory pathways for grizzly bears, elk, deer, pronghorn, wolves, bighorn sheep, cougars, and other fauna, enabling them to seasonally move from the core national park lands to lower-elevation settings. As climate change takes firm hold, wilderness lands are becoming even more important for the region's wildlife. Conservation groups, recognizing key species' far-flung migration patterns and ongoing dispersal needs, have long sought to designate additional wilderness areas, which will enable wildlife to move securely throughout the ecosystem and connect with distant ecosystems, namely the central Idaho and Glacier-Bob Marshall wilderness complexes. These efforts, however, have yet to bear fruit in the form of new wilderness designations.[56]

The GYE national forest wilderness areas were established more than forty years ago beginning with the original 1964 Wilderness Act and encompass 3.7 million acres. No new GYE wilderness areas have been designated since 1984, when Congress passed the Wyoming Wilderness Act.

The expansive Bridger, Teton, Washakie, and North Absaroka wilderness areas date back to the original 1964 legislation, while the other GYE wilderness areas have been added since then in various bills, culminating in the 1984 Wyoming legislation. Although the GYE's national park lands are eligible for wilderness designation and both Yellowstone and Grand Teton parks have made official wilderness recommendations covering more than two million acres, Congress has not shown any interest in moving on these recommendations. The Forest Service, judging by existing forest plans, is also not keen about new wilderness designations. The 2015 Shoshone forest plan, noting that 55 percent of the forest already enjoys wilderness protection, did not recommend any new designations, even though more than 684,000 acres were considered roadless. The revised 2022 Custer Gallatin forest plan recommended merely 140,000 acres of new wilderness, yet the forest contains 847,000 acres covered by the agency's roadless rule. Similarly, the revised Caribou and Targhee forest plans proposed only paltry wilderness additions covering less than 200,000 acres within this 2.6-million-acre forest, where only 134,000 acres are presently designated as official wilderness. The 3.4-million-acre Bridger-Teton National Forest contains roughly 1.2 million acres of wilderness, but the existing forest plan makes no new wilderness recommendations. Whether the Forest Service will see its way clear to support significant additional wilderness as it continues to revise the GYE forest plans remains to be seen.[57]

In Montana, conservation groups have long pursued new GYE wilderness designations without success, despite several high-profile collaborative efforts designed to end the impasse. In the northwestern corner of the GYE, the Beaverhead-Deerlodge Partnership—a limited group composed of local conservation and timber company representatives—recommended wilderness protection for roughly 570,000 acres in these combined forests, substantially more than the 329,000 acres recommended for wilderness in the Forest Service's 2008 draft forest plan revisions. Although the group secured support from Montana senator Jon Tester who introduced legislation endorsing the proposal, it met strong political resistance from fervent wilderness advocates who objected to provisions guaranteeing a steady timber supply for forest restoration purposes and the exclusion of some wilderness-eligible lands. The Forest Service also objected to the bill, citing the lack of science behind the large-scale restoration cutting provision. Although introduced in Congress during three sessions, the bill failed to pass, which means the agency's roadless rule—covering 1.8 million acres in the forest—provides the primary means of legal protection. The Beaverhead-Deerlodge wilderness debate continues unresolved with little prospect of a resolution on the horizon.[58]

Elsewhere in Montana, the Custer Gallatin National Forest found itself deeply embroiled in the wilderness controversy during its six-year forest-plan revision process. Much of the controversy focused on the 155,000-acre Hyalite-Porcupine-Buffalo Horn Wilderness Study Area (HPBH WSA), which sits astride the Gallatin Range and runs northward from just outside Yellowstone's northwest corner for thirty-six miles. In the years following adoption of the Montana Wilderness Study Act of 1977, the HPBH WSA attracted mounting recreational activity, including ORVs, snowmobiles, and mountain bikes.[59] Concerned about the growing impacts on the area's wilderness and wildlife values, conservation organizations sued the Forest Service to curtail motorized uses and mountain biking. They succeeded in convincing the courts that the agency was legally obligated to maintain the WSA's wilderness character until Congress finally decided whether to include it in the national wilderness system. Soon thereafter, a local group calling itself the "Gallatin Community Collaborative" coalesced to negotiate a wilderness compromise, but its efforts proved unsuccessful.[60]

In 2018 the Forest Service released the draft Custer Gallatin forest plan setting forth its initial 70,000-acre wilderness proposal for the HPBH WSA, with additional backcountry and recreation designations where motorized and mechanical activities would be allowed. A new collaborative initiative—the Gallatin Forest Partnership composed of conservation and mountain-biking groups—surfaced and eventually endorsed a revised proposal slightly increasing the wilderness acreage while providing a range of mountain-biking opportunities. The proposal pleased neither committed wilderness advocates, who argued the entire HPBH WSA should be protected, nor the motorized-recreation groups, who wanted little or no additional wilderness.

Impressed by the Gallatin Forest Partnership's work, which had gained important political support from nearby county commissions and other groups, forest supervisor Mary Erickson endorsed a modified version of its proposal. Her decision, released in 2022, recommended 78,071 acres of the HPBH WSA be set aside as wilderness, with an additional 45,910 acres designated as backcountry areas and another 12,500 acres designated as recreation emphasis areas. Erickson's decision, in her own words, essentially endorsed the agency's existing approach to the forest's WSA lands: "those areas . . . are manageable, [and] currently have few or no uses that would be inconsistent with wilderness designation."[61]

The Custer Gallatin National Forest contains more than 740,000 acres of designated roadless lands, much of which potentially qualify for wilderness protection. The revised forest plan, however, only recommended 140,000 acres for wilderness designation, consisting mostly of separate patches not exceeding 14,500 acres except the HPBH WSA recommendation. Nonetheless, the 140,000-acre figure far surpassed the 33,700 acres recommended for protection in the original forest plans finalized during the 1980s. But the Forest Service's final acreage figure fell far short of the 704,000-acre proposal advanced by the Gallatin Yellowstone Wilderness Alliance, which argued extensive wilderness lands were necessary to protect the forest's rich wildlife assemblages. Most conservation groups, while disappointed that the Forest Service's final recommendation lagged their own proposals, largely accepted the outcome. Mountain-biking organizations were pleased with the overall result, which retained most existing biking trails. Motorized-recreation groups viewed the wilderness recommendation differently, generally opposing any new wilderness designations that would further foreclose ORV, snowmobile, and motorcycle activities. Staunch wilderness-wildlife advocates lamented the plan's wilderness recommendation as a missed opportunity to fully safeguard wildlife habitat and corridors north of Yellowstone.[62]

Congress has the final word on the Forest Service's Custer Gallatin wilderness recommendations, and the Montana delegation has yet to introduce legislation. For the most part, the forest plan proposal maintains the status quo, basically tracking how the forest's wilderness-eligible lands are being managed. In fact, the forest supervisor candidly observed that her wilderness recommendations "represent[ed] little change in the overall forestwide suitability of motorized and mechanized recreation opportunities from current conditions."[63] Without question, in the Custer Gallatin wilderness debate, recreation has become a priority forest resource, reflective of the larger changes occurring throughout the GYE national forests. Conversely, wildlife conservation concerns proved not as dominant a factor in the debate, presaging continued wildlife-recreation controversies across the region. It is not surprising, then, that a new 102,000-acre wilderness proposal for the Gallatin Range has drawn criticism from wilderness-wildlife advocates.[64]

In the southern GYE, the much-touted Wyoming Public Lands Initiative (WPI) failed to resolve lingering wilderness questions. The WPI, conceived in 2015 by the Wyoming County Commissioners Association, invited the state's counties to address the fate of their national forest and BLM WSA lands through a transparent and inclusive process with the goal of producing a statewide wilderness bill. In the GYE the process focused on the Palisades and Shoal Creek WSAs in the Bridger-Teton National Forest along with several BLM WSAs. The 136,000-acre Palisades WSA spans Wyoming's Teton and Lincoln counties, with 53,000 additional acres situated on adjacent Idaho national forest lands. In Teton County, despite the efforts of a broadly representative advisory committee, wilderness advocates and motorized-recreation groups proved unable to reach agreement, while neighboring Lincoln County, with an economy resting on ranching and mining, opposed any wilderness protection and refused to even participate in the process. Elsewhere, Teton and Sublette counties initially agreed to discuss wilderness protection for the shared Shoal Creek WSA, but to no avail in the end.[65]

Deep-seated opposition to wilderness across much of Wyoming ultimately doomed the WPI effort, at least as it related to the GYE. Congressional legislation introduced in 2018 by US House representative Liz Cheney as the "Restoring Local Input and Access to Public Lands Act" reflected this sentiment and was plainly designed to kill the WPI effort. But when several Wyoming counties unexpectedly managed to achieve a level of consensus over wilderness, US senator John Barrasso introduced legislation that ignores the GYE WSAs and remains pending. The WPI episode further highlights the political and jurisdictional complexities overlaying the GYE landscape and public land management issues. It proved impossible to bridge the divide between neighboring Teton and Lincoln counties over the Palisades WSA. But even had the two counties reached agreement, the Idaho portion of the Palisades WSA would have remained unresolved. Meantime, ORVs and snowmobiles continue to surreptitiously invade the area, eroding its wilderness character and diminishing its quality wildlife habitat.[66]

Although the Idaho portion of the GYE contains little formal wilderness acreage, the Caribou-Targhee National Forest lands offer wilderness designation opportunities. Covering 2.6 million acres, these combined forests manage only 134,000 acres of official wilderness, namely the Jedediah Smith and Winegar Hole wilderness areas adjacent to Grand Teton National Park. The Targhee's 1985 forest plan recommended a mere 65,000 acres for wilderness protection, which then grew to 171,000 acres in its 1997 revised plan, while the Caribou only recommended two small areas for wilderness designation in its revised plan. Though the Targhee shares the Palisades WSA with Wyoming's Bridger-Teton National Forest, there was no apparent effort to review these lands in tandem with Wyoming's WPI effort—representing a missed opportunity to coordinate across boundary lines for conservation purposes. Further north in the Targhee, the Centennial Mountain Range is regarded as a key connective corridor for grizzly bears and other animals from Yellowstone, but the Forest Service has not identified any of these ecologically important lands for wilderness consideration. When a proposal surfaced in 2014 to establish a Caldera National Monument in the Targhee forest adjacent to Yellowstone's western border, it was promptly condemned and soon faded from public view. Given the level of local opposition, additional wilderness protection seems unlikely in the Idaho GYE national forests, which have long been associated with logging, mining, grazing, and motorized recreation.[67]

In sum, wilderness proposals involving GYE national forest lands have consistently languished during the forty years that have elapsed since Congress passed the 1984 Wyoming Wilderness Act. Any wilderness proposal generates passionate feelings on all sides of the matter, presenting manifold political challenges that have yet to be surmounted in the region. Too many hardened and diverse interests—wilderness proponents, the extractive industries, county officials, and motorized recreationists, as well as factions within the conservation community itself—must be brought together in order to pass congressional wilderness legislation.[68] Even when collaborative groups, such as the Beaverhead-Deerlodge Partnership or the Gallatin Forest Partnership, have managed to coalesce around a proposal, they have met stiff resistance that has stifled congressional action. Moreover, it has proven near impossible to bring the affected states, counties, and local communities together behind any shared wilderness proposal, while the GYE national forests have shown little interest in coordinating any cross-boundary wilderness review process. Though the existing GYE wilderness areas are widely acknowledged as central to the region's identity and play a critical wildlife conservation role, there is little prospect in this politically fragmented area for additional wilderness acreage in the foreseeable future.

Notwithstanding the prolonged wilderness stalemate, the GYE's undisturbed national forest lands enjoy some degree of legal protection. In 2000 the Forest Service adopted a nationwide roadless area rule that protected 58.5 million acres of national forest lands—nearly one-third of the entire national forest system acreage—from most industrial activity. Due to their wilderness characteristics, these un-roaded forest lands had long generated intense controversy usually ending in litigation whenever a development proposal surfaced that threatened to eliminate the wilderness designation option for these lands. The rule prohibited new road building and timber harvesting on the designated roadless lands with some exceptions. According to the Forest Service, its roadless lands served as "biological strongholds for terrestrial and aquatic plants and wildlife and as sources of clean water," and also "provide large, relatively undisturbed landscapes that are important to biological diversity and the long-term survival of many at risk species."[69] In the GYE, national forest roadless lands blanket nearly six million acres across the five national forests; the Bridger-Teton contains 1.4 million inventoried roadless acres, the Beaverhead-Deerlodge has 1.65 million acres, and the other forests have lesser amounts.[70] Researchers have confirmed that national forest roadless areas enhance wildlife and other conservation values, particularly when the area is located nearby to national parks and wilderness areas, which is the case in the GYE.[71]

Not everyone was pleased when the Forest Service prohibited intensive development activity on its roadless lands. Wyoming and Idaho led the way among displeased states and challenged the roadless rule in two separate lawsuits, one of which dragged on for ten years before the courts finally confirmed the rule. Arguing that the Forest Service's action amounted to the creation of new wilderness areas—a decision that Congress reserved for itself in the Wilderness Act—Wyoming secured an injunction in its federal district court blocking the rule from taking effect. The Tenth Circuit Court of Appeals reversed, however, ruling that the agency acted well within its statutory authority and that the roadless rule did not establish de facto wilderness lands.[72] Meanwhile, the George W. Bush administration sought to alter the roadless rule to allow individual states to decide national forest roadless designations for themselves, but the courts twice blocked this revisionary effort. Along the way, Idaho successfully petitioned the Forest Service to alter more than thirty roadless area designations in the GYE's Caribou-Targhee National Forest. Although conservation groups challenged Idaho's actions in court, the state prevailed, leaving portions of the Caribou-Targhee's roadless lands with reduced protection. Barring another assault on the 2000 roadless area rule, the GYE's revised national forest roadless acreage seems relatively secure, bringing significant ecological benefits to the region's national forests and resident wildlife.[73]

The Recreation Enigma: Is the GYE Being Loved to Death?

By nearly any measure, recreation has burgeoned across the GYE and now represents a primary use on the region's national forest lands. In fact, recreational activity has ballooned throughout the national forest system, growing from 46 million visits to national forests in 1955 to more than 148 million visits in 2016, while contributing $9 billion and 148,000 jobs to the nation's economy.[74] The revised Custer Gallatin forest plan illustrates the outsized economic role of recreation in the GYE: In this forest alone, recreation generates $85 million in local income and 2,900 jobs, compared to timber, which generates $19 million in local income and 410 jobs, and livestock grazing, which produces $12 million in local income and 384 jobs.[75]

Across the GYE, more people are engaging in more different activities with attendant ecological and social consequences. Owing to new technology, ORVs and snowmobiles have become more powerful and can penetrate ever deeper into once-remote backcountry, while mountain bikes are now ubiquitous on national forest trails, where they too are bringing riders into areas that were once inaccessible. Electric bikes further magnify these problems. The inevitable consequences of this explosion in recreation on the national forests are evident throughout the GYE: wildlife displaced from important habitat and growing conflicts among recreationists, often between those seeking speed and thrills and those seeking solitude and quiet. Indeed, knowledgeable, longtime observers view recreation as one of the GYE's most challenging problems, characterizing it as "a ticking time bomb" and the "third rail of conservation."[76]

Although one might assume that designated WSAs prohibit intensive recreation activity and provide refuge for wildlife, that assumption is not accurate in the GYE. Under a WSA designation, preexisting activities, including ORV use, snowmobiling, and mountain biking, are allowed to continue so long as they do not alter the wilderness characteristics that existed at the time of the protective designation. In 2005 a conflict surfaced involving the Bridger-Teton National Forest's Palisades WSA when the Forest Service renewed a commercial helicopter ski operator's special-use permit to allow a tenfold increase in permitted skier days. Conservation groups objected to the impact increased helicopter traffic would have on the WSA and went to court, arguing that, under the 1984 Wyoming Wilderness Act, the agency was obliged to maintain the area's existing wilderness character for potential wilderness designation. The court interpreted the term "wilderness character" to mean maintaining "outstanding opportunities for solitude or a primitive and unconfined type of recreation," and concluded that the increased volume of helicopter traffic violated this standard. The matter was settled when the operator agreed to phase out the increased flights over a five-year period. Regardless, the court ruling imposed important limits on the type and amount of recreational activity allowed in designated WSAs, which were intended to protect the area's preexisting wilderness character.[77]

Further north, a similar controversy arose on Montana's Gallatin National Forest in response to mounting motorized and mechanical recreational pressures in the Hyalite-Porcupine-Buffalo Horn (HPBH) WSA. When the Forest Service completed a travel plan that ignored the explosive growth in motorized activity across the HPBH WSA, citing a lack of pre-1977 data, conservation groups turned to the courts. And the courts agreed that the agency had violated its legal obligation to protect the WSA: "The Service entirely failed to explain how the travel plan provides current study area users with opportunities for solitude comparable to those that existed in 1977 despite increased volume of motorized and mechanical use."[78] The practical effect of these court rulings is to safeguard the GYE's WSAs from being overrun with new recreational users while the wilderness designation debate plays out in the political arena. This buys time for wilderness advocates to build support for permanent protection of these lands, while also providing a modicum of protection for wildlife dependent on these contested lands.

Recreation controversies extend well beyond the GYE's WSAs, infecting the region's multiple-use national forest lands too. A principal point of conflict, not surprisingly, involves motorized uses, both the environmental impacts ORVs visit on the landscape and the potential for conflict with other recreationists seeking a quieter outdoor experience. By all measures, ORV use has grown exponentially during the twenty-first century, as reflected in figures from Montana where the number of registered ORVs increased from 14,000 in 1995 to 77,200 in 2013.[79] Early on, the problems presented by burgeoning ORV use on the public lands drew presidential attention in the form of two executive orders. In 1972 President Richard Nixon directed the federal land management agencies to zone their lands to minimize environmental and wildlife damage as well as user conflicts linked to ORV use. And in 1977 President Jimmy Carter prohibited ORVs from areas where they might cause "considerable adverse effects."[80] The courts have not hesitated to enforce these executive orders and related Forest Service regulations in the GYE. In a case involving the Beaverhead-Deerlodge National Forest, for example, the courts ordered the Forest Service to limit snowmobile use to protect at-risk wildlife habitat.[81] Other cases have upheld the Forest Service's authority to close undeveloped areas to mountain-bike use.[82] As recreation pressures mount in the GYE and proponents garner additional political and economic clout, the emerging question is whether the Forest Service has the fortitude to assert its legal authority to limit potentially damaging motorized and mechanical recreational activities to protect the GYE's ecological integrity.

Another long-standing GYE recreation issue involves public access to national forest lands interspersed with privately owned lands, particularly involving checkerboard ownership patterns. This is not only a problem in the GYE but across the public lands, and one that is becoming more challenging as private ranchlands are changing hands and being subdivided. Recreational access concerns surfaced more than fifty years ago in the Custer Gallatin National Forest's Crazy Mountains, where private landowners and recreationists have come into conflict over use of local trails and roads that cross privately owned lands. After lengthy negotiations, the Forest Service has endorsed a land exchange to consolidate federal ownership in the area as well as miles away in the Big Sky area. The complicated multiparty exchange—which must satisfy legal equal value and public interest requirements—involves exchanging 3,855 acres of national forest land for 6,110 acres of privately owned lands. According to the Forest Service, the Crazy Mountains dimension of the exchange would resolve long-standing backcountry recreation access problems, increase wildlife habitat, ensure tribal access to sacred sites, and preclude possible development in the area. The Big Sky portion of the exchange would have similar recreation and wildlife benefits by transferring 425 acres of forestland protected by a conservation easement to the Yellowstone Club for downhill skiing purposes in exchange for 605 nonfederal acres. Individual landowners have indicated a willingness to allow continued access across their lands and to negotiate conservation easements in some cases. Although many conservation and sportsmen groups support the proposal, others are skeptical of its wildlife benefits as well as the conservation easement promises. They also fear potential water pollution emanating from the Yellowstone Club's snowmaking practices. The matter highlights the legal and other complexities involved in resolving recreational access problems in the region's national forests—a problem that will only worsen as outdoor recreation continues to grow, conflicts with landowners mount, and private lands continue changing hands.[83]

The ascendancy of recreation as a principal use of the GYE's national forests is further manifested in the region's expanding ski resorts. A prime example is the ongoing transformation of the Grand Targhee Resort, which is located in Wyoming on national forest lands, perched just outside Grand Teton National Park and adjacent to the Jedediah Smith

Wilderness Area. Established in 1969 as a modest local ski area with a couple chairlifts, successive resort owners have gradually converted the area into more of a destination location. The resort crowns Idaho's Teton Valley, formerly a sleepy farming community that has evolved into a hub for outdoor enthusiasts and vacation homes. In the process agricultural lands abutting the small towns of Driggs and Victor have been rapidly subdivided to accommodate the influx of new residents and visitors, outpacing the ability of local officials to keep up with the development pressures. Many visitors to the resort, which has limited mountainside accommodations, stay at the base of the mountain, putting them in Idaho while the ski area sits in Wyoming.[84]

In 2018 the Grand Targhee Resort submitted a master development plan to the Forest Service, seeking a special-use permit to add another 600 acres of skiable terrain, new chair lifts, a mountain-top restaurant, and summer-season mountain-bike trails. Additional lodging was also contemplated on the base area private lands to accommodate more overnight visitors. Problems surfaced immediately, however. The proposed new skiing areas—intended to double the resort's skiable acreage—would extend into critical bighorn sheep habitat, raising alarm bells with Grand Teton park officials as well as Wyoming's Game and Fish Department over potential impacts on the area's small sheep population. The project would, moreover, bring additional visitors and development pressure to the nearby Idaho communities, requiring additional infrastructure to accommodate the growth. While these costs would be borne by Idaho's Teton County, Wyoming's adjoining Teton County would legally oversee the base area development on private land and thus reap the tax and other revenues associated with it. Although the plan has been modified to reduce the skiable acreage and number of new chair lifts, bighorn sheep habitat remains at risk. And the two counties have yet to resolve their differences—another example of the cross-boundary challenges embedded in the GYE's multistate jurisdictional complexities. If the Forest Service approves the modified expansion plans, as seems likely, the enlarged Grand Targhee Resort will further confirm the powerful role outdoor recreation now plays in the region's national forests.[85]

Across the Teton mountain range, the Jackson Hole Mountain Resort (JHMR) is situated twelve miles west of the town of Jackson, Wyoming, and represents the largest ski area in the southern GYE. Its steep, challenging slopes attract skiers from throughout the world, many of whom fly into the nearby Jackson Hole Airport. Built during the early 1960s, the JHMR ski area operates under a Forest Service special-use permit that covers 2,500 acres in the Bridger-Teton National Forest. Now a major year-round destination, the JHMR includes the Teton Village development at the base of the ski area. In 2022 the resort submitted a revised master development plan to the Forest Service seeking to secure an additional 620 acres to provide more summer recreation options in the form of mountain-biking trails and a rock-climbing site. The forest supervisor has accepted the proposal as consistent with the existing forest plan.

But before proceeding with any new projects, the resort will have to secure further agency approval, which will require NEPA review and public involvement. While the proposal does not involve new ski lifts, conservation groups and backcountry skiers fear such a request is inevitable, which could negatively affect the area's dwindling bighorn sheep population and displace powder-hungry backcountry skiers. Meantime, the JHMR remains a major attraction for skiers during the winter months, adding to the welter of recreation and tourism impacts cumulatively altering the town of Jackson and its surroundings to the dismay of many local residents.[86]

The Big Sky resort complex—located south of Bozeman, Montana, in the scenic Gallatin Canyon—personifies the marriage occurring in the GYE between recreation, tourism, and real estate development. The Big Sky development, established in the early 1970s on former national forest lands consolidated into private ownership through land exchanges, has transformed a bucolic mountain valley. Covering more than 76,000 acres and boasting 5,680 acres of skiable terrain, Big Sky is now dotted with upscale hotels, condominiums, and private homes, and crowned by the exclusive, members-only Yellowstone Club. No longer a wildlife haven in the Custer Gallatin National Forest, Big Sky has become a major destination resort setting with a local population that has grown from a few hundred in the 1970s to more than 3,300 permanent residents today, many of whom find the cost of living there increasingly unaffordable. Early on, poorly constructed water and sewer lines contaminated the local aquifer along with the nearby Gallatin River, adding to the area's extensive environmental impacts.[87]

Nonetheless, under its "2025 Vision," the resort continues to expand its skier capacity and to construct pricey new residential units. The plan contemplates adding 5,700 new units to the 4,300 units already in place, along with upgraded chair lifts and remodeled guest facilities. Recent additions include the Montage Big Sky luxury resort—billed as "Montana's biggest building"—with nightly rooms starting at $1500, and the One and Only Moonlight Basin development featuring a new lodge, seventy-three hotel rooms, nineteen villas, and sixty-two private residences for sale at $8.8 million each. Since the Big Sky area is privately owned, it is primarily state and local law that governs zoning and land use. Not only has this ongoing build-out converted the valley into a small city in the midst of what once were wildlife-rich forest slopes, but the area's growing popularity is bringing mountain biking and other recreational pressures onto the nearby Gallatin Range's wilderness quality lands. The resort's outsized presence, as we shall see, speaks to the notable absence of meaningful zoning and other legal constraints on much of the GYE's private lands, as well as the accumulating environmental costs associated with industrial-scale recreation facilities.[88]

* * *

Today, with notable exceptions, resource-management priorities in the GYE national forests reflect the region's emergent conservation values and recreation economy. Controversies over wilderness designation, mountain-biking trails, and ski-area expansion have largely replaced the battles of yesterday over clear-cut logging, mining projects, and oil rigs abutting national park lands. To be sure, significant industrial activity—including the Pinedale Anticline oil field, the "Phosphate Patch," and the Stillwater complex—still persists on GYE public lands, but mostly on the outer edges of the ecosystem. Timber management is now focused on restoring forest ecosystems to address the worsening wildfire situation rather than on commercial harvest quotas, though questions still surface about the environmental impact of individual project proposals. New mine proposals are being met with staunch opposition that has succeeded in blocking most such projects. Cattle grazing continues on the region's forest lands as do predator depredation problems, but sheep allotments are a distant memory in deference to the grizzly bear recovery effort. And owing to the roadless area rule, major portions of the GYE national forest lands are legally off-limits to logging and road construction. All of which has generally proved beneficial for the region's highly regarded wildlife populations.

Yet, the transition is neither complete nor without challenges. Despite the importance of the GYE's hallmark national parks and wilderness areas, politics have doomed the prospects for additional wilderness designations. While recreation activity initially appeared relatively benign, the sheer volume and scale of today's increasingly diverse activities is cumulatively putting new pressures on the region's backcountry lands, rivers, and wildlife, the very attributes that make the area so unique.

The region's growing ski resorts are only adding to these recreation pressures. Moreover, even as priorities have evolved in the national forests and even with conservation commitments evident, serious problems loom beyond the national forest boundary line. With more people flooding into the region, development has surged on the privately owned lands, creating devilish problems for wildlife and cluttering once-open spaces with new homes and fences. As more people are drawn to the GYE's natural attributes and once-uncrowded wild landscapes, these ever more important, nature-based qualities remain at risk.

CHAPTER SIX

A Fragmenting Landscape
The Private Land Conservation Challenge

An aerial view of the GYE exposes several distinct images of the underlying landscape. One image captures the region's expansive, largely undeveloped federal lands. It shows mostly unbroken forests and soaring mountain ranges punctuated by sparkling lakes, open meadows, and meandering rivers. Another image highlights the region's privately owned lands as well as unmistakable development pressures. It shows new subdivisions, homes, and multi-acre ranchettes spreading across the valley floors, steadily consuming the remaining open spaces. This GYE growth pattern—a testament to the region's popularity as an attractive place to live, work, and play—has fomented significant ecological, economic, and social changes, which in turn present interrelated conservation challenges. These challenges are mainly being addressed at the state and local levels, where the federal laws and policies applicable to the region's public lands are not available.

As originally conceived during the 1980s, the GYE concept and related ecosystem conservation strategies were primarily directed toward federal land management policies, seeking to protect the region's natural attributes and improve coordination between the Park Service and Forest Service. Because private lands occupy about a quarter of the GYE acreage, environmental advocates initially focused on the federal lands to control logging, mining, drilling, and other industrial-scale activities that threatened important wildlife habitat, water quality, and recreation opportunities. At the time, the region's relatively small communities faced minimal growth pressures, while nearby ranchlands provided abundant open space. Located at lower elevations near water sources, the GYE's extensive ranchlands seasonally supported wildlife seeking food and shelter during the harsh winter months.

Of course, some astute observers early on perceived the ecological importance of the region's private lands and anticipated the need to integrate them into conservation efforts. That time has plainly arrived, given escalating population pressures across much of the GYE and related subdivision activity with associated environmental and aesthetic impacts. Not only are open space and winter habitat evaporating in the region's valleys, but also migration corridors are being severed by new structures, roads, and fences while water quality is a mounting concern. Efforts are underway to address these increasingly urgent problems, including numerous scientific studies on wildlife use and behavior patterns, updated zoning and planning requirements, conservation easement transactions, livestock-management reforms, and other creative solutions. The federal government, though occupying a reduced role in these growing private land conservation efforts, is still an important player and one with money. But the region's state and local governments, along with an array of conservation organizations—all guided by state property, land-use, and wildlife laws—are key to these efforts, which rely heavily upon incentive-based strategies that property owners generally welcome.[1]

Private Land Development: A Deepening Problem

The GYE's privately owned lands are now widely recognized as a critical concern in sustaining the region's ecological integrity and natural values. Upon arrival, the early settlers claimed the lower-elevation riparian lands to support their agricultural and ranching activities. Today, as then, these lands provide critical wildlife habitat, movement corridors, and ecosystem services, which are now better understood owing to a wealth of scientific studies. Because the GYE's protected federal lands are mostly located at high elevations, the region's elk, deer, bison, bighorn sheep, and pronghorn populations regularly migrate onto lower-elevation lands in their quest for food and shelter during the winter months. Increasingly, they encounter new homes, roads, fences, and dogs, making it difficult for them to utilize their traditional wintering grounds. And as the GYE's grizzly bear and wolf populations have increased so too has their range, which can bring them into conflict with landowners as well as livestock and pets.[2]

Unparalleled growth across much of the GYE has fueled profound changes in land use and development. Owing to its attractive natural amenities, along with modern technology that makes working from afar possible, the region's populace more than doubled between 1970 and 2015, when it approached 450,000—a growth pattern that only intensified when the COVID-19 pandemic arrived. During the same period, the number of homes tripled, going from 79,000 to 228,000, which has significantly altered the landscape in several communities. The impact of this sprawl is stunning: From 1970 to 1999, while the region's population grew by a robust 58 percent, the amount of rural land devoted to exurban housing increased by a whopping 350 percent with corresponding impacts on wildlife, water and air quality. By 2010, due to rising population and development pressures, the GYE's developed lands—namely agricultural, exurban, suburban, and commercial lands, as well as roadways—stretched across more than a quarter of the area. The predictable result of all this development was a 50 percent decline in all habitat types on the region's private lands.[3]

No end is in sight as people continue to flood the region, filling its open spaces with new houses, structures, and fences. Indeed, current growth patterns put the GYE among the nation's fastest growing areas, and far eclipse what is happening elsewhere in the three GYE states. Communities like Bozeman and Jackson—situated near protected public lands with educated workforces and ready access to air travel—have been heavily impacted by in-migration. Other attractive GYE communities—Red Lodge, Livingston, and Driggs, for example—are also experiencing an influx of new residents and second-home buyers. Under current projections, Montana's Gallatin County (including Bozeman) will balloon from 110,000 to 220,000 people in less than twenty years, reaching the size of Salt Lake City by the 2040s, then hitting 450,000 people by the 2060s, equal in size to Minneapolis. Knowledgeable observers expect another 100,000 homes will be added to the GYE landscape by 2040, many in the form of large-lot, rural subdivisions.[4]

These GYE growth patterns have major implications for the region's federal lands as well as local communities. In the fifty years from 1950 to 2000, the number of rural homes bordering federal lands increased by 302 percent, presenting the responsible agencies with significant new wildfire control, wildlife management, and other challenges. As new homes proliferate adjacent to the GYE national forest lands, the Forest Service and its state counterparts face a growing wildfire threat that requires additional fuel treatments and corresponding expenditures, which have grown exponentially in recent years. With more homes sprouting on the region's lower-elevation lands and nearby hillsides, important wildlife winter range is being lost along with migration corridors. Conflicts with grizzly bears, wolves, and other predators have increased, and the same holds true for bison, elk, and other large mammals that are prone to destroy fencing, carry disease, and otherwise present a nuisance to homeowners. Moreover, the three major GYE ski areas have regularly pressed to expand onto adjacent national forest lands, encroaching onto wildlife habitat and further altering the natural character of these mountainsides. Although confronted with this array of external impacts, the federal land management agencies have little control over the region's escalating private land development pressures.[5]

The three GYE resort counties—Gallatin, Teton (Wyoming), and Teton (Idaho)—have drawn new residents at a near-record pace for several decades. These newcomers are attracted to the area by proximity to the two national parks, nearby airports, and ski areas that have become all-season destinations. Real estate in these counties, which host the Jackson Hole, Big Sky, and Grand Targhee ski resorts, is at a premium while subdivision and construction activity continues unabated. In Jackson, Wyoming, a single-family home sold for more than $3 million in 2022, a price that would make it nearly impossible for average workers to find affordable housing; rather, they must daily travel to work across the Teton Mountains from Idaho or northward from Star Valley, where their presence is putting new housing pressures on these landscapes. At Big Sky, Montana, a previously unoccupied mountain valley has filled with second homes, roads, condominiums, and ski lifts, displacing the native elk, grizzly bears, and other wildlife while imperiling the area's water resources. Big Sky's growth pressures extend to Gallatin County's Highway 191 corridor between Bozeman and Big Sky, where new commercial development and spiraling vehicle traffic are proving deadly to the local elk herd, which also faces loss of its migration pathway. In Idaho's Teton Valley escalating subdivision activity cooled during the Great Recession but then accelerated again, converting once-open farmland into new subdivisions, condominiums, and golf courses—all of which is straining county planners as well as local public services.[6]

Although GYE counties are not all experiencing the same intense growth pains, the problem is nonetheless pervasive throughout the region and increasingly controversial. Places like Montana's Paradise Valley and Madison River corridor, as well as Wyoming's Sublette and Park counties, are seeing new homes sprout across once-open ranchlands along with new subdivision plats. In Wyoming's Park County, subdivision permits in the Cody area increased by 500 percent between 2019 and 2020.[7] Even in the GYE's more remote pockets development pressures are seemingly ever present. In Sublette County, when a transplant billionaire acquired 1500 acres in the Upper Hoback Basin with plans for a high-end resort, his rezoning requests met stiff local opposition due to its impact on the area's rural character and "the world's longest known mule deer migration corridor." When the county commission, after approving the project with wildlife restrictions, refused to waive the restrictions, the owner abruptly terminated the partially completed project. Two other Sublette County development proposals that intrude upon existing migration corridors have also provoked controversy. One involves a 614-acre parcel slated to house a therapy center for traumatized young women that would adjoin the same mule deer pathway; the other involves a fifty-one-lot subdivision that would bisect the renowned Path of the Pronghorn corridor. Although the county planning commission recommended against both developments, it was overridden by the elected county commissioners, who viewed the projects as contributing to necessary local growth.[8]

By any measure the GYE's communities and privately owned lands are much different places today than a few decades ago. There are more people and more homes amid sprawling subdivisions and growing resort communities, which is carving up the landscape with attendant impacts on the area's wildlife, river corridors, and scenic character. Wildfire dangers have only mounted as homes materialize in the region's wildland-urban interface zones. More people also means more visitors and recreational activity on the GYE's public lands, compounding wildlife and environmental impacts in areas previously undisturbed. Moreover, with the proliferation of second homes and the arrival of Airbnbs, the character of GYE communities is changing as more newcomers appear on the scene without any effort to integrate into the local culture. By all indications, these development patterns will continue, bringing even greater impacts to the region's natural and cultural attributes, as well as mounting pressures on the GYE's ranchlands and ranching heritage.

Preserving Ranchlands and Ranching

Ranches and ranching are synonymous with the history of the three GYE states, which still retain and cherish that heritage. Release of the popular television series *Yellowstone* has firmly embedded that connection in the public imagination. But ranch ownership in the GYE is changing as longtime ranchers retire, and their children choose not to continue working cattle or sheep for a living. The accumulating sales of traditional ranches have raised concerns that the properties will be retired from ranching and subdivided to the detriment of wildlife and the GYE's open-space appearance. But that has not occurred, at least not yet, across much of the region. Although wealthy individuals from elsewhere are purchasing area ranches for their amenity values, traditional ranchers are also buying neighboring properties to enhance their livestock operations. And many amenity buyers are retaining their properties as open ranchland. The impact on conservation values is mixed thus far, but changes are afoot in the region's traditional ranching operations.

The pattern of ranch ownership in the GYE has clearly evolved as the sale of ranching properties has accelerated. According to a detailed 2003 study of ranch sale transactions in ten GYE counties, nearly a quarter of the large ranching properties (covering roughly 1.5 million acres) sold between 1990 to 2001. Significantly, most of these properties were not acquired by real estate developers, at least this was true in areas outside the three GYE resort counties. Amenity purchasers bought 39 percent of the ranches and 43 percent of the acreage sold, traditional ranchers bought 26 percent of the ranches and 25 percent of the acreage, and developers purchased only 6 percent of the ranches sold. Amenity purchases predominated in Montana's Madison and Park counties, and in Wyoming's Sublette and Park counties. Outside of these four counties, the bulk of sales were to traditional ranchers seeking to expand their operations.[9]

Amenity purchasers in the GYE appear to have mixed motives for acquiring their new properties, but generally no intention to convert the land from its existing appearance. In fact, these amenity-based transactions are often facilitated by a growing cadre of local "conservation brokers," who specialize in matching nature-oriented buyers with available ranch properties. Few amenity buyers seem interested in running livestock as an economic matter, and many have been willing to relinquish federal grazing permits, providing the federal land management agencies an opportunity to retire allotments. Some amenity buyers come with strong conservation values, and are consciously committed to preserving wildlife habitat and open space. They even may be amenable to placing a conservation easement on the property, thus limiting future development options. One prominent example is billionaire Ted Turner, who has placed conservation easements on his 113,000-acre Flying D Ranch outside Bozeman, where he husbands bison rather than cattle, welcomes grizzly bears, and undertakes ecological restoration projects.

Other hobby ranchers may have purchased the property for recreational or privacy purposes, concerned less about wildlife or conservation. But as amenity owners become more familiar with the local landscape and culture, original conservation plans for their new place have occasionally evolved, sometimes to the point of reassessing the role of livestock on the land.[10]

Over the long term, absent a conservation easement arrangement, it is unclear what will eventually happen to these amenity properties. Will the original amenity buyers keep the land within the family and, if so, will the next generation choose to keep the land as is? Or will the land ultimately be disposed of for financial or other reasons? If so, who will then purchase the land and for what purpose? As population pressures continue to mount across the GYE, will the subdivision market value of ranches situated near growing communities prove irresistible, further diminishing open space and wildlife habitat? And with the region's open spaces filling in, will state or local officials be moved to revise zoning and land-use laws to better control sprawl, protect wildlife habitat, and preserve the area's open spaces? The answers to these questions are urgent and rest with the regional populace, their vision for the future, and the political leaders they elect.

Of course, traditional ranching continues across the GYE, along with conflicts between livestock and wildlife. Such conflicts can and do occur on both private and public lands. Grizzly bears, wolves and other predators occasionally appear on ranching properties, where they have preyed upon grazing cattle and sheep. Beyond their base ranch property, most GYE ranchers hold federal grazing permits, entitling them seasonally to release livestock onto the public lands, and in the case of cattle often without supervision. Once turned loose to graze on federal lands, domestic cattle and sheep can fall prey to grizzlies and wolves and come into contact with elk, bison, and bighorn sheep, raising the possibility of a brucellosis infection or a pneumonia outbreak, which is almost always fatal to bighorn sheep. These wildlife depredation and disease transmission events can prove costly to ranchers, even when compensated for their losses.

As a practical matter, very few domestic sheep are still present in the GYE, and almost none on GYE national forest grazing allotments. Grizzlies and wolves, both of which have enjoyed federal protection under the Endangered Species Act, often view sheep as easy prey. Federal law, however, has prohibited livestock owners and herders from killing marauding bears or wolves, sowing the seeds for conflict and nurturing antifederal sentiments within the ranching community. In 2006, to eliminate the depredation problem and reduce ensuing grizzly bear mortalities—with the expectation that the bears would soon be removed from the federal endangered species list—the GYE national forests collectively amended their forest plans to phase out sheep allotments within the primary grizzly bear conservation area. Since then, the Forest Service—proceeding on a voluntary basis—has retired most sheep grazing allotments while transferring others to cattle, where the potential for conflict is much less. Conservation organizations have joined in this effort by raising funds to purchase sheep allotments from willing ranchers and then transferring the grazing permit to the agency with the understanding that it would be permanently retired.[11]

The facts speak for themselves. According to a 2020 Interagency Grizzly Bear Committee Study Team report: "Since 1998, there has been a 98% reduction in the acreage grazed by sheep on public lands inside the [Grizzly Bear Recovery Zone and] . . . no domestic sheep grazing on public lands inside the [zone] for the past 13 years." Even outside the Grizzly Bear Recovery Zone, the GYE national forests have significantly reduced sheep grazing. For example, the Shoshone National Forest's detailed records reveal that, from 1986 to 2010, the permitted sheep animal unit months (AUMs) declined from 13,700 to 600, and have since been further reduced. Besides helping safeguard the GYE's grizzly bears, the absence of sheep also minimizes the risk of a pneumonia outbreak within the region's bighorn sheep populations. In short, the Forest Service's ecosystem-wide approach to wildlife–sheep conflicts in the GYE has succeeded in virtually eliminating the problem, aided by conservation organizations willing to buy out sheep grazing permits. This collaborative, incentive-based approach therefore stands as a proven model for pursuing GYE nature conservation objectives.[12]

While domestic sheep depredation problems have faded, cattle–wildlife conflicts persist in the GYE. Grizzly bears still seize opportunities to prey on cattle, igniting controversy among ranchers and bear advocates. Also, several cases of brucellosis transmission from elk to cattle have occurred in recent years. But the Forest Service, rather than retire cattle allotments, continues to maintain an active cattle-grazing program across the GYE national forests. In an effort to reduce depredation and disease incidents, the National Wildlife Federation (NWF) and other conservation groups have actively sought to purchase and retire cattle-grazing permits. In 2009 the NWF helped retire the Royal Teton Ranch's 6,000-acre allotment on Yellowstone's northern border, thus safeguarding a migration corridor for the park's bison that were otherwise being killed upon exiting the park. Overall, the NWF has retired more than 700,000 acres of critical habitat in the GYE since 2002 through its Adopt-a-Wildlife-Acre program. Moreover, conservation groups have extended an array of monetary incentives—paid range riders, guard dogs, damage payments, fladry, and even "reverse" bounty payments for allowing wolves to den on private lands—to local ranchers and landowners in an effort to protect bears, wolves, and other wildlife. Advocates of these incentive-based strategies have consciously eschewed alternative, often controversial regulatory strategies in favor of market-oriented strategies designed to promote wildlife conservation on the GYE's public and private lands. It is clearly succeeding in some instances in reducing the potential for conflict between wildlife and cattle, but has not eliminated all controversy.[13]

Indeed, litigation flares periodically over grizzly and cattle conflicts in different corners of the GYE. A major trouble spot is the Upper Green River area in the Bridger-Teton National Forest, where Wyoming ranchers seasonally graze more than 8,000 cattle in prime grizzly bear country. The area has long been regarded as a "mortality sink" for grizzly bears due to ongoing livestock depredation incidents that have prompted ranchers to surreptitiously dispatch marauding bears. In 2019 the Forest Service, after securing approval from the US Fish and Wildlife Service, adopted a new range-management plan that allowed 8,819 cattle to graze six allotments during summer months. Conservation groups objected to the plan and sued both agencies in an effort to overturn it. Their challenge primarily focused on the FWS's biological opinion that permitted up to seventy-two grizzly bears (including reproductive females) to be "taken"—that is, killed—to protect cattle during the plan's ten-year lifespan.

Although a Wyoming federal judge dismissed the lawsuit, the Tenth Circuit Court of Appeals ruled that the FWS's failure to consider limiting female grizzly mortalities was unreasonable, also noting the agency must address the area's "mortality sink" problem. Despite these legal shortcomings, the court did not prevent the grazing plan from taking effect. A similar lawsuit is pending over the Custer Gallatin National Forest's decision to open six allotments covering 21,000 acres in Paradise Valley to livestock grazing, citing few grizzly bear depredation incidents during recent years. Although three of the allotments will remain vacant, conservation groups fear the presence of livestock for an extended grazing season will attract bears and wolves, putting them in undue jeopardy notwithstanding the Forest Service's and FWS's contrary conclusion.[14]

At bottom, these wildlife–cattle controversies highlight the ongoing tensions over resource-management priorities on the GYE national forests. Long accustomed to grazing unattended livestock on national forest lands, can the region's ranchers and traditional ranching practices coexist with robust grizzly bear populations along with other wild animals that epitomize the region's unique wildland character? Thus far, the answer has been a reluctant "yes," as ranchers have continued adapting to new, on-the-ground realities, sometimes working collaboratively with conservation organizations to address wildlife-related problems. But if the answer becomes "no," then what will happen with the base ranch properties? Some ranches would likely be purchased by amenity buyers open to the presence of bears, but other ranches could be broken up or subdivided, creating new habitat fragmentation problems and further changing the area's open appearance. In some respects the ultimate fate of the GYE's critically important ranchlands rests in the hands of agency officials, local politicians, judges, and others, rather than the ranchers themselves. In other respects, the marketplace may well determine their fate.

Private Land Conservation: Assessing the Available Tools

Federal and state laws contain an assortment of tools that can be employed to promote conservation on the GYE's privately owned lands. At the federal level regulatory options are limited due to the traditional state role overseeing property ownership and use. But federal funding has come to play an increasingly larger role in nature conservation efforts on private lands in the GYE and elsewhere. Although the states have legal authority over private lands, the land-use planning and zoning laws in the three GYE states are relatively weak, reflecting a strong regional commitment to personal autonomy and property rights. State and local officials have regularly proven reluctant to assert their available authority aggressively, while local governments often lack the resources to develop meaningful land-use plans. Instead, the conservation easement has emerged as the principal tool for addressing wildlife and other conservation concerns on the region's private lands. The challenge has been—and remains—how to employ these tools locally with a vision of the landscape as a whole.

The federal government possesses a modicum of regulatory authority over private land use in the GYE, but it has relied primarily on monetary incentives to promote conservation values. The Endangered Species Act prohibits anyone from "taking" federally listed species, which includes shooting grizzly bears to protect livestock as well as destroying critical habitat. Under this provision, a recalcitrant resident who killed a protected wolf during the early days of the Yellowstone wolf reintroduction was successfully prosecuted, but the habitat protection requirement has not been employed regionally as a land-use control device. When conservation organizations sought to invoke the Clean Water Act to block a golf resort development in the Snake River floodplain, the courts rejected their arguments, allowing the project to proceed on secluded and previously undeveloped ranchland.

During the late 1980s, when Congress considered limiting geothermal development on private lands quite near Yellowstone to protect the park's iconic geothermal features, the bill's proponents were unable to move the Old Faithful Protection Act proposal, stymied by arguments invoking private property rights and state sovereignty. Ultimately, this troubling threat to the park's geothermal plumbing was resolved through negotiations between federal and state officials over federal reserved water rights, which established a fifteen-mile zone prohibiting the development of hot water found within the area's underground aquifers. Simply put, federal regulatory laws—even the powerful Endangered Species Act—have had little evident impact on private land use across the GYE.[15]

The story is proving quite different, however, in the GYE for federal incentive-based conservation programs designed to protect wildlife habitat and ecological systems. The 1965 Land and Water Conservation Act (LWCA) has long been the principal source of federal funding for conservation purposes. Now permanently funded at $900 million annually following passage of the 2020 Great American Outdoors Act, the LWCA makes these dollars available to federal, state, and tribal entities to purchase privately owned lands for wildlife and recreation purposes. In the GYE these funds have been used repeatedly over the years to secure key private parcels, including lands at a migratory bottleneck point in the High Divide region west of Yellowstone National Park and a state parcel situated in the Path of the Pronghorn migration corridor where it enters Grand Teton National Park. In 2003 Congress adopted the Healthy Forests Restoration Act (HFRA) in an effort to reduce the wildfire threat to homes situated in the growing wildland-urban interface—an increasingly serious problem in the GYE where amenity buyers are routinely building next to public lands in fire-prone areas. The HFRA dispenses federal funds for fuel-reduction projects in the wildland-urban interface zone and to assist communities in developing wildfire protection plans.[16]

Given the importance of open agricultural lands to wildlife, the US

Department of Agriculture's (USDA) conservation funding programs are geared toward incentivizing landowners to dedicate portions of their acreage to wildlife conservation. Under the USDA, the Natural Resources Conservation Service Working Lands for Wildlife initiative includes several different programs to voluntarily enlist private landowners in conservation efforts. These programs employ various funding mechanisms designed to protect wildlife habitat, migration corridors, and ecological processes, while supporting ranching operations. In 2022 the USDA announced a significant new pilot conservation initiative in partnership with the state of Wyoming that makes more than $15 million available to ranchers through the agency's Agricultural Conservation Easements Program and its Environmental Quality Incentives Program. Built upon USDA's voluntary, incentive-based approach to conservation, this pilot program funds ranchers to protect wildlife habitat and corridors while enabling them to continue ranching without having to consider selling all or portions of their property holdings. Now underway in Wyoming, this politically palatable program has spawned similar federally funded conservation programs in Montana and Idaho, which should further protect wildlife habitat on the region's private ranchlands.[17]

While federal funding is playing an important conservation role on the GYE's private lands, the three GYE states hold the key to land use and development policy in the region. By long tradition, state and local governmental entities oversee land planning and use matters, doing so through an array of state laws, local codes, and zoning regulations. In the GYE states these laws grant state and local authorities considerable discretionary authority over conservation concerns on private lands, but they have not been aggressively implemented or enforced in most communities. For example, although two elk herds rely heavily during winter months on private lands in the Cody area, the county's aging land-use plan is silent about elk habitat needs, and it is unclear if that will change in pending revisions. This is explained, in large part, by the region's attachment to individual property rights, along with evident antigovernment sentiments—all of which permeates local land-use planning and zoning efforts. As one GYE political official has observed, "taxes and zoning are four letter words" across the region.[18]

Nonetheless, the three GYE states have each adopted land-use planning, zoning, and subdivision laws that empower local governments to direct and control growth and related development activity. These laws require counties to prepare land-use plans that are enforced through zoning standards and subdivision requirements. Several provisions in these laws identify conservation and wildlife as relevant factors to consider in local land-use decisions, providing a legal foothold for officials to give nature conservation a prominent role in planning and zoning matters.[19] Reflecting these concerns, the Teton County (Wyoming) land-use plan is "organized around stewardship of . . . ecological resources" and highlights wildlife, natural, and scenic resources as "the core of [its] heritage, culture, and economy." The Wyoming Supreme Court has sustained the county's goal of preserving the area's "rural western character" against the argument that it violated the state's constitution and land-use planning laws. Similarly, the Montana Supreme Court has ruled that wildlife impacts were a legitimate consideration in overturning a county's decision to rezone an undeveloped parcel near Yellowstone National Park. The court decision blocked a zoning change that would have allowed nearly 1,000 new residences on a tract originally permitted for thirty-two single-family residences. But politically, when local officials incorporate wildlife and conservation values into land-use decisions, they risk losing the next election.[20]

These same land-use planning and zoning laws provide communities with the authority to address the region's growing rural sprawl and affordable housing problems. The sprawl problem is glaringly evident in Montana's Gallatin County, where Bozeman and the surrounding areas are growing at a breakneck pace, making it the state's unrivaled fastest-growing county. By 2022 the county's population had swelled to 122,000 people, treasured open space was rapidly disappearing, and three-quarters of the county's residents expressed deep concern about these unsustainable trends. With 53 percent of the county's lands in private hands and only 16 percent of these lands protected by conservation easements, a clear sense of urgency prevailed. County and city officials responded by convening a broad-based working group that included federal and state agency representatives, local officials, conservation groups, agricultural organizations, and land trusts. The group's charge was to develop a science-based plan consistent with local values that would identify sensitive lands and recommend how to protect them.[21]

The result, following a community engagement process and comprehensive review of existing land-use plans, is the Gallatin Valley Sensitive Lands Protection Plan. According to the working group: "continuing the current course will result in significant habitat fragmentation, displacement of wildlife, and loss of quality of life—ultimately diminishing the sense of place that makes the Gallatin Valley a distinct and valuable landscape." Focusing on four broad themes—wildlife and biodiversity;

agricultural heritage; connectivity; water quality and quantity—the working group amassed extensive data on these concerns, and employed GIS modeling technology to identify "hot spots," trends, and constraints to guide future land-use planning. The plan concludes with specific recommendations addressing land-use regulation reform, incentive-based conservation strategies, subdivision regulation reforms, density bonuses, and revisions to the environmental assessment process.[22]

If implemented, this farsighted Sensitive Lands plan would go far toward protecting wildlife habitat, environmental values, and open space on the county's private lands. During the 1990s, however, a similar conservation-oriented planning process died stillborn, which set the stage for the sprawling development occurring today. Whether the political forces that doomed that earlier effort are still extant remains unclear, as does the appetite of local officials for the recommended land-use reforms. Regardless, the growth impacts across the Gallatin County landscape are undeniable, and they will only intensify absent the adoption of meaningful regulatory reforms and additional incentive-based conservation strategies.[23]

The related affordable housing problem is mounting across the GYE. With more people attracted to the area, the affordable housing stock—at least in several popular locations—has dwindled as newcomers bid up home prices, effectively pricing longtime residents and lower-wage workers—often nonwhite immigrants—out of the market. The problem is particularly acute in Teton County, Wyoming, where essential workers are now commuting across Teton Pass from Idaho and from the Star Valley area some fifty miles distant, and even from Idaho Falls ninety miles away. In Bozeman a booming influx of Spanish-speaking immigrants have found attractive work opportunities but few housing options due to the red-hot real estate market. Similar problems are evident in the Driggs area, as well as the gateway town of Gardiner where nonresident Airbnb purchasers are cutting into the limited local housing stock.[24]

One answer appears straightforward: Promote higher density, cluster-style development to address affordability concerns while reducing sprawl impacts. But whether that approach is politically feasible is open to debate, given the NIMBY syndrome that is alive and well in the region. Wyoming has sought to address the problem by authorizing higher-density development. The governing law establishes a "conservation design process" that enables counties to protect wildlife habitat through cluster development and density bonuses, which permit developers to increase the maximum allowable development on a tract. Although the

Teton County land-use plan identifies affordable housing as an important consideration in its planning processes, it continues struggling to meet this need in the face of political headwinds. In fact, tensions have surfaced between Teton County and the state, which is considering legislation that would cap mitigation fees paid by developers to offset, among other things, the cost of new housing. Tensions are also evident between county and city officials over appropriate strategies to construct affordable housing without jeopardizing important local values, such as neighborhood parks, or promoting commercial development that will require more workers and hence more housing.[25]

In several respects, Teton County, Idaho, exemplifies the intense development pressures confronting many of the GYE's smaller, rural communities. Forty years ago, the area was a homogenous agricultural community with a population of 2,900 residents and a small ski area that attracted a few die-hard ski bums. Since then, the county—particularly the towns of Driggs and Victor—has become a bedroom community for Jackson, Wyoming, as well as a flourishing second-home destination with its own luxury homes and resort-like facilities, including three world-class golf courses and an expanded airport with private jet hangars. Once the area was discovered during the 1990s, local developers and outside speculators descended on the county purchasing open farmland to subdivide, abetted by a lax comprehensive plan permitting unlimited 2.5-acre plats on the valley floor's agricultural lands. From 2000 to 2009, the county was the second-fastest-growing rural county in the nation. Local farm families, attracted by the purchase prices offered by developers, sold out, creating a major real estate boom cycle.[26]

Then came the 2008 Great Recession and falling real estate prices, leaving the county with an abundance of "zombie subdivisions"—platted lands with bankrupt developers and few buyers in sight. As the "bust" deepened, the county developed and then adopted in 2012 a new comprehensive plan better attuned to local growth realities, conservation issues, and quality of life concerns confronting the area. "Our vision," according to the plan, "is for a Valley with a vibrant economy and high quality of life. This requires educational opportunities, recreational opportunities, cultural amenities, public land access, and protection of natural resources and scenic vistas." Owing to contentious local politics as well as embedded antigovernment sentiment, another ten years elapsed before the county finalized a zoning code that imposed meaningful subdivision and permitting standards. To avoid the new zoning standards, however, developers quickly filed more than 100 subdivision proposals, seeking to grandfather projects under the permissive old code, many of which have been denied.[27]

The boom cycle is back. With a population of 12,500 Teton County now ranks third among Idaho counties in per capita income, a reflection of recent growth and a transformed economy. Real estate, insurance, construction, and professional services make up the county's top job categories, while farming, logging, and mining are at the bottom of the list. Expansion at the Grand Targhee Ski Area has attracted new residents and put additional strains on county services. The hot real estate market, the ski area, and other recreational amenities have enticed Airbnb investors as well as speculators, who have bid up prices while generally not engaging with the local community. The booming market—single-family homes in Driggs and Victor fetch a median sale price around $750,000—has increasingly put home ownership out of reach for local residents. Meantime, the Idaho legislature has limited the regulatory authority county and local officials can exercise over short-term rentals—another example of the jurisdictional complexity bedeviling the GYE region.[28]

Moreover, amid the overheated real estate market, the county has dealt with an array of large-scale development proposals that would further alter the valley's character. A Florida real estate venture firm, for example, has sought approval for the 562-acre High Noon Ranch development, a mix of commercial and residential units—ostensibly a dude ranch—on undeveloped agricultural land at the foot of the Big Hole Mountains. Bordered by farmlands under conservation easements and national forest lands, the development would sever an important migration route, and also raise wildfire and water issues. Following a public hearing, the county planning commission voted to reject the concept plan, placing the project in limbo. Notwithstanding efforts by local officials to control these rampant growth impacts, unrelenting market forces are plainly transforming this once-placid farming area with unmistakable environmental and quality-of-life consequences.[29]

Beyond the two Teton counties and Gallatin County, most other GYE counties are not well positioned to address the gathering real estate development storm that threatens much of the region. Generally lacking substantial financial resources, these smaller, largely rural counties often lack the staff resources and expertise to conceive and implement progressive land-use and zoning plans. Local politics is a factor, too. The counties tend to lean conservative, and thus generally support private property rights and regularly oppose stringent planning and zoning regulations as government overreach. No surprise, then, that these political sentiments were on full display in Park County, Montana, in early 2024, when a voter-initiated referendum seeking to repeal the county's growth-management policy generated heated political controversy before being defeated. Nonetheless, as the consequences of rampant growth and development within the GYE become ever more evident in the region's rural communities, local pressures may—or may not—prompt proactive efforts to preserve the natural and open space values that most residents have long held dear. With much at risk, the clock is clearly ticking.[30]

Conservation Easements and Habitat Leases

State law also governs the establishment of conservation easements, which have become the principal legal device for protecting wildlife habitat and other natural values on the GYE's private lands. More than a dozen national and local land trusts—The Nature Conservancy, Jackson Hole Land Trust, Montana Land Reliance, Rocky Mountain Elk Foundation, Wyoming Stock Growers Land Trust, and others—are active in the area, as well as several federal and state agencies that also hold conservation easements. By 2024, although difficult to calculate precisely, around 1.3 million acres (or roughly 20 percent) of the six million acres of privately owned land in the GYE were under conservation easement protection. Roughly 2,400 easements were spread across the twenty counties considered part of the GYE, with nonprofit land trusts holding about 75 percent of the easements (covering around one million acres), and the remaining easements in governmental hands. Most of the protected acreage is situated in Montana—about one million acres—with Wyoming at around 250,000 acres followed by Idaho at 50,000 acres.

An assortment of governmental and private funding sources are in place, providing the money necessary to consummate GYE conservation easement purchases. Federal and state laws extending income tax deductions to conservation easement donors have helped immeasurably to facilitate these transactions. Yet, with real estate prices soaring across the GYE, conservation easement purchases are becoming ever more expensive, potentially curtailing future transactions while also heightening fundraising pressures on easement purchasers.[31]

Conservation easements, a purely voluntary arrangement, have emerged as a popular incentive-based approach to achieve nature conservation goals on private lands. Each of the three GYE states has conservation easement laws in place, though Wyoming did not adopt its law until 2005, much later than most other states. The Wyoming and Idaho laws generally track the Uniform Conservation Easement Act with local variations, while Montana's law is not modeled on the uniform act and contains several unique provisions. By law, conservation easement agreements are treated as private transactions between the property owner and the easement holder. Such agreements impose permanent or "in perpetuity" limitations on future land uses, which is also required by federal law in order to qualify for favorable tax treatment. The courts have proven willing to enforce stringent easement provisions, as the Wyoming Supreme Court did when it prohibited a new Jackson Hole property owner from disregarding a building-size limitation in a preexisting easement agreement.[32]

The private nature of conservation easement transactions, while an attractive feature for many landowners, can create challenging transparency and enforcement problems. A private transaction in the nature of a contract, conservation easement agreements afford the public no role in either negotiating or enforcing the agreement. The easement holder, typically a land trust, is responsible for monitoring and policing its conservation easement agreements—an obligation that continues indefinitely no matter how often the property may change hands. The easement holder must regularly check the property to protect against disallowed uses and guard against improper amendments that undermine the transaction's initial conservation purposes. According to the Wyoming Supreme Court, neither neighboring landowners nor the general public has legal standing to enforce conservation easement commitments in a court action. For large land-trust organizations with professional staff and adequate resources, these ongoing monitoring and enforcement obligations may not prove particularly onerous, but for smaller land trusts lacking substantial resources such obligations can prove infeasible.[33]

Indeed, considerable controversy erupted in the GYE when the new owners of the Carney Ranch sought to alter an existing conservation easement. Beginning in 1995 and extending over several years, different land trusts acquired conservation easements on the 5,500-acre Carney Ranch in the Upper Green River Valley adjacent to Wyoming's Bridger-Teton National Forest, ultimately putting 2,571 acres in protected status. Collectively, the easements were designed to protect wildlife habitat, including a "bottleneck" stretch on the high-profile Path of the Pronghorn migration corridor as well as sage grouse, elk, and moose habitat. After the ranch changed hands in 2014, the new owners began constructing a cabin in the protected pronghorn pathway in violation of the easement and without consulting the responsible land trust. Once the illegal cabin project surfaced, the ranch owner and land trust officials met and agreed to eliminate 15 acres from the original easement, allowing the cabin to stand in exchange for putting 115 acres elsewhere on the property in a conservation easement.[34]

By then, however, news of the easement violation and proposed amendment had broken, triggering a vociferous public response calling for enforcement of the original easement in order to protect the pronghorn migratory corridor. After the land trust's board of directors became involved, the parties resolved the matter by agreeing to maintain the original easement agreement and remove the cabin. Although the incident ultimately left the conservation easement in place, it revealed some of the problems inherent in these private, nontransparent easement transactions. But for the news media coverage of the matter, the easement may have been amended and the cabin constructed without any public scrutiny or input, let alone judicial oversight. Such a change would have been inconsistent with the intent behind the original easement, which was designed to benefit the public interest in wildlife by protecting this critical, highly visible, and popular migratory corridor—an interest subsidized in part by public funds and tax benefits. Thus, to obtain the full benefit of such conservation easements, the arrangement requires careful, science-based planning and constant vigilance, as well as the willingness and resources to enforce the easement terms.[35]

Alternatives to the permanent conservation easement agreement have emerged in recent years, most notably federal and state habitat lease programs. Limited duration habitat leases between a property owner and a governmental body or conservation organization offer landowners regular lease payments in exchange for maintaining the property in a largely undeveloped condition suitable for wildlife use over a set term of years. The US Department of Agriculture's Farm Service Agency offers ranchers ten- to fifteen-year habitat leases through its Grassland Conservation Reserve Program, which is designed to protect grassland habitat while enabling haying and grazing to continue. Habitat leases are also part of the 2022 Wyoming–US Department of Agriculture private land conservation pilot program that is focused on protecting wildlife-migration corridors. Montana's Fish, Wildlife, and Parks agency has also developed a Habitat Conservation Lease program that utilizes the revenue from nonresident hunting license sales to protect important, seriously threatened habitat; the program envisions thirty- to forty-year lease arrangements that also obligate enrolling landowners to open their property for limited public recreation use. Moreover, conservation organizations are pioneering similar habitat-leasing programs in the GYE area, such as the elk occupancy agreement between two nonprofit groups and a Paradise Valley rancher who has agreed to segregate and open part of his private lands to elk during winter months in exchange for compensation payments to offset his costs.[36]

These emergent habitat lease programs contain several attractive features. Being of limited duration, they provide landowners with a degree of flexibility over time not available under a permanent conservation easement. In areas like Jackson Hole with exorbitant property values, habitat lease arrangements may prove more financially feasible as a means to stretch limited funds and protect more acreage. Elsewhere, flexible lease agreements can serve as an entry point for landowners into a discussion about a more enduring conservation easement agreement. But unlike conservation easement transactions, limited duration habitat lease arrangements do not qualify for federal or state income tax deductions.

By most accounts, conservation easements and habitat leases occupy an important role in ongoing efforts within the GYE to preserve vital wildlife habitat and corridors on the region's ecologically critical private lands before it is too late. These incentive-based approaches to nature conservation have proven both politically feasible and increasingly popular with landowners—a fact confirmed when the conservative Montana legislature, during its 2022–23 session, abruptly rejected a bill that would have put a forty-year time limit on conservation easements. The same legislature also adopted a bill (and overrode a gubernatorial veto) that directed a portion of the state's marijuana tax receipts to habitat conservation, consistent with an earlier voter-endorsed initiative approving this funding arrangement. In the ongoing quest to safeguard the region's ecological integrity, it is essential to sustain this political momentum as well as the funding sources necessary to enlist landowners in these critical nature conservation initiatives.[37]

Landscape-Scale Conservation: The High Divide Model

As the GYE concept and associated acreage have expanded over time, so too have regional conservation efforts. In fact, GYE conservation projects are increasingly focused on the larger landscape, with private lands playing a pivotal role in these efforts. Science-based considerations are a driving force in this notable expansion in regional conservation activity. Climate change is altering GYE ecological conditions, making it imperative that climate-stressed species have the opportunity to move across the landscape to meet their habitat needs. For isolated wildlife populations like the Yellowstone grizzly bear, a landscape-level strategy is essential to preserve genetic diversity by connecting the GYE bears with their northern cousins in the Glacier-Bob Marshall region. New digital technology has enabled scientists to document and better understand elk, pronghorn, mule deer, and other migration patterns, which are now understood to extend across and often well beyond the generally accepted GYE boundaries. The GYE's waterways likewise flow across the region and interconnect with distant waters—all of which are vitally important to the region's wildlife, fisheries, and human communities. In short, to preserve the GYE's ecological integrity and natural features, conservation efforts must also embrace the surrounding landscape, including both public and privately owned lands.[38]

The most visible GYE landscape conservation initiative involves the High Divide region and the namesake High Divide Collaborative (HDC). The High Divide region extends west and north from Yellowstone National Park across several Montana and Idaho counties, occupying a critical role connecting Yellowstone with central Idaho's expansive wilderness areas and the Crown of the Continent Ecosystem. The HDC focuses its attention on nearly four million acres, which includes the Centennial, Madison, and Pioneer mountain ranges; the Madison, Centennial, Beaverhead, and Big Hole valleys; the Henry's Fork and Salmon-Lemhi country; and other nearby lands. The High Divide country—rural in character, dotted with small towns, and home to numerous family-owned large ranches—is representative of areas widely referred to as "the Old West." It encompasses federal national forest and BLM lands along with extensive private ranchlands mostly situated at lower elevations that contain important riparian habitat.[39]

Although the region has long looked to mining, logging, and agriculture for its economic sustenance, these traditional High Divide industries have declined in recent years. At the same time, the region's service sectors, including recreation and tourism, have shown considerable growth. New home construction has also accelerated in the region, raising wildlife-habitat fragmentation and wildfire-management concerns. During the early 1990s, recognizing the High Divide country's superb conservation values, land trusts and local conservation organizations began focusing on the area, seeking to acquire private ranchlands through outright purchases or conservation easements. With financial support from the federal Land and Water Conservation Fund as well as private foundations, these groups succeeded over the next couple decades in se-

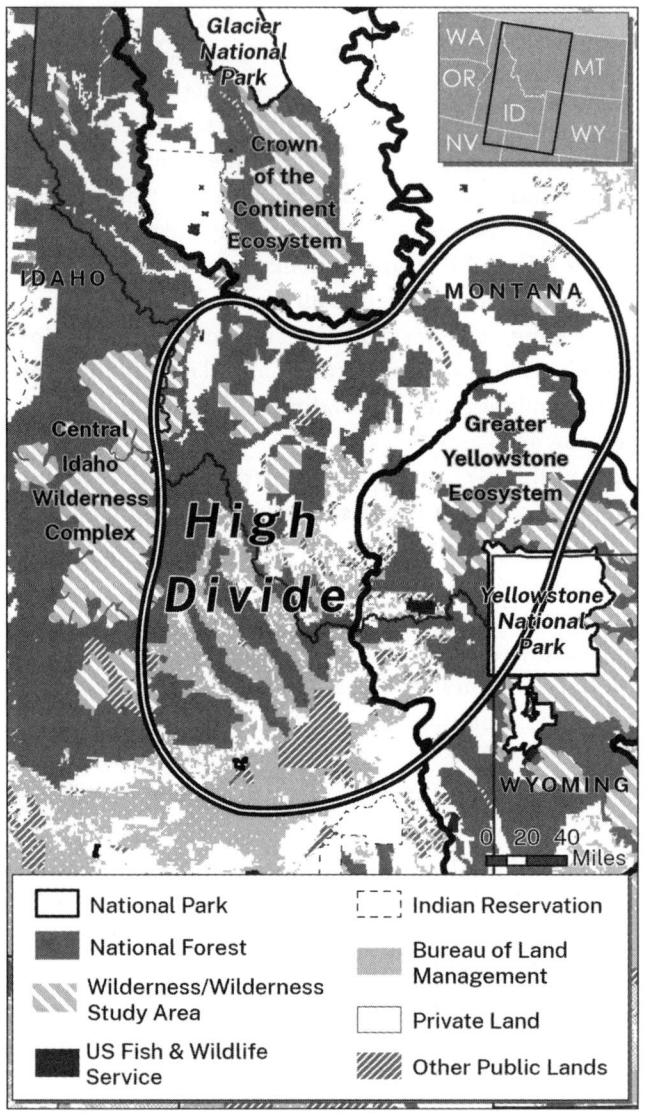

The High Divide. Situated on the northwestern edge of the GYE and beyond, the lightly populated High Divide straddles the Idaho–Montana border, encompassing both public and private lands. Because the area embraces critical wildlife-movement corridors, the High Divide Collaborative is working to protect it with conservation easements and other tools. These efforts are designed to help GYE grizzly bears to connect with bears in the Crown of the Continent Ecosystem, to enable other species to relocate in response to climate change, and to safeguard critical habitat. The High Divide initiative is expanding GYE nature conservation efforts to the broader landscape. Map by the University of Utah DIGIT Lab.

curing $437 million that enabled them to protect piecemeal more than 750,000 acres.[40]

To build upon these accomplishments, the HDC was established in 2012 in an effort to better coordinate existing conservation initiatives while more fully engaging local residents to address persistent community concerns. Today, the HDC boasts a broad membership of more than eighty different entities, including federal land management and wildlife agencies, state agriculture and wildlife agencies, land trusts, conservation groups, community officials, local groups, private landowners, scientists, and others—all overseen by an intentionally diverse High Divide Collaborative Coordinating Committee. The HDC's basic strategy is to build trust and relationships with local landowners and residents by employing high-quality science to identify key parcels, emphasizing the economic advantages of its conservation efforts, and proceeding with sensitivity to local economic and cultural concerns. The ultimate goal is to secure local support for its conservation initiatives, which includes overcoming a legacy of distrust toward federal officials and outsiders. To move forward, the HDC has identified several key features critical to the region's ecological health and cultural future: working ranchlands, healthy forests, wildlife connectivity, community economic resilience, recreation, wildfire management, aquatic resources, and sagebrush ecosystems. And recognizing that change already grips the area, the HDC has embraced a collaborative approach to conservation in an effort to preserve the region's natural values, traditional ranchlands, and community spirit.[41]

As an entity, the HDC has obtained substantial support from federal as well as private sources to purchase conservation easements in the region. In 2016–17, the HDC received $30.5 million from the federal Land and Water Conservation Fund, which it used to protect 23,000 acres. A 2019 study of the High Divide region concluded: "conservation easements, though only a small part of the regional land base, contribute substantially to landscape-scale representation of ecosystems and provide complementary conservation value to public land at the regional scale." The study found easements more prevalent at lower elevations with more productive soils, but also determined that existing easements only slightly improved connectivity. Another study revealed that High Divide ranchers harbored diverse attitudes toward grizzly bears dependent in part on their experience with conservation easements, noting that ranchers in the region's southwestern reaches were least likely to be accepting toward bears. In fact, resistance persists from some residents fearful the HDC's agenda is driven by outside interest groups hostile to private property rights and intent on turning the region into a large nature reserve. Nonetheless, the HDC has garnered significant attention as a workable model for landscape-scale conservation efforts in the GYE and beyond.[42]

Landscape-level conservation efforts involving private lands are also evident elsewhere in the GYE. One prominent example involves the 160-mile Red Desert to Hoback (RD2H) mule deer migration corridor, which the state of Wyoming has designated an official migration corridor. Summering west of Grand Teton National Park, the RD2H mule deer herd travels across a mosaic of land ownerships during its seasonal migration that skirts the western flank of the Wind River mountain range in route to lower-elevation winter habitat in the Red Desert area just north of Interstate 80. To eliminate a potential bottleneck along the migration route due to an impending subdivision development near Fremont Lake, the Conservation Fund purchased the 364-acre parcel and then transferred it to the Wyoming Game and Fish Department, which maintains it as a Wildlife Habitat Management Area. Elsewhere along the RD2H corridor, government officials, conservation groups, and private landowners have worked together to eliminate problematic fencing on private lands to open passageways for the migrating herd.[43]

* * *

As concern mounts within the GYE over intensifying development pressures as well as climate change impacts, conservation efforts have expanded across the ecosystem and beyond to the surrounding landscape. Given their ecological importance, the region's lower-elevation private lands have assumed an ever more critical role in the success of these efforts. Numerous entities—the relevant federal and state agencies, local governmental bodies, land trusts, conservation organizations, informed landowners, and others—are engaged in an assortment of voluntary initiatives designed to preserve vital wildlife habitat and historic migration corridors emanating outward from the region's national parks and wilderness areas. Scientists are involved too, providing critical data concerning wildlife ranges, winter habitat, migratory corridors, riparian zones, and other ecological factors, enabling the involved parties to target and expand their land acquisition and conservation easement purchases. State and local laws governing planning, zoning, and easements are also critical to these conservation efforts, though they have not always been implemented aggressively in the face of powerful local political opposition tied to property rights, antiregulatory sentiments, and distrust of outside influences. The challenge, moving forward, is to address and surmount these obstacles while more fully engaging the region's private landowners in wildlife and other conservation efforts to preserve the GYE's unique natural attributes and appearance.

Given the diverse conditions prevailing on the GYE landscape and the range of attitudes prevalent within the region's landowners, an amalgam of approaches is necessary to achieve an ecologically meaningful level of private land conservation. Incentive-based approaches have proven particularly effective at enlisting ranchers and other landowners in regional conservation initiatives. Working locally to meet landowners where they live, establishing trust between conservation advocates and landowners, and demonstrating sensitivity to each other's needs and knowledge are commonsensical keys to advancing these efforts. Improved collaboration and coordination among the various entities is also necessary to develop a joint vision for the region's future and to implement that vision at the necessary landscape scale. Money is essential, too, and is available from a variety of federal, state, and private sources to engage potentially interested landowners in these efforts and to ensure all parties can participate in charting the course ahead.

Due to local sensitivities, however, the federal role is more circumscribed than on the region's public lands, but a federal presence remains important for coordinating management of the region's ecologically significant public and private lands. Moreover, given ever-increasing development activity and inevitable funding shortfalls, the existing state planning, zoning, and other laws must be employed determinedly and strengthened to advance GYE conservation goals. As development pressures continue mounting relentlessly throughout the GYE, private land conservation efforts are now central to preserving the region's ecological integrity and well-being.

CHAPTER SEVEN

Ecosystem Conservation Revisited

The GYE is a much different place today than it was sixty years ago, and it will be a different place again in the years ahead. The region's formal boundary lines remain unchanged, but the landscape and its occupants have changed in significant and sometimes unanticipated ways. Yellowstone and Grand Teton national parks still sit at the core of an expansive federal land complex awash in charismatic wildlife, unique thermal features, soaring mountain peaks, and other natural attributes that have seized and held the public imagination. The GYE national forests still adjoin the parks, providing critical habitat as well as commercial, recreational, and wilderness opportunities. The same communities still ring the ecosystem, though much more heavily populated and subdivided as nearby open space dwindles. Meantime, the laws and policies governing the GYE landscape have evolved, reflecting broader changes in public values that have elevated protection of the region's natural character and appearance under the greater ecosystem rubric.

What then is the current state of the GYE? What role has the ecosystem management strategy played in safeguarding the GYE's ecological integrity and natural resources? What forces account for existing conservation advances, and corresponding setbacks? What challenges lie ahead for the GYE as we move through the Anthropocene where climate change and other human impacts are omnipresent? And how does the GYE experience fit into the larger conservation narrative unfolding across the nation? Clearly, much has transpired during the past sixty years that holds important lessons for today and tomorrow. This chapter seeks to answer these questions in order to put the GYE experience into proper perspective and learn from it.

The GYE Today: Taking Stock

The logical point to begin assessing the GYE and related ecological management efforts is the post–World War II era, which set the stage for today's world. Remarkable changes have occurred since the 1950s. Nationally, these sweeping changes include unparalleled economic prosperity, explosive population growth, increased personal mobility, additional leisure time, and improved outdoor recreational equipment—all of which have had significant implications for the nature conservation movement. At the same time, manifold scientific advances and technological changes have reshaped our understanding of both the natural world and the fields of ecological, environmental, and climate science. By the early 1970s these changes, combined with other developments, had spawned a new environmental movement that secured a spate of new laws extending legal protection to our air, water, land, and wildlife. All of these changes have profoundly affected the GYE, helping account for the emergence of compelling new conservation priorities and policies.[1]

Several bedrock ideas have shaped the GYE and related resource-management policies as the region has evolved during this period. First, scientists have determined that nature is dynamic and in a constant state of flux or change, which has put the long-standing "balance of nature" shibboleth to rest and elevated the ecosystem as a conservation target. Second, science has confirmed that the natural world is comprised of interconnected ecosystems that extend across the manmade boundaries we have imposed on the landscape, thus expanding the scope of our conservation efforts. Third, increasingly evident climate- and biodiversity-related impacts are further expanding our understanding of the necessary scope of conservation efforts. Fourth, recognizing the natural world has both intrinsic and instrumental value, we have committed to leaving it undisturbed to take its own course in national park, wilderness, and other select locations, determined to impose some limits on the increasingly ubiquitous human imprint. Finally, we have come to realize that these nature conservation efforts regularly bring significant economic and social value to nearby communities, though negative impacts can occur too.[2]

Since the 1960s, National Park management policy has evolved in both Yellowstone and Grand Teton in response to the challenges facing each park. While the Park Service's mandate—to conserve unimpaired and provide for public enjoyment—has not changed during the past 100 years, the agency's interpretation of that mandate has changed profoundly. Following the 1963 Leopold Report, the Park Service has consciously injected ecological science into its management policies and practices. In Yellowstone and elsewhere, the agency has embraced ecological integrity as its primary resource-management goal, largely employing natural regulation—let nature take its course—policies to achieve this goal while also engaging in active ecological restoration efforts. Further, the Park Service has clarified and affirmed that its conservation responsibilities take precedence over conflicting recreation and visitor enjoyment policies in order to leave these special places unimpaired for future generations. And the agency has acknowledged the need for individual parks to address external developments that threaten park resources and their conservation obligations. This science-driven approach to park resource management has not escaped controversy, yet continues to govern park policy, albeit with some modifications.[3]

Both Yellowstone and Grand Teton have implemented and generally adhered to these ecologically based management standards. Wildfires have been allowed to burn in the backcountry despite the controversial 1988 fires, wildlife wander about free from human intervention, exotic species—most notably lake trout and mountain goats—are being aggressively managed, snowmobiles are finally under the agency's firm control, whitewater kayaking and other intrusive recreational activities have been held at bay. Outside the front-country hotels, roads, and popular visitor attractions, nature still dominates in both parks, reflecting the agency's commitment to light-handed, science-based resource-management policies. Yet, increasing visitation as well as related economic and political pressures emanating from nearby communities are putting additional strains on the natural setting. These pressures have confirmed the fundamental reality that the GYE national parks do not exist in isolation and cannot be managed isolated from their surroundings. With both parks situated at the palpable core of the GYE, park managers face the formidable challenge of preserving ecosystem integrity within their enclaves and beyond, notwithstanding contrary outside pressures and demands.[4]

This reality is starkly evident in the case of the GYE's wildlife populations, which serve as an essential barometer of the ecosystem's health and present some of the most daunting management challenges. Bears, wolves, elk, and bison are principal visitor attractions in the two GYE parks, where they are protected (except for seasonal elk hunting in Grand Teton) and play vital biological roles. But these animals regularly cross park boundaries onto adjacent lands where different wildlife management policies prevail, including state hunting and predator control laws as well as tribal treaty rights and supplemental feeding on Wyoming's winter feedgrounds. However, the federal Endangered Species Act has displaced state law in the case of the GYE's wide-ranging grizzly bears and for wolves, too, during the decades-long reintroduction project, providing a high level of protection not available under state law. Consequently, wildlife management in the GYE is enmeshed in jurisdictional complexities at all governmental levels, both within and among the responsible federal, state, and tribal entities.[5]

Despite this complex jurisdictional setting, GYE wildlife populations are in generally good shape at present, though intense controversies persist over federal and state management policies as well as the long-term outlook for several species. The region's isolated grizzly bear population, after dropping precipitously when Yellowstone closed its garbage dumps during the 1960s, has rebounded under federal endangered species protection. In fact, the Yellowstone grizzlies—with a population estimated at nearly 1,000 bears—could soon lose federal protection, which would then subject them to state management and trophy hunting. Long-absent wolves were reintroduced in Yellowstone during the mid-1990s under a special Endangered Species Act provision, and they soon thrived, notably altering predator-prey relationships across the landscape while occasionally predating on local livestock. After dispersing throughout the region and reaching specified population recovery levels, the Yellowstone wolves were returned to state management following pitched judicial and political battles. As the states have approved more aggressive wolf hunting and trapping polices, the wolf recovery accomplishment stands on even shakier ground while grizzly bear delisting appears more problematic due to these state policy changes. Regardless, the existence of presently viable grizzly bear and wolf populations stands as a signal conservation achievement, one that defines the GYE's wild character and perpetuates important ecological relationships.[6]

The GYE's bison, elk and other ungulates are protected when inside the region's core national parks but come under different state management policies when they migrate onto adjacent national forest and private lands, where they are subject to hunting and progressively face the loss of important habitat. Yellowstone's free-ranging bison population has proliferated during the past several decades despite carrying brucellosis and facing extreme persecution by the ranching community. While Montana has gradually moderated its tolerance level, the park's bison are still rigorously controlled upon leaving the park through hunting activity that often resembles a shooting gallery, a demanding quarantine, test, and slaughter protocol, and strict range limitations. This dismal situation is improving as larger numbers of healthy bison are being transferred to Native American reservations, where they are welcome as historical, cultural, and spiritual icons. The new Yellowstone National Park Bison Management Plan should help further improve matters.

GYE elk populations similarly roam between the region's national parks and adjacent lands, where they too are subject to different management regimes. Although documented purveyors of brucellosis to domestic cattle, elk have escaped the same harsh management protocols attached to the region's bison, largely because they are prized by hunters and support lucrative outfitting businesses. Scientists have recently documented regional long-distance elk, pronghorn, and mule deer migration patterns, capturing popular attention and prompting collaborative efforts to protect vital migration corridors. These new scientific insights have notably expanded regional wildlife conservation efforts to a large landscape level that extends beyond the de facto GYE boundary line. As these wildlife corridor protection efforts unfold, ungulate conservation in the GYE is writing a new chapter in the region's rich history of wildlife conservation.[7]

The GYE national forests have long occupied a critical role in the region's ecological health and related conservation efforts, as well as its economy. During the past sixty years that role has changed as Forest Service policies and priorities have shifted away from industrial-scale timber, mining, and energy projects toward wildlife conservation, ecological restoration, and recreation. In the GYE today, timber harvest levels are a mere shadow of prior levels; most cutting is for ecological restoration and wildfire-management purposes. Oil and gas activity is mostly absent from the region's national forests, while the BLM hosts an expansive gasfield development in Wyoming's Upper Green River Basin at the GYE's outer reaches. New mining project proposals involving GYE national forest lands have been routinely contested and mostly rejected, though two long-standing mining ventures continue operating on the ecosystem's periphery where they are local economic mainstays. The Forest Service has eliminated most domestic sheep grazing permits to accommodate the region's burgeoning grizzly bear population, and cattle-grazing permits have also declined over time. Consequently, the GYE national forests are no longer managed principally as commercial settings defined by an extractive mindset and supporting policies.[8]

Indeed, the region's national forests have assumed a new role, one focused on ecological sustainability, wildlife conservation, and recreational opportunity. Since the mid-1960s Congress has designated 3.7 million acres of wilderness and 338,000 acres as wilderness study areas in the five GYE national forests, while the Forest Service has administratively set aside nearly 6 million acres as roadless areas—representing an expansive space where nature largely endures. According to the 1976 National Forest Management Act, the Forest Service is obligated to meet biodiversity goals, which includes maintaining habitat for the GYE's ungulate populations and protecting the grizzly bear, wolverine, and other endangered species. Under the NFMA rules, moreover, forest-management decisions must be based on the best available science, and ecological sustainability is denominated a primary management goal—mandatory provisions consistent with fundamental ecosystem-management principles. By any measure, recreational activity has skyrocketed across the GYE's national forests, reflecting significant population growth as well as newly ascendant public values and economic realities. Though often regarded as benign compared to logging and mining, outdoor recreation is not without environmental impacts that can displace wildlife, erode soils, and harm delicate ecosystem features. In short, the GYE national forests are quite different places today than sixty years ago, but remain quite central to the region's ecological and economic health.[9]

While management of the GYE's public lands has perceptibly shifted toward ecological sustainability, the same cannot be said for the region's private lands, which have long occupied a critically important—but often taken for granted—role in wildlife conservation. Situated at lower elevations and generally near water sources, privately owned ranch property and other lands cover nearly one-third of the GYE, offering winter habitat and migratory pathways for elk, bison, deer, pronghorn, and other animals during their life cycles—habitat that can mean the difference between survival or death during the rugged winter months. With rapidly escalating population pressures evident across much of the region, ranchlands are being subdivided and developed at a breakneck pace to accommodate more people and more homes. The resulting incremental fragmentation is displacing wildlife from traditional habitat and migratory routes while the region's defining open-space appearance is simultaneously disappearing. The future, from a biological, cultural, aesthetic, and even economic perspective, is not promising. Some see it as an impending disaster for the ecosystem.[10]

Although various legal and other tools are available to address the GYE's private land quandary, it has become a veritable race against time in a growing number of places. The federal laws tempering development on the region's public lands simply do not apply on privately owned lands. State and local laws instead govern these lands, and they are relatively weak. Even when elected officials suggest new zoning measures, the political pushback has often been intense given the local commitment to property rights and antigovernment sentiments. The most promising solutions, therefore, have proven to be incentive-based approaches, namely direct governmental payments to landowners, philanthropically financed buyout arrangements, and private conservation easement transactions—all of which are taking center stage in efforts to forestall loss of the region's vital open spaces.

While federally funded conservation programs are relatively new to the region, land trusts have been active for some time. As the land trusts have acquired more knowledge about migration patterns and potential climate change impacts, they are developing thoughtful, science-driven strategies to guide their investment efforts toward such critical areas as the High Divide and other historic migration routes. The question, however, is whether sufficient public or philanthropic dollars are available to forestall serious damage given the influx of new people and the powerful market forces at work. If not, then the burden falls on state land use and zoning laws, and those who implement and enforce them.[11]

In sum, the GYE federal lands and charismatic wildlife populations are in relatively good shape at present, a testament to the ecological conservation progress that has taken place during the past sixty years. For the most part grizzly bear, wolf, bison, and elk populations are at high levels, with grizzlies, wolves, and bison pushing social tolerance bounds. Likewise, wildlife habitat across the GYE federal lands is in generally good shape. The GYE national park, refuge, and wilderness lands provide secure space for these animals, while the multiple-use national forest lands are also relatively secure today as intensive commercial activities have dwindled on these lands except in a few locations. However, the same is not true for the GYE's ecologically important BLM and private lands, which have long provided critical lower-elevation habitat for regional wildlife. With development and subdivision pressures continuing to mount on these lands, valuable habitat has already been lost and more remains at risk. Such rampant development leaves the GYE's migratory species fewer route options during their regular journey, and also blocks dispersal options for climate-stressed animals. Thus, despite current wildlife population levels, the future is far from certain for GYE wildlife absent more urgent efforts to protect critical habitat and migration routes on the region's BLM and privately owned lands.

These GYE conservation achievements, however, come with distinct economic, social, and other costs. As noted, the influx of often-wealthy newcomers attracted to the GYE for its natural attributes is altering the region's communities, prompting a runaway real estate market, related housing shortages, and shifting local values, which collectively are changing the region's character. Local residents, service workers, and others have increasingly found themselves unable to afford housing in the GYE's high-priced gateway communities, forcing them to leave the area or travel long distances to workplaces. Service workers—many of whom are members of minority groups—are attracted to the area by available employment opportunities, but then face the high cost of living and surging rental costs. In Wyoming's Teton County and Montana's Gallatin County a growing Hispanic immigrant population fills local, low-paying service jobs, but cannot afford to live in these areas and barely manages to make ends meet financially, standing in stark contrast to the wealthy elite who daily depend on these workers.

Although local officials are working to address the affordable housing crisis, the affluence divide raises evident environmental justice concerns. Moreover, as wealthy newcomers continue to displace local residents within the GYE, long-standing norms concerning community engagement, neighborly courtesies, recreation access on private lands, and the like are fading. Thus, while the widely shared commitment to protect the GYE's wildlife and natural character is attracting legions of new residents and businesses to the region, their arrival has brought challenging economic and social changes with troubling social equity implications.[12]

Ecosystem Conservation at Work

Nearly fifty years ago, it was thought that the new "Greater Yellowstone Ecosystem" moniker and related ecosystem management concepts would take hold and help save the area's imperiled natural attributes. The fundamental idea was to knit together the region's diverse national park, national forest, and private lands as part of a large, interconnected ecological entity, and to coordinate planning and management decisions across this expansive region renowned for its wildlife, scenery, and wilderness character. For an area encompassing millions of acres and spanning three states, these notions were, frankly, audacious yet alluring given the natural values at stake. With the benefit of hindsight, what can be said about the current state of affairs within the GYE and the ecosystem management idea? What worked and what has not worked? Why?

The "Greater Yellowstone Ecosystem" terminology—coined by the

Craigheads and adopted by conservation organizations—has clearly taken hold across the region within governmental circles and among the public. The term, which sometimes appears as the "Greater Yellowstone Area" (GYA), is regularly used in official documents to describe the area and to acknowledge its invaluable wildlife populations and other natural attributes that cohere this expansive landscape. It shows up in local real estate and other promotional business literature that extol the region's natural attractions, recreational opportunities, and quality lifestyle. It has enabled the federal land management agencies, working together through the GYCC and otherwise, to focus beyond their individual boundary lines, thus expanding their efforts to sustain and restore the region's ecological integrity. It is commonly used by the state wildlife agencies and other governmental entities throughout the region, also broadening their perspective as well as public understanding.

Moreover, the term undergirds the related ecosystem-management concept that has improved coordination efforts among the responsible federal and state agencies. It both reflects and legitimizes the role that science now occupies in agency planning and decision processes. Nevertheless, the related GYE and ecosystem-management concepts have not brought about an integrated, legally binding vision or plan for the region. Instead, GYE conservation accomplishments have come about in a piecemeal fashion, reflecting a functionally pragmatic approach to the region's natural-resource-management challenges and its jurisdictional complexity.[13]

The principal threats confronting Yellowstone and the GYE from the region's federal lands during the past sixty years have been mostly curtailed or mitigated, as we have seen. Where commercial logging was once rampant on the Targhee, Gallatin, and other GYE national forests, it is now largely quiescent across these same forests. Oil and gas development activity that once seemed imminent in several GYE national forests has not materialized, while Congress has effectively put new leasing in the Bridger-Teton's Wyoming Range off-limits. Mining proposals have been consistently blocked, while Congress has withdrawn 30,000 acres north of Yellowstone from any new mining projects. Domestic sheep are absent from the forests, and cattle numbers have declined. To be sure, drilling continues in the BLM's Pinedale Anticline gas field, phosphate mining persists in southeast Idaho on the Caribou-Targhee, and the Stillwater mining complex on the Custer Gallatin continues to operate, but these intensive development activities are removed from the GYE core. Otherwise, industrial-scale activity is now largely absent from the GYE national forests, representing a fundamental reorientation of federal resource-management priorities toward maintaining and restoring the region's natural attributes.[14]

However, Congress has made no new wilderness designations on the GYE national forests since the mid-1980s, despite regular efforts to protect additional acreage. The fact that none of the region's congressionally designated wilderness study areas (WSAs) within the region's forests have yet received formal wilderness protection stands as an unambiguous testament to the intense political passions that wilderness proposals generate in each of the three GYE states. Serious wilderness designation efforts involving Beaverhead-Deerlodge National Forest lands and the Wyoming Public Lands Initiative both foundered despite considerable local support. It remains to be seen whether the Custer Gallatin National Forest's wilderness recommendations, which garnered support as well as opposition from both pro- and anti- wilderness groups during the forest planning process, can be translated into law. Nonetheless, nearly six million acres in the GYE national forests are covered by the Forest Service's roadless-area rule, providing an important measure of protection against most industrial activity and new road construction. The 2009 Wyoming Range Legacy Act protects these Bridger-Teton forest lands from potential energy development, while three GYE river segments—the Clarks Fork, Upper Snake, and East Rosebud—have been designated wild and scenic rivers, with congressional proposals pending to protect additional river corridors. Thus, even as Congress has been stymied from extending formal wilderness protection within the GYE national forests, these same forest lands have gained a significant degree of legal protection that largely safeguards their ecological and natural attributes.[15]

This evolution in GYE public land policies and ecological thinking has benefited the region's wildlife while also improving relations and cooperation among the responsible agencies. It is common knowledge that growing concern during the 1980s over the region's at-risk wildlife populations—particularly as grizzly bear numbers spiraled downward— played a pivotal role in defining the original GYE boundaries and in convincing agency officials to better coordinate their resource-management efforts. The most prominent example is the Interagency Grizzly Bear Committee, which has overseen a joint federal-state grizzly bear recovery effort that has succeeded in growing the GYE bear population from a couple hundred bears to a figure approaching 1,000 bears. As more has been learned about regional wildlife-migration patterns, the GYE agencies have begun collaborating in efforts to safeguard vital migration corridors, as occurred with the Path of the Pronghorn initiative that is designed to enable this pronghorn herd to continue seasonally navigating its historical migration route. To address the perceived bison-brucellosis problem, the agencies collaboratively established an interagency working group that, after imposing strict geographical limits on migrating bison, has loosened these restrictions, opening the door for greater seasonal bison movement as well as an expanded bison translocation program reuniting Native American tribes with culturally important buffalo. Related efforts to enlist the GYE's private landowners in these evolving wildlife-management and restoration efforts have met some success through conservation easement transactions, private philanthropic initiatives, and federal incentive-based programs, perhaps most notably in the High Divide area. Although controversy persists over wildlife-management policies (particularly the federally administered Endangered Species Act), the GYE boasts the full complement of species present at the time of Euro-American settlement, the region's wildlife populations are generally healthy, and habitat on the GYE national forests is more secure today than it was before.[16]

These GYE conservation accomplishments have been achieved on a piecemeal or issue-by-issue basis, not through any formal, region-wide vision or integrated policy agenda. Put differently, GYE ecological sustainability and restoration goals are being pursued individually not as a result of any overarching law or policy statement. After local politics derailed the GYCC's early 1990s Vision document initiative, the federal agencies have shown no appetite for another comprehensive, region-wide planning or visioning process. Instead, the agencies have addressed GYE wildlife and other cross-boundary conservation issues on a case-by-case basis. Prime examples of this approach are the joint federal-state interagency grizzly bear recovery effort and the interagency bison-management plan.

This issue-based conservation strategy acknowledges that the Park Service, Forest Service, and the other GYE federal and state agencies are each governed by different laws with different management priorities and planning procedures. The strategy also recognizes that the GYE federal agencies answer to different national and local constituencies, while the relevant state agencies operate in a more parochial and conservative political environment. As a practical matter, it has proven challenging enough to coordinate management strategy among the responsible GYE agencies for a single species or issue—particularly controversial species like grizzlies, wolves, and bison—let alone to achieve consensus on broader conservation priorities or practices that extend across the jurisdictionally fractionated GYE landscape. Put simply, no single entity is in charge of the entire GYE or has the authority to impose its will on the landscape or the people living there.[17]

The GYE resource-management coordination problem—and the corresponding issue-by-issue approach—is also attributable to the fact that the GYE falls within three different states, each with its own politics, laws, and priorities. Federal officials may have the final say on the GYE public lands, but the states have jurisdictional authority over the region's privately owned lands and its wildlife outside the national parks and refuges. Although the three GYE states have generally agreed among themselves on certain issues, such as grizzly bear delisting and wolf management, significant differences are evident over other conservation matters. Wyoming's elk feedground policy stands in sharp contrast to Montana and Idaho elk-management policies, and is regularly decried by these state's wildlife managers. While Idaho has long supported phosphate mining in the Caribou-Targhee National Forest, Montana residents have strongly contested new mining proposals outside Yellowstone National Park. Wyoming's effort, through its Public Lands Initiative, to resolve the wilderness issue for the Palisades WSA gained no traction in Idaho even though the WSA straddles the two states. Recent legislative changes to Montana and Idaho laws permitting controversial wolf hunting and trapping practices have no counterpart in Wyoming law and prompted a federal review of state management of the GYE wolves, which could have adversely affected Wyoming if the wolves were relisted under the ESA.

Striking differences are also evident among the GYE's counties and communities, despite their increasingly shared economic reliance on tourism and recreation. Sublette County in Wyoming has generally welcomed development on private lands in the Hoback area and elsewhere despite evident wildlife impacts, while nearby Teton County tightly regulates land-use decisions to protect wildlife. Lincoln County flatly refused to engage in the Wyoming Public Lands Initiative even though adjacent Teton County embraced the opportunity to try to resolve management of a WSA straddling the two counties. The adjacent Idaho and Wyoming Teton counties are not in full agreement over expansion of the Grand Targhee Resort. The Big Sky complex, which spans two Montana counties, has presented both counties with significant environmental concerns. Further, several GYE counties have found themselves at odds with their state legislatures and officials, perhaps most notably Wyoming's Teton County over its affordable-housing issues. Other examples abound, highlighting the formidable obstacles to any coordinated effort to articulate and implement a comprehensive vision or plan for the jurisdictionally fragmented GYE landscape.[18]

Notwithstanding the absence of a formal regional vision or plan, nature conservation has assumed a prominent role across the GYE. Indeed, protecting the region's wildlife and other natural attributes has organically become a de facto priority that extends across the GYE's public lands and even spills onto its private lands. Likewise, the ecosystem-management approach has been absorbed into agency policy within the Park Service, Forest Service, and US Fish and Wildlife Service, which each give wildlife and biodiversity conservation a priority position in their planning and management decisions. To a lesser degree, the state agencies are also invested in sustaining the region's ecological integrity, though constrained by state and local political realities. To be sure, controversy persists over regional resource-management issues and appropriate conservation strategies, but an ecological management approach geared toward conserving the GYE's natural amenities is plainly evident and now reaching across the broader landscape. In sum, much has changed during the past sixty years, including an unmistakable region-wide embrace of ecological conservation.[19]

The Forces at Play

This momentous shift toward ecosystem conservation in regional policies and priorities is attributable to several interrelated forces and factors. These agents of change, while connected, have each played distinctive roles in the transformation that has occurred across the GYE. What follows seeks to identify and describe the forces that have brought an ecological management approach to conservation in the GYE.

The region's conservation groups along with their funders occupy a prominent role in this transformation. Individually and collectively, they have stalwartly promoted the GYE concept and ecosystem-management principles, regularly extolled the region's deep and long-standing connection to nature conservation, passionately defended—through political advocacy, strategic litigation, and other tactics—the area and its wildlife against ill-advised development proposals, and articulated a compelling nature-first vision for the region. Many of these same groups, along with allied land-trust organizations, have also been deeply involved with the region's large landowners through conservation sales, easement transactions and otherwise, a reflection of their shared interest in preserving the GYE's natural qualities, wildlife, and open spaces. Moreover, private philanthropy has been instrumental in GYE conservation efforts, not only financially supporting the region's various environmental groups but also providing the funds necessary to pursue incentive-based solutions to livestock grazing concerns, development pressures, and other specific problems.[20]

Scientists, too, have played a major role advancing the GYE nature conservation agenda. Over the years, they have intensively studied the region and its wildlife, documented its inherent ecological connections, identified impending risks, and conveyed this information to the agencies and the public. Their work, for example, accounts in large measure for the ongoing efforts to protect wildlife migration corridors and for the wolf-reintroduction decision that has restored vital predator-prey relationships. Agency officials have regularly drawn upon the scientific literature to fortify their resource-management decisions, including the shift toward more ecologically sensitive policies. Conversely, when agency officials have ignored relevant science, their decisions have proven vulnerable in the courts, as reflected in the judicial decisions overturning the FWS's grizzly bear delisting decisions. Sound scientific information will be more important than ever as the GYE agencies grapple with climate change and its ecological impacts.[21]

The federal land management agencies have plainly evolved in their approach to the GYE. Today, Yellowstone and Grand Teton park officials regularly employ science in their resource-management decisions and work consciously to establish relationships outside park boundaries to address wildlife, fire, and other resource issues. The Forest Service's shift from a commodity production agenda to an ecological management approach is now widely acknowledged. In the GYE it is reflected in the agency's second-generation forest plans as well as its management decisions that now mostly avoid intensive development activities across the region's national forests. The FWS has similarly shifted toward an explicit ecological planning and management approach to its wildlife conservation responsibilities, though the National Elk Refuge still struggles with its artificial winter-feeding program.[22]

Over time the GYE's communities and residents have come to recognize that their economic and social well-being is linked to the area's natural attributes. Tourism and recreation are now indisputably big business across much of the region, as reflected in park visitation numbers, hotel occupancy figures, proliferation in outdoor recreation retail stores, and the sheer volume of traffic during summer months. The GYE ski areas bring people to the region in winter as well as summer, while also providing a recreational outlet for residents. With few exceptions, local residents, political leaders, and business owners are acutely aware that the region's natural attributes are what has drawn visitors as well as new residents and businesses to the area. The result of this obvious amenity-based transformation is evident in the level of local support for conservation initiatives such as the Path of the Pronghorn, Wyoming

Range Legacy Act, Yellowstone Gateway Protection Act, and new wild and scenic river designations. And it is reflected in the Park County Business Council's opposition to the Emigrant and Crevice mine projects. In each instance, these legislative accomplishments and related initiatives were born out of widespread local opposition to troublesome development proposals. All of which represents, by several accounts, the arrival of the "New West."[23]

During the past decades law and litigation have also played a prominent role in preserving the GYE's ecological integrity and advancing conservation values. Indeed, it is hard not to view the law as a pivotal factor in virtually all of the consequential resource-management issues that have confronted the region during the past forty years. Federal laws like the Endangered Species Act, NEPA, National Parks Organic Act, National Forest Management Act, and Wilderness Act have not only driven the federal agencies toward more ecologically enlightened decisions, but also provided conservation groups with the legal ammunition to challenge environmentally problematic decisions in court. Although these federal laws do not expressly require the GYE federal agencies to engage in ecosystem management, they have collectively nudged the agencies toward more ecologically sensitive and better coordinated resource-management decisions. When the agencies have ignored their legal obligations, the courts have proven willing to intervene and overturn questionable decisions that threaten key resources or ecosystem integrity. But with state law now taking center stage in GYE wildlife and private land development issues, it remains to be seen whether these quite different laws will serve as a meaningful restraint on ecologically troublesome decisions.[24]

Without question, the ESA has loomed large over the GYE since the grizzly bear was listed in 1975 as a federally protected threatened species. Grizzly habitat needs were originally used to define the GYE's informal boundaries, and the bear has since figured prominently in litigation to stop rampant timber harvesting on the Targhee, curtail oil and gas leasing on the Bridger-Teton and Shoshone forests, and reduce national forest livestock-grazing activity, as well as in related allotment buyout efforts throughout the region. The ESA-protected bear spawned the region's most far-reaching interagency coordination effort—the Interagency Grizzly Bear Committee and the related Yellowstone Ecosystem Subcommittee, which have been instrumental in prioritizing the grizzly bear and its habitat in agency decisions region-wide. Indeed, agency officials regard the grizzly bear as a "defining part of our management" and the "driver for ecosystem management." Concern over the isolated

Yellowstone grizzly bear population's long-term welfare has underpinned recent landscape conservation initiatives, specifically ongoing efforts to connect the GYE bears with their northern cousins in the Crown of the Continent Ecosystem—a conservation initiative the courts have interpreted the ESA to require. It is no wonder, given the grizzly bear's considerable impact on GYE resource-management decisions, that conservation groups have aggressively sought to prevent it from being "delisted" under the ESA and removed from federal protection. In short, the grizzly bear remains a central legal concern as well as a critical biological presence within the GYE.[25]

The ESA was also key to Yellowstone wolf restoration, which has reshaped the GYE ecology and stands as a landmark ecological restoration accomplishment. In 1994, employing the ESA's experimental population provision, the FWS and Park Service reintroduced wolves to the park where they have flourished while preying upon the park's abundant elk herds, thus restoring important ecological processes and components. Though the wolves brought throngs of visitors to the park, they also wandered far beyond park boundaries, provoking local ranchers and hunters to decry the damage they were doing to domestic livestock and the region's prized elk herds. Eventually, Congress entered the fray and, after a bitter legal fight, adopted legislation removing the wolves from federal protection and turning their management over to the states. Since then, Montana and Idaho have separately loosened limitations on wolf hunting, which has been directly responsible for more Yellowstone wolves killed upon venturing outside the park, despite Park Service objections and public protests. While the ESA initially carried the day to bring wolves back to the GYE, politics has since triumphed and given the states control over the wolves. The ongoing wolf controversy illustrates the limits of this powerful but politically vulnerable law.[26]

Other federal laws have figured prominently in GYE resource-management matters. The National Environmental Policy Act (NEPA), with its rigorous environmental analysis and public participation requirements, has regularly been deployed—often successfully—by conservation groups to challenge environmentally troublesome federal agency decisions, such as timber sales, oil and gas leases, drilling projects, new mine proposals, livestock grazing, and recreational uses. During the 1980s, for example, NEPA claims sent the Forest Service back to the drawing board when it sought to lease portions of the Bridger-Teton and Gallatin national forests for oil and gas exploration. More recently, NEPA litigation has forced the Forest Service to reassess its decisions granting

Wyoming a permit for its elk feedgrounds on national forest lands and allowing livestock grazing in prime grizzly habitat. The National Parks Organic Act's nonimpairment mandate provided snowmobile opponents with a persuasive legal argument in the battle over winter use in Yellowstone, but the same law has failed to protect the park's bison when they seasonally migrate outside the park. The Wilderness Act's strict management requirements were successfully invoked to protect WSAs in the GYE national forests from potentially damaging motorized-recreation activities. The revised National Wildlife Refuge Administration Act's "biological integrity" provision may yet force the FWS to eliminate the National Elk Refuge's supplemental winter-feeding program, thus restoring more natural processes on the landscape. Though conservation groups have not prevailed in every legal challenge invoking these powerful federal laws, their litigation successes have cumulatively stopped an array of ecologically ill-advised decisions and compelled agency officials to rethink other questionable decisions. The upshot, by any measure, is a more intact GYE where nature has obtained greater priority.[27]

These federal laws, however, have not succeeded in bringing the GYE land-management agencies to formally embrace an integrated ecosystem-management approach to the region. To be sure, the GYCC serves as a consultative body that brings the GYE land managers together to discuss regional conservation matters, and the agencies have undertaken specific cross-boundary coordination efforts on behalf of the grizzly bear and bison. But, in the aftermath of the failed early 1990s Vision process, the GYE federal agencies have avoided any comprehensive planning initiatives, clearly sensitive to political realities within the three GYE states and the power of their congressional delegations. Nonetheless, federal law provides ready opportunities to coordinate planning processes across boundary lines. The National Forest Management Act requires the Forest Service to coordinate its planning efforts with other federal, state, and tribal agencies, while the NFMA planning regulations obligate forest planners to prepare a landscape assessment before revising forest plans. The Park Service, US Fish and Wildlife Service, and BLM operate under similar planning coordination obligations in their governing laws and policies. NEPA also provides for interagency consultation and coordination when agencies face planning and project decisions, including the opportunity to participate as a cooperating agency in the process. Although these legal provisions implicitly acknowledge the cross-boundary ecological realities in the GYE and elsewhere, the law offers few enforcement mechanisms and none that have yet compelled region-wide planning or decision-making. Moving forward, coordinated planning will likely continue to occur on individual resource-management issues and among acquainted individual managers, who today better understand and appreciate the GYE as an interconnected ecological entity.[28]

Where federal law has long dominated the GYE legal landscape, state law is assuming equal—or perhaps even greater—importance for conservation purposes in this jurisdictionally complex setting. Outside the national parks and wildlife refuges, state law governs wildlife management except for endangered species, leaving the grizzly bear (and the recently listed wolverine) the only major GYE species under federal protection. Elk, deer, bison, pronghorn, and even wolves now fall under state management. The three GYE states follow the North American Model of Wildlife Management, which is built upon hunting and fishing license revenues and thus emphasizes game animals and consumptive uses. Unsurprisingly, the states are no friend to the region's wolves or bison, and they are poised to implement a trophy grizzly bear hunt once the bear is removed from federal control. Only Montana among the GYE states has a state environmental policy act, which means development proposals involving valuable wildlife habitat in Wyoming and Idaho are not subject to careful environmental review or public comment. State law also governs land-use planning, zoning, and conservation easement transactions on the region's privately owned lands, where intense development pressures are strikingly evident. These state land-use laws are notoriously weak and generally not rigorously implemented or enforced, which has enabled subdivision and other development proposals to proceed with minimal scrutiny. As the Carney Ranch conservation easement controversy illustrates, even these transactional arrangements can be fraught with troublesome oversight and enforcement issues under existing law.[29]

To be sure, state law is not toothless and has been invoked successfully in several instances to advance GYE conservation goals. In the part of the Emigrant Mine proposal that involved development on private lands, the Montana Supreme Court found violations of the state's environmental review requirements and blocked further mining. The same court approved the transfer of bison within the state, opening the door for a nonlethal solution to Yellowstone's growing bison population problem that also benefits the state's Native American tribes interested in restoring bison on their reservation lands. Under its governing statutes, the Wyoming Game and Fish Department embraced the idea of designating wildlife-migration corridors, a promising initial move toward protecting these vitally important routes, though the governor later assumed responsibility for such designations. The Wyoming Supreme Court's decision rejecting a rancher's wildlife damage claim emanating from a brucellosis outbreak in his cattle herd represents an instance where the court refused to second-guess the existing policy allowing wildlife to range freely in the GYE. On occasion, the state courts have also enforced specific conservation-oriented provisions in local land-use plans as well as subdivision requirements in various GYE cases. While these results are heartening, state law imposes few of the strict standards and procedural requirements commonly found within the federal laws governing environmental and conservation-related matters on public lands.[30]

Politics has also been a defining presence in GYE conservation efforts. The high-profile GYE, with its charismatic wildlife and world-renowned natural features, attracts both national and local political attention, as reflected in the diverse constituencies engaged in the region's conservation controversies. Under the US Constitution's property clause, Congress is vested with authority over the public lands, giving it the final say on matters involving the GYE's national parks, forests, wildlife refuges, and BLM lands. When confronted with compelling conservation arguments, Congress has been moved to protect GYE lands, adopting laws like the Gallatin Land Exchange legislation, Wyoming Range Legacy Act, and Yellowstone Gateway Protection Act. Under the Wilderness Act Congress has retained for itself the authority to designate wilderness areas, which it did in 1984 with passage of the Wyoming Wilderness Act following lengthy statewide negotiations between wilderness advocates, industry groups, and local officials. But since then, Congress has not been persuaded to designate additional GYE wilderness areas, bowing to local opposition from the various sides in the ongoing wilderness debates. Though stymied on wilderness, Congress has added three GYE river segments to the nation's wild and scenic river inventory. Initially resistant to Yellowstone wolf-reintroduction proposals due to strong local opposition, Congress eventually approved the reintroduction in response to growing national pressure. Then, as the wolf population proliferated over the course of seventeen years, Congress was convinced by a growing consensus within the region's ranching and hunting communities to intervene and remove the wolves from federal management under the ESA.

Conversely, despite local support for grizzly bear delisting, recreational snowmobiling in Yellowstone, and whitewater boating in the park, Congress has not intervened in these matters, giving its nod to conservation concerns. Nor has Congress revised several key laws—including the ESA, NEPA, and the General Mining Law—that have figured prominently in

GYE conservation controversies. At the same time, Congress has shown no appetite for comprehensive, GYE-specific legislation that would address regional resource-management issues at the tri-state level.[31]

Political considerations have also prompted presidents and cabinet officials to intervene in GYE conservation matters, further highlighting the great degree of national interest in these issues. President Bill Clinton involved himself in the New World Mine controversy, even traveling to Yellowstone to announce a settlement removing the threat. Interior Secretaries Sally Jewell and Ryan Zinke both entered land withdrawal orders to stop mining on national forest lands north of Yellowstone, protecting the area from further development until Congress could adopt the Yellowstone Gateway Protection Act. Consistent with the Republican Party's long-standing political agenda promoting domestic energy production and with substantial local support, Interior secretaries during the Reagan, Bush I, and Bush II administrations leased large acreages across the GYE for oil and gas exploration, giving rise to the expansive Pinedale Anticline gas-field development. But the Obama administration, citing local opposition to more drilling, reached a contrary decision for the Wyoming Range, deciding not to proceed with a pending lease offer. Clearly, some GYE conservation issues will garner national attention and will thus be resolved at that level, especially when local support is also evident. Even when the matter may not rise to the nationally significant level, the issue may still merit congressional or cabinet-level intervention, especially when the state congressional delegation can point to a strong local consensus on the path forward.[32]

State-level politics suffuse nearly every GYE conservation issue, driven by powerful state sovereignty and simmering antifederal sentiments evident within the three GYE states. This is particularly true in the case of federal grizzly bear and wolf management, as well as with bison in Montana. Indeed, the GYE states have sued the federal government for control over endangered species, migrating bison, and National Elk Refuge management policies. The states have also sued when federal officials have limited development or recreational opportunities on the GYE public lands, for example, when the Bridger-Teton placed forest lands off-limits for energy leasing or when Yellowstone curtailed snowmobile opportunities in the park. The three GYE states have each also flirted with asserting ownership claims to the federal lands within their borders, and Wyoming has succeeded in securing jurisdiction over nonfederal lands within Grand Teton National Park. But the states are not always in an adversarial position. State and federal political interests have aligned on several GYE conservation matters, such as safeguarding the Wyoming Range from energy development and prohibiting mining on national forest lands north of Yellowstone. Federalism is plainly alive and well in GYE politics, as reflected in the shifting federal-state alignments over regional conservation issues.[33]

Equally noteworthy as a political matter is the relationship between the three GYE states. In fact, interstate relationships are particularly important in the GYE given the need for collaboration and coordination on the region's cross-boundary wildlife and other resource-management issues. The three states have generally seemed to agree among themselves on several GYE conservation matters, such as grizzly bear delisting and wolf management. This is not the case, however, with elk management, where Wyoming's winter feedgrounds have been roundly criticized by Idaho and Montana, concerned that these concentrated feeding operations will facilitate the spread of brucellosis and chronic wasting disease across the region. It is also clear that Montana and Idaho view mining in the GYE quite differently; phosphate mines in Idaho's Caribou-Targhee National Forest are continuously expanding, while Montana has steadfastly rejected new mining proposals north of Yellowstone. Earlier, Wyoming took issue with the Noranda mine proposal, which would have burdened the state with mine waste while the benefits would have gone to Montana. For the most part, local economic and cultural concerns have driven state politics on these matters as well as new wilderness designations, the federal wolf reintroduction effort, and visitor limitation proposals. Given these manifest political realities and divergent state interests, it is difficult to conceive the GYE states coming together, or joining with the federal agencies, in any comprehensive planning exercise to establish overarching regional ecological management standards.[34]

In recent years, the region's Native American tribes have become more engaged in GYE-related conservation matters, bringing another political player to the table. In fact, tribes have begun broadly asserting themselves in National Park and public land issues across the West, reminding everyone that these places were their traditional homelands prior to Euro-American westward expansion and the late nineteenth century treaties that dispossessed them. In the GYE, Yellowstone National Park now recognizes twenty-seven tribes historically connected to the park's land and has pledged to inform and collaborate with them on issues of mutual concern. Already the tribes are deeply involved in the park's bison-transfer program, enabling them to restore buffalo on their reservation lands and thus reconnect with an important aspect of their history and culture. The program has not escaped controversy as the state of Montana and ranching groups have sought, unsuccessfully, to block the movement of bison within the state. Public concern is mounting, however, over treaty-based bison hunting outside Yellowstone, where some tribal hunters are shooting migrating bison in large numbers and then butchering them haphazardly.

In western Wyoming, the Wind River Indian Reservation lies within the informal GYE domain, and its expansive wilderness-like lands provide important habitat for wildlife, including the expanding grizzly bear population and a recently restored bison herd. The Greater Yellowstone Coalition and other conservation organizations have added tribal members to their boards and staff, recognizing their shared interests as well as the potential political benefits of such an alliance. In many respects, the mounting tribal presence in GYE matters, perhaps most notably the bison-transfer project, represents an important marriage between conservation and culture with elements of social justice also evident.[35]

Taken together, these agents of change have ushered in a new era for the GYE—one that views the region as an interconnected ecological entity and prizes its wildlife populations, natural attributes, and recreational assets. Along the way, faced with new economic realities and opportunities, most GYE communities have recognized these realities and embraced tourism, recreation, and real estate development—long perceived by many to be more compatible with the region's emerging nature-first agenda than logging, mining, and other resource production activities. As this new GYE era unfolds, this assumption of an alignment between nature conservation and these new economic realities is being tested as outdoor enthusiasts flood the GYE public lands and rampant subdivision activity carves up the region's privately owned open spaces. The question, then, is whether the same agents of change that succeeded in embedding the GYE concept and related ecological conservation advances across the region will prove able to harness the economic forces now besetting the GYE.

The Conservation Challenges Ahead

What does the future hold for conservation in the GYE in the coming years as the Anthropocene unfolds? Several impending challenges are evident and already garnering attention as they continue to defy easy resolution. These matters pit powerful interests against one another, making consensus political answers hard to find in the jurisdictionally fragmented GYE setting. The impending challenges include the ever more apparent climate change crisis, mounting private land development activity, escalating recreation and visitation pressures, and persistent predator management, wildlife disease, and migration corridor issues. Whether the same nature conservation and ecosystem management approaches employed during the past decades will be sufficient in the future remains to be seen.

It is already quite clear, however, that GYE conservation efforts must expand beyond the ecosystem to the broader landscape scale, requiring greater coordination efforts among the responsible agencies, governmental entities, advocacy groups, and the public. Given the magnitude of the looming challenges, science will continue to be essential for understanding and responding to the changing environment, which may entail more active management strategies. Based on recent history, conservation advances will also continue to require an assortment of advocacy strategies ranging from public education and political persuasion to litigation and market-based incentive solutions. Failure to acknowledge and focus on these matters puts the existing GYE conservation achievements at risk, jeopardizing once again the region's ecological integrity, its renowned wild character, and quality of life for its citizens.

Certainly, climate change casts the largest shadow over the GYE's human and wildlife communities. Where the term "climate change" was nonexistent in the popular lexicon forty years ago, it is today a central concern throughout the GYE region and beyond, with its impacts being felt on the ground as well as in the region's communities. Due to global-warming trends attributable primarily to human-induced greenhouse gas emissions into the atmosphere, scientists are predicting profound environmental changes. In the GYE both air and water temperatures are expected to rise and more precipitation is expected to fall as rain rather than snow. These climate-induced changes will reduce the annual snowpack more rapidly, accelerate the spring runoff in mountain streams, and dry out area vegetation earlier. Scientists predict these environmental changes will, in turn, modify animal migration patterns, impact native trout habitat, alter forest conditions, increase and intensify wildfire events, and endanger habitat-specific wildlife, such as the wolverine and pika. Some of these impacts are already occurring. The widespread pine bark beetle epidemic, triggered by warmer winter temperatures, has decimated the region's whitebark pine trees that provide an important seasonal food source for grizzly bears. In response, the FWS has listed the whitebark pine tree as a threatened species, while the federal courts blocked the FWS's initial grizzly bear delisting proposal due to concern over the future availability of this critical food source.[36]

Other consequences also attach to these climate-related changes. Although the warmer temperatures are expected to increase the agricultural growing season in the GYE, the spring runoff will come earlier, which will affect irrigation practices and mean less moisture later in the season. Warmer stream temperatures will reduce cold-water trout habitat, impacting recreational anglers as well as the region's thriving fly-fishing and guiding industry. A shorter snow season will affect winter recreation, contracting the ski season at the GYE ski areas and otherwise reducing seasonal recreational opportunities. More frequent and intense wildfire events will undoubtedly affect homeowners in the WUI zone, while the likelihood of such fires could induce communities to consider stricter zoning codes, deter future construction proximate to the region's flammable forests, further strain the Forest Service's firefighting capacity, and prompt the agency to cut more trees to lessen fuel loads. Warmer temperatures are also expected to reduce home and business energy needs due to a shortened winter season. Although climate change impacts will vary across the geographically diverse GYE, the region's wildlife, water, and vegetation will undergo change as will the nearby communities that depend on the region's natural attributes for their economic well-being and cultural identity.[37]

"For the most part," according to the 2021 *Greater Yellowstone Climate Assessment*, "meaningful policy to address and adapt to climate change is lacking in the GYA." To be sure, federal climate change policy has vacillated in the twenty-first century, pinballing from one presidential administration to another, while the three conservative-leaning GYE states have been slow even to acknowledge the problem. Nonetheless, the *Climate Assessment* and related studies provide guidance regarding useful adaptation measures. To build public support for the necessary policy adjustments, the report's science-trained authors support additional research and public education about expected climate impacts and necessary responses. Mirroring basic ecosystem management-based adaptation principles, they also call for rigorous monitoring of actual and projected environmental changes in order to frame effective management policies and responses to the impending changes.[38]

Because the climate crisis knows no manmade boundaries, scientists generally agree that a landscape-scale approach is essential to promote ecological resiliency and to implement effective adaptive management responses. In the GYE a landscape-level management approach would focus on the region's all-important watersheds and wildlife, both of which are at severe risk as temperatures warm. The GYE watersheds are not only the region's lifeblood but they are vital lifelines for downstream communities and industries situated outside the GYE. Several GYE wildlife species must be able to move or migrate to more suitable up-elevation habitat as warmer temperatures alter their existing habitat. Prime examples are the wolverine that depends on deep snow, and the pika that relies on cooler, high-elevation habitat. A landscape approach is also regarded as essential to ensure the grizzly bear's long-term survival in the GYE through genetic interchange with more northern bear populations. The nascent movement toward identifying and protecting ungulate migration corridors across the GYE likewise represents a landscape-conservation strategy that will only become more important as hotter temperatures further alter existing habitat. Should native wildlife be unable to move or adapt to the region's changing environmental conditions, many scientists endorse a more interventionist management approach that could entail human-assisted translocation to a more suitable environment. In the jurisdictionally complex GYE, any new landscape-scale planning or management initiative will require an unprecedented level of collaboration among the regional federal, state, local, and tribal players, as well as private landowners. It would, in short, represent a paradigm shift in GYE resource-management practices, albeit an essential shift to preserve the region's ecological integrity and wild attributes.[39]

There is another dimension to the GYE climate-conservation challenge—one that involves mitigating the region's carbon footprint as a matter of energy policy, individual responsibility, and institutional leadership. The overarching question is how climate change should factor into future GYE energy development debates, recognizing that the region's growing population will require ever more energy in the modern world. Should climate change concerns deter future fossil-fuel development in the Pinedale Anticline or other locations? Is there a role for alternative energy development—solar or wind power—on the GYE multiple-use lands, recognizing that alternative energy infrastructure can have substantial environmental impacts, including on migratory corridors, bird life, and water consumption? Are hydro or geothermal power viable options in the GYE? Or should the GYE communities just continue to rely upon distant energy sources rather than provide for their own needs? Institutionally, who should take the lead in addressing these matters in the GYE: the federal land management agencies, the states, local communities, industry, conservation groups, others? Finally, at the individual level, what responsibilities do residents have to curb their own energy consumption? This final question is unavoidable in light of the number of private jets regularly ferrying wealthy residents to Jackson, Big Sky, and even Pinedale. It is also relevant given the number of extravagantly large homes built on large rural lots distant from the region's towns and basic services, which necessitates more carbon-emitting auto travel.

Development pressures on the GYE's private lands show no sign of abating. Indeed, the cumulative impacts associated with private land development have become a primary concern across much of the GYE as open space continues to disappear and wildlife are squeezed from critical habitat. Even before but certainly after the COVID-19 pandemic, the allure of the GYE has been breathtaking, attracting new residents, businesses, and construction to the region. The market forces at work with this influx are enormous, propelled by demand for housing and modern rural, ranch-style living in an idyllic setting. Any effort to regulate or constrain these forces brings forth arguments about unbridled property rights and individual freedom. Most of the region's politicians are noticeably reluctant to tackle this issue absent a palpable crisis. Although state and local laws contain various land-use, zoning, and subdivision provisions, the laws also vest local officials with considerable discretion, which often leaves their implementation and enforcement subject to short-term political considerations. But in several instances, when conservation concerns have carried the day locally, the courts have sustained wildlife-friendly zoning and subdivision decisions. This suggests that GYE land-use laws, though weaker than they should be in today's world, can be employed as a basis for conservation-oriented planning and decision-making. Nonetheless, it remains incumbent upon state and local officials to strengthen these laws.[40]

Absent action by local governmental officials, however, conservation groups and land trusts have assumed what has become a key role in addressing the GYE's mounting private land development concerns. In doing so, these groups are relying upon economic incentives to entice private landowners to enter conservation easement and other land-control transactions. The federal government has weighed in, too, in the form of Department of Agriculture conservation programs that provide cash grants to willing landowners who agree to participate in these programs. What this means in many instances, absent a change in local attitudes toward land-use regulation, is that self-appointed, nongovernmental players—namely private individuals and nonprofit entities—will play a significant role in future decisions about the location and manner of conservation commitments on the region's ecologically vital private lands.

But are there enough private sector funds to meet future conservation needs as the GYE landscape fills in? The answer is almost certainly "no," so federal as well as state funding will continue to be critical in addressing the private lands conservation challenge. Given this financial reality and underlying accountability concerns, government officials and land trust personnel must coordinate their private land conservation efforts and utilize the best available science to maximize the ecological returns.[41]

For decades, the conservation community extolled the economic benefits attached to outdoor recreation and tourism as a panacea for eliminating the extractive industry activities severely damaging the GYE's national forest lands. That day has now arrived in the GYE. Logging, mining, drilling, and livestock grazing—often referred to as the "Lords of Yesterday"—are no longer dominant economic players in the region, except in a few areas. Instead, tourism, recreation, and the related businesses serving them have become driving forces in many GYE communities, but not without environmental and other troubling impacts that are only accelerating.

During both summer and winter months, hordes of outdoor enthusiasts are now afield on GYE public lands engaged in a plethora of activities: hiking, mountain biking, fishing, hunting, backpacking, climbing, skiing, rafting, kayaking, ORVing, snowmobiling, and the list goes on. Their mounting presence, including the use of motorized equipment and mountain bikes to penetrate deeper and faster into the backcountry, is displacing wildlife and disrupting life-cycle patterns. More recreational use means more trail erosion, more unauthorized trails, and growing sanitation concerns, not to mention conflicts between users seeking solitude and others intent on thrills and speed. During summer months, prized GYE trout fishing rivers are often packed with float boats and anglers, leading one newspaper to proclaim "Rivergeddon" on Montana's storied Madison River. Meanwhile, the principal GYE ski areas have become all-season resorts, expanding to accommodate more visitors and thus attracting more development and private housing. The visitation season at Yellowstone and Grand Teton is extending into what used to be the quiet spring and fall shoulder seasons, while park visitors increasingly find it difficult to land a parking spot at popular attractions. According to one knowledgeable observer, "the GYE front country is now filled up, and the back country is becoming increasingly crowded." And there is no end in sight.[42]

Potential solutions to what amounts to an industrial-scale recreation onslaught in the GYE are readily identifiable but not particularly popular on public lands that have traditionally been open to recreationists and largely unregulated. Dedicated GYE conservationists have described the region's burgeoning recreation problems as "thorny," "a ticking time bomb," and the "third rail of conservation." The Custer Gallatin forest planning process illustrated the challenges involved in finding common ground between different conservation and recreation groups—a challenge exacerbated by clashes over new wilderness designation proposals, ORV use, and mountain-biking opportunities. The Park Service and Forest Service have the legal authority to impose limits on recreational activity through reservation requirements, route designations, closure orders, and the like, but have been reluctant to take these steps. Yet, the courts have upheld such decisions when implemented and even compelled the agencies to do so in a few instances, most notably during the Yellowstone snowmobile imbroglio and the WSA mechanized recreation controversies. Both agencies currently collect entrance or access fees and maintain "leave no trace" educational programs, but this has done little to shift visitation patterns or use practices.[43]

The dispersed nature of backcountry recreation activities makes oversight or rule enforcement difficult. Given the expansive GYE terrain and the array of activities available on these lands, greater coordination among the federal land management agencies along with local communities, businesses, and involved user groups and conservation organizations could help to better disperse recreational users and activities across the region, relieving some of the pressure. Alternatively, it may be time for Congress, which has long been silent regarding recreation on the multiple-use national forest and BLM lands, to finally engage with the issue and provide the agencies with clear guidance on their authority and responsibility along with adequate budgetary resources. That could, in turn, embolden the agencies to confront the issue. Otherwise, those who derive so much individual pleasure or profit from recreation on the GYE public lands may succeed in "loving it to death," permanently altering the landscape's ecological and natural character.[44]

Wildlife management continues to generate controversy throughout the GYE even as the responsible agencies have made strides toward working together to address ongoing predator, disease, and migration concerns. These wildlife management issues plainly transcend the conventional boundaries we have imposed on the GYE landscape and require a meaningful level of federal-state collaboration to resolve. Recent scientific studies detailing elk, deer, and other ungulate migration patterns has further highlighted the need for transboundary management approaches. Moreover, neither brucellosis nor chronic wasting disease recognizes the GYE's federal or state boundaries. The GYCC's decision to extend membership status to the state wildlife agencies in 2020 formally acknowledges these realities and provides a new, more inclusive forum for addressing GYE wildlife issues. Since federal and state officials have worked together over the years on specific wildlife issues, including grizzly bear recovery and bison management, they have already established important relationships that can be built upon. Yet, because these wildlife conservation issues present different management challenges, they will require individual coordinated responses that take account of the ecosystem and related local concerns.

In the case of predator management, namely grizzly bears and wolves, the issue revolves around the respective management regimes within the GYE national parks and outside them. Although the grizzly bear, as a federal threatened species, remains fully protected against hunting and otherwise, the GYE wolves are now under state management when outside the national parks and subject to state-controlled hunting and trapping. The same would be true for grizzly bears if delisted. Where Yellowstone and Grand Teton protect both species from hunting, the three GYE states have implemented aggressive hunting policies for wolves, which have seen entire packs eliminated when they have left National Park confines. The states are also poised to hunt grizzly bears outside the parks once they regain control over them. While the GYE states have not set coordinated wolf-hunting quotas, they have preemptively agreed on grizzly mortality quotas for hunting purposes upon delisting with an eye toward maintaining a viable population to avoid relisting. A similar coordinated approach makes sense in the case of the GYE wolves; otherwise, one state's aggressive wolf quotas could tip the GYE wolf population numbers over the acceptable mortality threshold and trigger federal relisting for all three states.

Given the importance of grizzly bears and wolves for National Park visitors as well as local businesses serving them, improved coordination is clearly necessary between the Park Service and the states on hunting policies adjacent to the parks. Or, recognizing the costly fifty-year effort involved in recovering the GYE grizzly population and the public's distaste for trophy hunting of bears, perhaps it is time to consider a grizzly bear conservation act modeled on the Bald and Golden Eagle Protection Act that would give the federal government some degree of control over the hunting and killing of delisted bears. In fact, Secretary of the Interior Deb Haaland co-sponsored such legislation when serving in Congress before her cabinet appointment.[45]

Beyond hunting, other predator-management issues and policies require attention in order to ensure viable GYE grizzly and wolf populations. In the case of the grizzly bear, an interagency landscape-level management approach is essential to enable the isolated Yellowstone bears to connect on their own with bears in the Northern Continental Divide Ecosystem to ensure genetic diversity and long-term survival, notwithstanding Montana's recent bear translocation announcement. Already the GYE wolves have derived some genetic benefit from connecting with their cousins in nearby ecosystems. Outside the national parks, the Forest Service occupies a critical role in GYE predator management through its livestock-grazing policies, which the agency should consider revising to curtail the level of cattle grazing in prime grizzly bear habitat, as it has for sheep. Governmental and private funding has proven helpful in reducing depredation risks for ranchers through range riders, guard dogs, and other deterrent strategies, which merit additional support. Likewise, adequate compensation is essential when depredation incidents occur. The goal must be to secure a high level of social tolerance for the GYE's large predators by reducing the disproportionate burden borne by the ranching community with grizzlies and wolves roaming about the ecosystem.[46]

Wildlife disease concerns involving brucellosis and chronic wasting disease present the GYE with an array of scientific, political, and coordination challenges. The brucellosis problem affects the region's bison and elk herds, both of which are infected with the bacteria and could transmit the disease to local cattle herds with significant economic costs. The responsible agencies have adopted a disease control policy designed to minimize the risk of transmission, which has treated Yellowstone's bison harshly while largely ignoring elk that have been the actual disease vector in every reported case. Although bison have gradually gained access to ground outside the park, they remain subject to troublesome hunting practices as well as problematic capture, test and slaughter policies when exiting the park, though Yellowstone's new Bison Management Plan should help improve this situation. The expanded quarantine facilities along with the Native American bison translocation program have relieved some of the pressure, but these strategies clearly need additional political as well as financial support at both the federal and state levels.

Meantime, the elk brucellosis problem persists without any comprehensive solution in sight. The National Elk Refuge has yet to begin phasing out its winter-feeding program while Wyoming shows little interest in closing its winter feedgrounds. With chronic wasting disease now afield in the Wyoming portion of the GYE, these feedgrounds, where the animals congregate in close quarters, represent ticking time bombs for a massive infection level that could devastate the region's elk and deer herds. The need for a coordinated response among the GYE states is obvious, as is the need for more science and less politics in this matter. It may be time, through an orderly feedground phase-out process, to let nature take its course in this portion of the GYE, grim as that prospect may be.[47]

The importance of migratory corridors and the need to protect them has emerged as a central concern in the GYE, which hosts several of the nation's longest wildlife migrations. It has become apparent—based on ungulate migration data as well as grizzly bear and other biodiversity concerns—that the original conception of the GYE as the target for ecological management has proven scientifically insufficient. Moving forward, conservation efforts must focus on the larger landscape, extending beyond the GYE to embrace the area's expansive federal lands as well as its ecologically important private lands. Wyoming has assumed a leading role in migration corridor conservation, though politics has intruded to the point that the heralded Path of the Pronghorn has yet to receive official state recognition. The officially designated Red Desert to Hoback mule deer migratory corridor represents a promising start, but it has been disrupted in several locations to accommodate problematic commercial activity.

While well aware, under the ESA, of the need to connect the Yellowstone and Glacier grizzly populations on the ground, Montana has yet to do so, choosing instead to truck a couple bears into the GYE. That politically motivated action intended to prompt the delisting of grizzlies under the ESA highlights the fact there is nothing in state law to mandate such connectivity or the establishment of protected migration corridors. In the High Divide area and elsewhere, federal and state agencies are working together, along with conservation groups, local officials, ranchers, and others to secure migratory passageways on critical private lands, setting an important precedent that is expanding GYE conservation efforts and related coordination strategies. Proceeding on a piecemeal basis, there is movement toward connecting the GYE with nearby ecosystems and toward safeguarding migratory corridors in order to maintain native species and vital ecological processes. These developments suggest landscape-scale conservation is organically displacing ecosystem management as the essential GYE conservation strategy in the years ahead.[48]

Beyond the GYE

The GYE concept and related ecosystem-management principles have inspired and spawned similar transformative nature conservation initiatives. Several of these initiatives have links to the GYE, while others have emerged to protect different ecological regions. In each instance, the overarching strategy involves employing scientific criteria to pursue nature conservation at an ecologically defined level to break down the legal boundaries that so often inhibit these efforts. While the origins of these ecological conservation initiatives vary (and some have faltered), the commitment to transboundary conservation is the same. Steady if uneven progress is plainly evident in several locations. And none too soon given the ongoing cumulative changes linked to climate warming, population growth, and inexorable development pressures. Indeed, the clock is ticking to preserve the nation's natural heritage, remaining wild places, and the creatures that inhabit them.

The GYE is included in two significant large-scale conservation initiatives centered north of the Yellowstone region. Established in 1993, the Yellowstone to Yukon (Y2Y) project seeks to safeguard a string of national parks, wilderness areas, and other protected lands as well as connective wildlife corridors that extends 2,100 miles through the Rocky Mountains from northern Canada to the Yellowstone country in the United States. Involving more than 450 conservation and other groups, the science-based Y2Y vision is to preserve this unique linear mountain range's ecological health to ensure its wide-ranging wildlife—grizzly bears, elk, and wolves—have adequate habitat and the ability to move safely. To date, the continental-scale Y2Y project has succeeded in increasing the region's protected acreage by 50 percent, while also protecting another half-million acres on private lands to benefit wildlife.

At a similarly expansive scale, the proposed Northern Rockies Ecosystem Protection Act (NREPA) would add twenty-three-million acres of national forest land in Montana, Idaho, Oregon, Washington, and Wyoming to the national wilderness system along with 1,800 river miles to the National Wild and Scenic Rivers System. It would also establish connective corridors to link the region's protected areas together with the goal of fully protecting native wildlife and fish species. Like Y2Y, NREPA embodies a multistate, landscape-scale conservation vision, but it has gained little political traction due to the region's conservative politics and local opposition to additional protected wilderness acreage. Both of these initiatives validate and seek to expand upon the GYE idea, invoking the latest science to extend what was initially a GYE-focused conservation effort to a much larger landscape-scale effort.[49]

Other ecosystem-oriented conservation initiatives have taken hold and attained a degree of legitimacy on the West's public lands. North of the GYE, US and Canadian land managers—operating as the Crown Manager's Partnership—have joined with nearby universities, conservation groups, and Native tribes in an effort to safeguard the 18-million-acre Crown of the Continent Ecosystem. The wildlife-rich "Crown," as it is known, encompasses Glacier and Waterton Lake national parks, adjacent national and provincial forest lands, several wilderness areas, Wild and Scenic Rivers segments, tribal lands, and local communities on both sides of the international border. Most of these communities are undergoing economic and social changes that attach them more firmly to the region's natural amenities and beauty.

On the Colorado Plateau, Grand Canyon National Park is the focal point of a sustained effort to protect the surrounding landscape and the Colorado River corridor. These efforts have involved expansion of the park itself, the establishment of several large nearby national monuments (Grand Canyon Parashant, Vermillion Cliffs, Grand Staircase-Escalante, Bears Ears, and Canyon of the Ancients), the presence of upstream and downstream national recreation areas, passage of the Grand Canyon Protection Act, and strategic mineral withdrawals on adjacent public lands. Native American interests and support have figured prominently in these conservation initiatives, adding an important cultural dimension to them. The common thread connecting the "Crown" and the Colorado Plateau initiatives is both the scale and cross-boundary nature of these conservation efforts, which mirror in several respects what has transpired in the GYE.[50]

Moving westward, related ecosystem-based initiatives are well established in the Pacific Northwest and California. In the Pacific Northwest, following a decade of acrimonious litigation, the region's expansive federal timberlands are managed since 1994 under the Northwest Forest Plan. The plan established a 24-million-acre ecosystem-management regime that involved three states (Washington, Oregon, and California), embraced nineteen national forests and seven BLM resource districts, and placed more than 18 million acres off-limits to commercial timber harvesting in order to protect the northern spotted owl, several salmon runs, and other species. Driven initially by court orders, the Northwest Forest Plan is aptly regarded as the seminal, official federal ecosystem-management project on public lands.

In California's Sierra Nevada mountain range, the Forest Service amended the region's eleven forest plans during the early 2000s to emphasize the ecological restoration of the damaged forests and watersheds and to promote nature conservation, including protections for the dwindling California spotted owl population. The so-called Sierra Nevada Framework covers 11.5 million acres, protects remaining old-growth ecosystems, emphasizes forest thinning to reduce the wildfire danger, and constrains commercial timber harvesting. Much like the Northwest Forest Plan, the Sierra Nevada Framework represents an agency-driven, landscape-scale plan that effectively links these eleven national forests together to promote ecosystem health and nature conservation objectives.[51]

Other cross-boundary landscape-scale conservation initiatives are underway elsewhere on western public lands and adjacent private lands. In southern California, the Desert Managers Group—an amalgam of the federal, state, and local agency officials—is working cooperatively to manage the roughly 30-million-acre California Desert landscape. The area contains 10 million acres of federally protected lands, including two national parks, several large national monuments, seventy-two designated wilderness areas, and numerous state parks, along with large military training bases, which retain conservation value. In southern Colorado's San Luis Valley, five national wildlife refuges along with the Great Sand Dunes National Park and Preserve as well as nearby ranchlands overlain with conservation easements anchor a complex of public and private lands covering several million acres. Known as the San Luis Valley Conservation Area, this complex of protected lands contains diverse ecosystems, which the responsible agencies manage primarily for wildlife, habitat restoration, and recreational use.

In Montana's Missouri Breaks country, the privately funded American Prairie reserve project is patching together an expansive wildlife reserve consisting of its own privately acquired lands alongside the Charles M. Russell National Wildlife Refuge and the Upper Missouri River Breaks National Monument as well as nearby undeveloped state, BLM, and tribal lands. Already dubbed "the American Serengeti," the project envisions a 3.5-million-acre public-private grassland conservation area dedicated to restoring plains buffalo and other native species. Regardless of whether these embedded large-scale conservation initiatives are directly traceable to the GYE concept and its related ecosystem management approach, the connections are unmistakable as is the commitment to promoting nature conservation at an ecologically relevant scale.[52]

Although federal public land management policies and priorities have vacillated during the past half century with the politics of the moment, related ecosystem management and landscape conservation concepts have secured a tangible foothold within the responsible agencies. This reality formally emerged during the Clinton administration with its Northwest Forest Plan, Florida Everglades Restoration Initiative, and Interior Columbia Basin Ecosystem Management Plan. Today, it is manifest in the Biden administration's America the Beautiful policy initiative, which is derived from the highly influential 30×30 biodiversity campaign supported by virtually all of the nation's conservation organizations. Both the Biden and Obama administrations endorsed far-reaching conservation policies in response to climate-change concerns. These included the Obama administration's creation of nationwide Landscape Conservation Cooperatives, the Park Service's *Revisiting Leopold* report and Director's Order 100, the Forest Service's revised NFMA planning rules, and the BLM's new public land conservation rule. Both administrations, drawing upon the Clinton model, established large-scale national monuments to protect ecological values and Native American cultural resources, which the courts have sustained.

Despite efforts by the Bush and Trump administrations to roll back many of these conservation advances, the basic ecosystem-focused conservation framework remains in place through agency rules and policy directives as well as the Northwest Forest Plan, Greater Yellowstone Coordinating Committee, Crown Managers Partnership, Desert Managers Group, Sierra Nevada Framework, and other place-based initiatives as noted above. That these rules, policies, plans, and initiatives generally involve some level of federal collaboration with state, local, and tribal officials, along with private landowners, reflects a robust commitment to conservation strategies that extend well beyond our conventional boundary lines.[53]

Congress, with some noteworthy exceptions, has been markedly absent as these ecosystem-based conservation policies have evolved, though it has consistently rejected efforts to reverse this trend. In a few instances, Congress has even adopted place-based legislation designed to conserve a larger regional landscape by superseding existing boundary lines. One option for the GYE, therefore, is to consider a federal, place-based bill that would confirm existing conservation achievements, create a formal institutional structure to oversee the area, and establish a coordinated, ecosystem-based planning process governed by conservation-oriented management standards. Two examples could serve as models for such legislation designed to treat the GYE as a legal whole. One example is the Columbia River Gorge Scenic Area Act, which created the multimember Columbia River Gorge Commission to oversee this stretch of the river that is bordered by two states and includes mostly national forest land along with some private landholdings. The act empowers the commission to prepare management plans for the area and enables the counties to adopt land-use plans consistent with the broader plan. Another example is the Tahoe Regional Planning Agency, established through a congressionally approved interstate compact between California and Nevada to manage the Lake Tahoe watershed by establishing environmental standards that apply in both states.

Of course, any such proposal for the GYE would generate considerable political opposition tied to state sovereignty, federal overreach, and the sanctity of private property arguments, and likely doom this approach. Moreover, the GYE differs noticeably from the Tahoe and Columbia River Gorge landscapes, where shared waters were a driving force for coordinated management, where only two states were involved, and where a single federal land management agency was the principal landowner. Nonetheless, as the Anthropocene unfolds, the conservation challenges confronting the GYE will continue to transcend the region's jurisdictional boundaries and will require yet more—and perhaps even legally mandated—coordination among the responsible entities.[54]

Therefore, given the growing acceptance of ecosystem conservation that transcends jurisdictional boundary lines, it may not be too soon to contemplate comprehensive legislation designed to formalize this quite evident trend toward landscape conservation. To be sure, today's hyper-divisive politics do not augur well for a bipartisan legislative agreement on this ecosystem-based approach to conservation on the public lands. But legislation that explicitly knits the various public lands together in specific ecologically important places while giving affected landowners the opportunity to participate in landscape-level conservation projects seems a logical next step in the overall scheme of public land and resource policy. The GYE experience and other related ecosystem-based initiatives provide working models for select, place-based conservation initiatives. Indeed, such legislation would go far toward institutionalizing landscape conservation on the GYE federal public lands as well as other ecologically significant locations.

One comprehensive option is what might be labeled a "National Conservation Network Act." Under such an act, Congress could empower the responsible federal land management agencies, acting jointly, to designate protected area complexes (PACs) consisting of existing national parks, wilderness areas, and other protected lands that would place these lands under a statutory umbrella for conservation management purposes. The act would require coordinated interagency planning (with state, tribal, and public participation) that included the identification and designation of connective wildlife corridors and ecological restoration areas for management purposes. It would afford private landowners and conservation easement holders the opportunity to affiliate with the network and, in turn, to receive economic and technical assistance for managing their own lands for conservation purposes. The goal would not be to standardize management of any particular landscape, but to treat each landscape or PAC individually. Simply put, the envisioned act would enhance and legitimize ongoing federal nature conservation efforts by formally extending them to the landscape level and by empowering federal agencies and nearby landowners to work together with a focus on the larger landscape to meet the challenges presented by a warming and ever more crowded world.[55]

* * *

Yellowstone National Park—the essential core of the GYE—still occupies its historical role as the principal model in our ongoing quest to preserve the nation's dwindling natural heritage. Much has transpired during the past sixty years in the wake of the Leopold Report and the Wilderness Act, both of which gave real credence to the idea of conserving and restoring wild nature. Since then, given the immense scope of the task, the GYE concept has taken center stage in the Yellowstone area along with related ecosystem-management strategies. While freighted with controversy at times, these concepts have expanded nature conservation efforts to the regional level, helped to restore wildlife population numbers, and brought a vital ecosystem perspective to these tasks. Although these concepts are now rooted within the federal agencies responsible for the GYE public lands, the same is not true when it comes to the region's private lands, which represent the next frontier in GYE nature conservation efforts.

To be sure, conflict persists and is likely inevitable on the region's public and private lands as we continue working to reconcile the needs of nature conservation with those of the diverse human communities that call the GYE home. But the conservation strides made thus far are undeniable, providing more than a glimmer of hope that we can meet the challenges presented by the Anthropocene and ongoing regional development pressures. It will continue to be a race against the clock, but we have a much better understanding of the scope of the task and the strategies required to preserve this iconic landscape.

Conclusion

The world's first national park, Yellowstone has stood as a towering beacon to nature conservation for more than 150 years. An extraordinary place where we now largely leave nature alone, the park represents both a powerful symbol and a working model. It symbolizes the nation's early and enduring connection to the natural world, a heritage that enriches us culturally, spiritually, and otherwise. Yellowstone also represents a working conservation model, a place where we are formally committed to preserving wild nature. For park managers, fulfilling these roles has proven an ongoing challenge in what has become an ever more crowded, developed, and contentious setting. To meet this challenge, we have come to realize that the park lies at the core of a much larger landscape—the Greater Yellowstone Ecosystem—that has appropriately become the actual conservation target.

From the outset, it was evident that the boundary lines encasing the park were insufficient to sustain its wildlife as well as its natural features and processes over the long term. As time passed, this stark reality became much clearer. The park was not an island and could not be effectively managed as such. Development on the lands outside the park—clearcuts, mines, oil fields, roads, subdivisions, and more—made plain that the park and its natural attributes were imperiled if these activities continued apace. Advances in ecological science helped too, providing us with a better understanding of the dynamic natural world as well as the toll peripheral development was taking on the park and regional wildlife.

The grizzly bear recovery effort and the 1988 fires further confirmed the region as an interconnected ecological entity. Drawing upon these critical insights, the GYE concept has proven a powerful idea that knits the region together, while the related ecosystem-management concept represents a strategic approach designed to sustain and restore the region's ecological integrity.

The GYE is a complex place, however, which makes prescribing a single resource-management regime a near impossibility. Spanning nearly 23 million acres of federal, state, tribal, and private land, the GYE is both ecologically complicated and jurisdictionally fragmented. No single entity has control over the GYE lands and resources. Rather, an amalgam of federal and state agencies along with twenty separate local governmental entities and nearby Native American tribes carry some degree of managerial responsibility for this remarkable landscape. While the Park Service manages the two GYE national parks under a distinct conservation mandate, three separate federal agencies oversee the other GYE federal lands, each operating under different legal mandates, albeit with distinct conservation responsibilities. The three surrounding states, local governmental entities, and nearby Native American tribes oversee the other GYE lands, each with their own laws and priorities within the region. All of which, in this jurisdictionally complex setting, makes a coordinated conservation approach challenging but plainly necessary.[1]

Once we acknowledged this jurisdictional fragmentation problem and its consequences, we have made significant if halting strides toward protecting nature in the GYE. We have come to recognize and value the irreplaceable wildlife and other natural attributes at risk, as well as the need for a meaningful managerial strategy. Consequently, we now view the region as an interconnected entity, one that requires a commitment to collaboration, coordination, and foresight among those responsible for its welfare. We have recognized, in some instances, the urgency of the matter, though it has been—and remains—a race against time if we are to sustain the GYE's ecological integrity and natural attributes. In short, the GYE story constitutes a vitally important and yet incomplete journey toward a new era in nature conservation.

* * *

Much has changed in the Yellowstone region during the past sixty years, following the introduction and promotion of the related "Greater Yellowstone Ecosystem" and ecosystem management concepts. Since then, nature conservation has emerged as what amounts to a de facto (or unofficial) resource-management goal across much of the GYE, reflecting the power of these ideas in this iconic setting. The federal and state agencies responsible for managing the region's public lands and wildlife have embraced the GYE concept, as have many of the region's communities, businesses, and citizens. Although the vaunted Vision process failed early during the transition toward ecosystem management, the agencies have nonetheless adopted a version of this integrated and coordinated approach to resource management, albeit rather quietly and focused on individual conservation issues. The multimember GYCC, since its inception in 1964, represents a conscious interagency effort to better communicate and coordinate across federal-state lines, while the Interagency Grizzly Bear Committee and similar entities do the same for specific resource-management issues.

The point is not that these coordination efforts work perfectly or comprehensively; rather, it is that the GYE agencies have moved in this direction, departed from their individual silos, and acknowledge the need to approach their management responsibilities at an ecosystem level. With the state wildlife agencies joining the GYCC in 2020, we might reasonably expect more coordination and collaboration on regional wildlife and related issues.

Conservation progress is evident on the ground in the GYE. Sixty years ago Yellowstone National Park was in danger of losing its grizzly bears, and the park's wolves had long ago passed into history. Park rangers were regularly shooting what agency officials perceived to be excess elk, and had only recently allowed bison to roam freely. The Park Service consistently snuffed wildfires, and its management decisions focused mainly on providing visitors with an aesthetically pleasing experience. That changed following the 1963 Leopold Report, which brought ecological science squarely into National Park resource-management decisions.

Today, the GYE grizzly bear population—nearly 1,000 bears by current estimates—fills the ecosystem; restored wolves prowl across the region; a bison population that approaches 5,000 animals has reestablished seasonal migratory patterns; elk continue to roam the park and migrate beyond it; and lightning-caused backcountry fires are allowed to burn subject to certain restrictions. Nonnative lake trout and mountain goats are unwelcome and systematically eliminated within the GYE parks to protect native species. Unrestrained snowmobiling activity is under control, while new recreational activity proposals are scrutinized to protect park resources. Serious efforts are afoot to safeguard migration and dispersal corridors that extend far outside the parks. In short, nature is generally receiving priority with management decisions focusing on the GYE rather than the park boundary line.[2]

The GYE national forests are also managed quite differently today than sixty years ago. Where commodity production activity, namely logging, mining, oil drilling, livestock grazing, and road construction, had the upper hand on the region's national forests, this is no longer true as these activities are a shadow of their former selves in most locations. Commercial timber production is largely quiescent, new mining proposals have been routinely forestalled, energy production is mostly limited to the GYE perimeter, and livestock-grazing levels are down in many locations.

Congress has abetted this transition, withdrawing select GYE national forest lands from mining and drilling. In deference to grizzly bear recovery, the Forest Service has closed roads and phased out domestic sheep grazing across the GYE, while also curtailing cattle grazing in problem areas. Recent forest plans have emphasized ecological restoration, wildlife conservation, migration corridors, and recreational opportunities. Taken together, Congress and the Forest Service have separately placed 60 percent of the GYE forestland into wilderness, wilderness study area, or roadless area categories—protective designations designed primarily to maintain the natural character of these lands. Simply put, the Forest Service is now giving wild nature a meaningful foothold on the GYE national forest lands.[3]

This reorientation in management priorities on the GYE's federal lands, while welcome in many quarters, has stirred controversy in others, complicating the search for collaborative solutions. Nowhere is this more evident than in the case of wildlife management. By several estimates, grizzly bears now fill the ecosystem, occasioning more conflicts with people and livestock. The GYE states are thus once again seeking to end federal control over the bears and to assume responsibility for managing them, principally through trophy hunting—a prospect that deeply troubles many scientists and bear advocates, who also perceive ongoing threats to grizzly habitat. Removing the bears from the endangered species list would eliminate a powerful legal tool constraining federal land managers—one that conservation groups have employed effectively to challenge problematic management decisions. Wolves, originally reintroduced to the GYE under strict federal control, are managed today by the states, which have ramped up hunting and trapping options while showing little regard for the Park Service's interest in maintaining wolf packs for research purposes and visitor enjoyment.

Bison, too, continue to generate controversy in Montana due principally to the largely unfounded fear of brucellosis transmission to cattle.

Although the state has provided park bison with additional space outside park boundaries, it has also created roadblocks to the tribal transfer program but has failed to stifle it. Elk management stirs interstate controversy, primarily over Wyoming's feedgrounds and its artificial winter-feeding policy. Nearly everyone is interested in safeguarding wildlife-migration opportunities, with the much-publicized Path of the Pronghorn serving as exhibit 1, but tensions are apparent over what official corridor designation will mean on the ground. These ongoing wildlife management controversies demand intergovernmental—federal, state, tribal, and local—collaboration with solutions aligned to the region's nature conservation priorities.[4]

Because the GYE's human population is growing rapidly and subdivision pressures are mounting, the region's private lands have assumed greater conservation importance for wildlife as critical habitat and migratory pathways. The strategies available to promote private land conservation are quite different, however, from those that have proven effective on the region's federal lands. Most federal environmental laws do not apply on privately owned lands; instead, state and local governmental officials generally have the final say over these matters. Across the GYE, state land-use planning and zoning laws are generally weak, and their wildlife protection provisions have not been consistently prioritized.

Rather than rely upon legal regulatory tools, private land conservation efforts center on incentive-based market approaches, including outright purchases, conservation easement transactions, and new habitat-leasing arrangements. The funds necessary for these transactions have come from such diverse sources as private foundations, individual philanthropists, the Land and Water Conservation Fund, the US Department of Agriculture conservation grant programs, and state programs like the Wyoming Wildlife and Natural Resource Trust Fund. The lesson is quite clear: different conservation strategies—ones sensitive to property rights and regulatory overreach sentiments—are required on the GYE private lands, which has primarily meant incentive-driven arrangements.[5]

With roughly 20 percent of the GYE's private lands now set aside for conservation purposes, the local interest in protecting these ecologically important lands from development is evident. But is this commitment and the underlying strategies sufficient to meet wildlife and other conservation needs given the rapidly accelerating pace of development and subdivision activity throughout the GYE? There is reason for concern. First, as property values escalate across the region, the amount of public and private funds available to underwrite conservation transactions may not be adequate to meet the opportunity, highlighting the need for additional funding options. Second, to the extent these are private transactions, there is no assurance they will benefit the public interest in wildlife. Conservation easement arrangements are privately negotiated transactions and nontransparent, as was evident during the Carney Ranch quandary, where the subsequent owner started building in the protected migratory corridor. Avoiding this type of problem requires an easement holder with adequate resources to monitor and enforce the easement terms. Third, to derive maximum conservation benefits from these private land transactions requires careful, science-driven planning based upon ecological knowledge as well as ample financial resources that are not always available in the conservation community.

These challenges are not insurmountable, however. They can be met through structured public-private partnerships that bring governmental entities together with property owners, land trusts, and other conservation organizations to collaboratively plan and jointly oversee private land conservation transactions—as appears to be happening in the High Divide area. The ultimate goal is to ensure these transactions add real and enduring conservation value on the landscape.

* * *

Several conclusions—resting upon deeply imbued constitutional federalism principles and the long-standing American commitment to pragmatism—follow from these diverse GYE conservation accomplishments and initiatives. First, there is no one-size-fits-all solution to the GYE's myriad conservation issues, given the region's diverse and jurisdictionally complex landscape. Rather, wildlife and other conservation challenges are being addressed on an issue-by-issue basis, as is true for grizzly bears, bison, and migration corridors. Second, given this issue-by-issue approach, both federal and state law have played a major role in framing and resolving transboundary GYE conservation problems. Federal law dictates management policy on the region's federal public lands and for endangered species, while state law controls other wildlife-management matters as well as private-land issues. Which suggests, given the nature of ongoing GYE conservation challenges, that state law is of growing importance within the region. Third, with federal-state tensions prevalent throughout the region, Congress can always intervene to advance an overriding national or local interest, as it has in the case of the Wyoming Range, Gallatin Land Exchange, mining north of Yellowstone, and state control over wolves. Fourth, given advances in scientific knowledge concerning migration patterns, genetic diversity, and climate change, the original ecosystem management approach to the GYE public lands is necessarily evolving into a landscape-scale conservation approach that includes the region's private lands and extends beyond the informal 23-million-acre GYE boundaries. This shift in thinking toward the landscape level is manifest in several initiatives, including the Red Desert to Hoback mule deer migration corridor, the High Divide project, efforts to promote genetic interchange between adjacent grizzly bear populations, and climate-related adaptation strategies.

Other noteworthy developments are emerging from these evolving GYE conservation initiatives. One is the growing importance of private philanthropy. During the late 1920s the Rockefeller family's philanthropic funds were critical in acquiring front country lands for the expansion of Grand Teton National Park, eventually bringing the park to its current size, though not without considerable controversy. Since then, private philanthropy has continued to play a major role in regional conservation efforts involving both public and private lands. Examples include the accumulating number of conservation-easement transactions, livestock-grazing lease buyout arrangements, the acquisition of troublesome mineral leases in the Wyoming Range, the Crevice Mine buyout, national park inholding purchases, and additional bison quarantine facilities. These ongoing philanthropy-supported initiatives highlight both the historic and continuing role market-based transactions are playing for conservation purposes across the region's private and public lands. Such transactions have proven to be a flexible and effective tool in responding to specific wildlife and land-use conservation issues. The caution, going forward, is to ensure these often privately initiated transactions provide meaningful conservation benefits.[6]

Another notable development is the emergence of "culture-related conservation." As employed here, the term "culture-related conservation" implies a marriage between nature conservation and cultural or heritage preservation, which may also contain overtones of environmental justice. In the GYE the merger between nature and culture for conservation purposes is discernible in two separate ways. First, ongoing efforts to incorporate Native American tribes in GYE bison-conservation policy as well as National Park management represents an important joinder between conservation and culture. Recognizing the cultural significance of bison to the region's tribes, the Yellowstone bison-transfer program is saving many of the park's wild bison from the indignity of being slaughtered like cattle. And it is bringing buffalo back to tribal lands under tribal management, not only restoring them within the broader landscape, but also reestablishing the vital linkage between Native peoples and this age-old animal. The same holds true for tribal treaty–based bison hunting outside Yellowstone, although the slaughter-like aspects of these hunting practices calls out for more tribal oversight. Ongoing discussions between the Park Service and the tribes concerning a Native role in park resource-management and interpretive programs, including the utilization of traditional ecological knowledge, constitutes another example of culture-related conservation rising to the fore. The fact that similar discussions and experiments are occurring elsewhere across the West between Native American tribes and federal land management agencies is yet further proof of an ongoing merger between culture and conservation. Such a marriage appears to bode well for nature conservation in the GYE and beyond, while also responding to long-standing social justice concerns.[7]

A second aspect of culture-related conservation is manifest in various regional private land conservation initiatives aimed toward maintaining the GYE's ranchlands and hence ranching heritage to conserve critical habitat and open space. Success in perpetuating traditional ranching operations and the related ranching lifestyle can also help to preserve long-standing migration patterns and other natural processes on the landscape. Examples of these initiatives include the High Divide Collaborative and innovative Paradise Valley (Montana) projects that offer workable models specifically designed to meet joint conservation and ranching needs. Of course, maintaining traditional, open-range livestock grazing practices can perpetuate wildlife–livestock conflicts, so such culture-related conservation initiatives are not without potential cost from a nature conservation perspective. But the fact that grizzly bears, wolves, and bison were on the landscape long before settlers arrived suggests ranchers can reasonably be charged with knowledge of their presence and thus to have accepted the risks involved running cattle among them. Moreover, these initiatives cannot be justified as an element of environmental justice as the region's ranchers have not faced a history of discrimination and persecution as is true for the nation's Native American populace. Nonetheless, the twin realities of legally protected property rights and long-standing political power effectively dictate financially based accommodations with wildlife on private lands in today's GYE, which in turn support conservation initiatives designed to sustain the region's ranching heritage.[8]

* * *

Although notable progress is plainly evident in GYE nature-conservation efforts, major challenges remain. Change is a constant throughout the ecosystem, and is only accelerating as we move into the Anthropocene. Climate change, of course, overhangs nearly everything in the GYE: water, wildlife, and wildfire, as well as the region's human communities, which must adapt to a near-certain warmer and drier future. To sustain the ecosystem and past conservation accomplishments in this milieu, most scientists articulate the essential resource-management goals in terms of ecological integrity, resilience, and connectivity. Additional current and long-term human welfare goals for the region are typically framed in quality of life terms taking into account economic, social, and cultural values, which increasingly include such concerns as affordability, crowding, and retaining the sense of place. As we move into the future, these overarching goals are potentially attainable, but doing so will require a well-conceived mixture of restraint and active management as well as creativity and tolerance. Having come this far in maintaining and restoring the GYE's natural attributes, we must recognize that the decisions made today will be critical to sustaining that progress and propelling it forward.

On the GYE's public lands, several critical resource-management challenges confront the region. These include climate adaptation, ongoing wildlife-management conundrums, and recreation issues, which present the most apparent problems. As the GYE climate warms, climate adaptation strategies must be implemented to promote ecological resilience, including landscape-scale planning, adaptive management protocols, additional protected areas embracing newly identified critical habitat, and designated connective corridors. Vexing wildlife management issues—grizzly bear and wolf management, ungulate disease control, and the protection of critical habitat and migration routes—continue to rankle managers and landowners alike. Solutions to these ongoing wildlife issues will require a deepened commitment to working together toward common-ground solutions that rest upon credible science and risk-management considerations. As recreational activity proliferates across the GYE public lands, the agencies may need to consider use, access, and timing limitations, as well as additional recreation fee sources in order to monitor and enforce applicable regulations to protect environmental values and the backcountry experience. The agencies should also explore ways to better disperse users across the region's public lands and to educate them about appropriate behavior. In each instance, a landscape-scale approach to the problem is essential, which means the solutions to these challenging issues do not rest solely in federal hands. Rather, the federal land management agencies must work with the state wildlife agencies, tribal governments, local communities, outdoor businesses, and private landowners. Simply put, a principal challenge remains one of collaboration and coordination, surmounting the jurisdictional boundaries that have too often put the various GYE entities at odds over the years.[9]

On the GYE private lands escalating development pressures continue fragmenting the landscape at an accelerating pace, impacting wildlife habitat and open space. The problem is both cumulative and urgent. At one level, the principal tools available involve money—primarily from federal and state government, land trusts, conservation groups, and supportive foundations—to compensate ranchers and other large landowners for conservation easements, habitat leases, livestock losses, and the like. At another level, however, regulatory controls are an essential tool, too. State and local governments are key players here through laws governing land use, planning, and zoning. These laws and local ordinances can and should be strengthened, and also implemented and enforced more rigorously at the local level. Improved coordination is important as well, entailing collaboration between the responsible federal and state agencies, local officials, conservation groups, and their funders to develop plans that identify and effectively protect critical habitat and migration corridors while also securing the funds necessary to do so. Better planning, coordination, and funding is also necessary to address the wildfire threat in the WUI zone as well as local housing affordability concerns. There is little time to lose, given warming trends, population growth, and home shortage pressures that are putting the region's critical lands, native wildlife, service workforce, and lower-income residents at heightened risk.[10]

Regardless, conflict seems inevitable over the GYE's impending nature conservation issues, which means the law and litigation will continue to play a central role in these matters. The sheer number and diversity of interest groups across the political spectrum operating in the GYE virtually ensures court challenges and administrative appeals. On the federal public lands, where conflict still simmers over the Forest Service's ecosystem restoration and thinning projects as well as mining proposals, energy development, livestock grazing, and road-building projects, federal laws such as the ESA, NEPA, NFMA, and FLPMA will continue to apply and constrain agency discretion both substantively and procedurally. Conversely, the GYE states, ranchers, and local officials have proven willing and able to challenge federal agency actions perceived to crimp their authority or rights; they will continue to invoke state sovereignty, property rights, and federal overreach arguments, as they have in the case of the grizzly bear. Should they prevail or should the FWS delist the GYE grizzly bear population—a decision that will surely provoke a legal challenge from conservation organizations and the tribes—then state wildlife management laws will assume even greater importance in dictating wildlife policy across the region's public and private lands.

On the GYE's private lands, the relevant laws are the state land-use, planning, and zoning statutes that lack firm mandates and mostly just require consideration of wildlife, open space, and other conservation concerns. Whether local planners and county commissioners will apply these laws more rigorously in the face of current development and political pressures remains to be seen, but the state courts have shown a willingness to sustain conservation-oriented zoning decisions. The alternatives to litigation over private land conservation issues remain incentive payments to induce responsible landowner behavior, which requires money, along with careful planning. And lurking in the background whenever conflict surfaces is the alternative of legislation either through Congress or state legislatures—a heavy but not impossible lift for nature conservation proposals.[11]

Institutionally, the sixty-year-old GYCC can and must assume an enhanced role in the quest for common ground on GYE conservation issues. Though the GYCC is not a formal legal entity, its federal members have managerial authority over nearly two-thirds of the GYE, while the recent addition of state wildlife agencies brings the region's wildlife squarely within its ambit. Over time, the GYCC's expanding role has been shaped by the evolution of the GYE concept, related ecosystem-management efforts, and the myriad controversies that have bedeviled the region. While often criticized for its unwillingness to tackle controversial issues, the GYCC has provided a useful forum for federal managers to discuss these matters and develop relationships that have informed progress on individual GYE conservation issues. The GYCC's commitment to supporting scientific research on specific cross-boundary matters has generated important objective information that can and has been used to support resource-management decisions. Its use of committees to address various conservation issues has proven effective in several instances. The addition of the BLM and state wildlife agencies to GYCC membership—long limited to just national park, forest, and refuge managers—should help to expand the group's perspective as well as its capacity to address a broader suite of issues, including climate change and wildlife migration, that extend beyond the federal domain. The addition of state officials to the GYCC's membership should also help to dissipate some of the federal-state tension that has accompanied wildlife management and disease issues.

While it is politically unrealistic and legally impermissible for the GYCC to assume an overt management role in the GYE, its enlarged membership augurs well for more coordination and collaboration at the landscape scale, which is essential to preserve the region's natural attributes and wild character. The challenge before the expanded GYCC is to demonstrate it is capable of overcoming the region's pervasive federal-state tensions as well as other jurisdictional complexities that have long dogged GYE conservation issues. If it can do so, the odds of meeting the forthcoming challenges will be much improved.[12]

* * *

The same visionary impulses that drove our ancestors to protect Yellowstone initially and then Grand Teton along with adjacent national forest and then wilderness lands are still plainly afoot in the GYE. However, the conservation target is even larger today than before if we are to preserve the region's unique wildlife populations and natural attributes into the future. The challenges of doing so are daunting to be sure, but so were the challenges that confronted our forebears who bequeathed this natural heritage to us. The GYE concept and related ecological management approaches have brought us to this point in time and enabled us to solidify the conservation impulse pervasive across the region. Yet the clock is ticking and there is much work still to do to consolidate the progress to date and to address the looming problems. In the face of a warming climate and unrelenting development pressures, we can—and must—build upon the approaches that have worked in the past, while also conceiving new strategies befitting this irreplaceable landscape.

The original bard of conservation, John Muir, observed, "When one tugs at a single thing in nature, he finds it attached to the rest of the world"—certainly an indisputable truth in the GYE today. While we cannot undo the jurisdictional lines that define the GYE, we can take Muir's observation to heart and continue breaking down these barriers to conserve the region's wild character and attributes—its wide-roaming grizzly bears, its shadowy wolf packs, its intrepid bison herds, its annual elk migrations, and the enlarged landscape these animals depend on.

Under the mantle of ecosystem management, much has already been accomplished to sustain and restore the GYE's natural character, yet much remains to be done to protect native wildlife and further knit the region together while extending these efforts to the larger landscape. With careful, science-driven planning, acute sensitivity to legitimate local concerns, and a shared commitment to meaningful coordination across boundary lines, we can meet nature conservation's imminent challenges. If not in the distinctive, irreplaceable GYE where wild nature still abounds and the region's citizens have come to revere it, then where else can we conserve the nation's extraordinary natural heritage? Our children and the generations ahead await our answer.

Postscript

As editing on this book concluded, events continued to unfold in the Greater Yellowstone Ecosystem. The US Fish and Wildlife Service, upon completing a status review under the Endangered Species Act, rejected Wyoming's petition to delist the GYE grizzly bear population from the endangered species registry. The GYE bears thus remain a "threatened" species and will not be considered a "distinct population segment." Rather, the FWS has proposed treating the GYE grizzly population as a contiguous part of the five other Northern Rockies grizzly populations, combining these formerly separate recovery zones into a single recovery zone that encompasses suitable habitat for bear recovery. Designed to promote connectivity and resiliency, the new recovery zone would extend across the state of Washington and portions of Idaho, Montana, and Wyoming. The FWS has also proposed revising the so-called 4(d) rule governing when problematic grizzly bears may be taken or removed, notably loosening the conditions when agencies, ranchers, or property owners may dispatch or harass bears.

Under the ESA, the FWS is required to employ the "best available science" when making delisting determinations. Doing so, the FWS concluded that the existing grizzly bear populations, including the more populous Yellowstone- and Glacier-area bears, should be treated as a single "distinct population segment." The agency explained that bears are dispersing well beyond the initial recovery zones, rendering them no longer separate and discrete populations. And, applying the ESA's listing criteria, this intermixed grizzly population remains "threatened," because it "is likely to become an endangered species" due to habitat destruction and modification, human-caused mortality, and the isolated nature of some populations. In short, despite significant progress toward recovering the GYE grizzly population, the grizzly population as a whole remains at risk, and the GYE bears have yet to connect with other grizzly populations.

Clearly frustrated, the three GYE states are contemplating litigation and will request the new Trump administration to review the FWS's delisting decision and proposed "distinct population segment" decision. Because the FWS's determinations are in the form of a proposal, the new administration will have a say in the matter. Whatever it decides, however, must comply with the ESA's "best available science" standard, and the explanation underlying any decision must prove reasonable under existing law. Conservation groups are unhappy too, dismayed over the proposed relaxation of the 4(d) restrictions on taking grizzly bears. As in the past, the courts will likely be called upon to review these decisions, and Congress could be enlisted in the matter as well.

This latest round in the GYE grizzly bear's recovery saga further manifests the inevitable controversies that accompany ongoing regional nature conservation efforts. It is no surprise that the GYE states are once again at odds with the FWS and conservation groups over grizzly bear recovery, or that the delisting question will likely end up in the courts, or that politics may creep into the matter. Nor is it surprising that science is playing such a prominent role, given the ESA and the widely accepted conservation biology principles underlying the proposed landscape-level approach to grizzly bear recovery. Nor should it surprise anyone that quite evident development pressures along with habitat loss and connectivity concerns are entwined with the bear's fate as well as other GYE nature conservation efforts.

Two other developments are noteworthy. In late 2024, Congress passed the EXPLORE Act (Expanding Public Lands Outdoor Recreation Experiences Act), which provides additional guidance to the federal agencies overseeing recreational activity in the GYE and elsewhere. Among other things, the act creates a federal interagency council on outdoor recreation; addresses motorized access, rock climbing, biking, campsites, internet connectivity, and other activities; streamlines permitting processes; assists gateway communities; and promotes public-private partnerships. The legislation not only reflects the growing importance of outdoor recreation on public lands and for local economies but also represents the first detailed congressional foray into the matter in decades. Further, the US Geological Service issued a report for the National Elk Refuge, assessing the likely impact of chronic wasting disease on the refuge's controversial supplemental winter-feeding program, projecting a significant decline in elk numbers under various future feeding scenarios with a corresponding impact on hunting-related revenues.

Even more changes and controversies surely lie ahead, an inevitable aspect of our enduring commitment to conserve nature in Greater Yellowstone.

ACKNOWLEDGMENTS

This book would not have been possible without the time, support, and assistance of many individuals along the way to whom I am indebted. More than forty years have passed since I first ventured into the world of national parks, greater ecosystems, and ecosystem management. As my attention increasingly focused on the Yellowstone country, I have had the pleasure of becoming acquainted with other like-minded individuals who have willingly shared their wisdom and experience. Over time, these professional and personal relationships have both shaped and reshaped my views about nature conservation, enabling me to grasp more fully the GYE region, the laws and policies governing it, and the forces at work determining its future.

I owe a debt of profound gratitude to the mentors and colleagues who early on shared their knowledge and insights while encouraging my own efforts. Ken Diem, who oversaw the University of Wyoming–National Park Service Research Station, provided me an initial research opportunity that sparked my nascent interest in conservation and law. Joe Sax took me under his expansive intellectual wing, graciously offering me the chance to collaborate on a national park research project that confirmed my path forward. Bill Lockhart did the same, opening additional doors along the way. Mark Boyce and Susan Clark helped enlighten me on the Yellowstone world and its challenges, while unlocking new science and publication realms. In the ensuing years, Charles Wilkinson and John Leshy helped me navigate that path while expanding my understanding of public land history, law, and policy.

I had the privilege of entering the Greater Yellowstone Ecosystem world as the idea was taking hold thanks to the efforts of a number of farsighted individuals. Rick Reese gave the idea a critical push forward with publication of his Greater Yellowstone piece in the *Montana Geographic* series, helped to establish the Greater Yellowstone Coalition, and remained a committed advocate throughout his life, willingly sharing his passion and knowledge with me over the years. In these early years, I benefited greatly from the insights, guidance, and counsel of several other individuals, who have played vital roles overseeing and safeguarding the Yellowstone region while also advancing the GYE idea. They include Hank Phibbs, Bob Barbee, John Varley, Paul Schullery, Norm Bishop, Jack Stark, Mike Finley, Brian Stout, Jim Caswell, Jack Troyer, Sandra Key, Ed Lewis, Louisa Willcox, Mike Clark, Michael Scott, Gretchen Long, Hank Fischer, and the list goes on. My thanks too to Todd Wilkinson, Angus Thuermer Jr., Mike Koshmrl, and the many other intrepid journalists who have kept the public well informed about happenings in the Greater Yellowstone region and its challenges.

This book is an outgrowth of several research projects over the years, most notably a 2020 *University of Colorado Law Review* article based in part upon a series of interviews that I conducted with key players in the region, who contributed significantly to my knowledge and understanding of the issues. For their time and insights, I extend my gratitude to Stephanie Adams, Peter Aengst, Bill Berg, Mike Brennan, Caroline Byrd, Franz Camenzind, Scott Christensen, David Diamond, Bob Ekey, Mary Erickson, Brian Glaspell, Dennis Glick, Andrew Hansen, Doug Honnold, Virginia Kelly, Bart Melton, Doug McWhorter, Tricia O'Connor, Tom Oliff, Tim Preso, Luther Propst, Ray Rasker, Kathy Rinaldi, Mary Gibson Scott, Liz Storer, Gary Tabor, Dan Wenk, and Michael Whitfield. Since then, I have continued learning about the region, its people, and natural attributes from the insightful work of Arthur Middleton, Jason Robison, Temple Stoellinger, Sam Kalen, Justin Farrell, and others. I eagerly await their future contributions to Yellowstone-related scholarship.

The book has benefited immeasurably from the assistance of my indispensable Behle and Quinney Student Research Fellows at the University of Utah S. J. Quinney College of Law. Over the years, they have tracked down obscure legal, scientific, and policy materials, often on short notice, retrieved vital government documents, deciphered statistical data, checked references, and generally helped me to clarify my own views on the issues. My profound appreciation goes to Spencer Williams, Jack Nunn, Ben Cilwick, Shannon Wolfe, Ian Girard, Adam Duncan, Sydney

Jo Sell, and Hannah Follender for their exceptional research assistance. My thanks too to Phoebe McNealy and Eric Goodwin at the University of Utah DIGIT Lab for their work preparing the maps that appear throughout the book. Susan Darais at the law school's Faust Library also provided timely research assistance, while my assistants Patti Beekhuizen, Amina Naveed, and Margaret Spight regularly solved my technical problems, helping me focus on the task. Therese Boyd copyedited the manuscript masterfully, while my editor at the University of Chicago Press, Joe Calamia, has offered both encouragement and guidance throughout this project. To all, I am extremely grateful.

On a personal level, my appreciation for and love of the Yellowstone country has only grown and intensified as a result of numerous backcountry hiking trips that have exposed me to the area's ecology, beauty, and solitude—qualities that we all should cherish and preserve. I have truly enjoyed the friendship and time together with my stalwart GYE hiking companions: Don Sharp, Jim Smith, Tom Mitchell, Naomi Cohen, Arielle Cohen, Ron Tipton, Barry Majorowicz, and Bob Adler, along with several others. I would be remiss not to also thank my lifelong friend, Dick Janis, for joining me on that fateful 1967 western road trip.

Finally, to my dear wife, Linda, words cannot express my gratitude for your continued support, understanding, and encouragement during this lengthy journey through yet another book project. In the end, I can only hope that the words on these pages enable others to understand and appreciate the extraordinary natural heritage that we have in the Yellowstone country and inspire additional efforts to ensure it remains a unique, special, and intact place for future generations to enjoy.

NOTES

Introduction

1. Thomas McNamee, *The Return of the Wolf to Yellowstone* (New York: Henry Holt, 1997), 239–44.

2. Hank Fischer, *Wolf Wars: The Remarkable Inside Story of the Restoration of Wolves to Yellowstone* (Helena, MT: Falcon Press, 1995), 108–57.

3. Associated Press, "Jury Convicts Montana Man for Killing Wolf," *Spokesman-Review*, October 26, 1995, https://www.spokesman.com/stories/1995/oct/26/jury-convicts-montana-man-for-killing-wolf/; United States v. McKittrick, 142 F.3d 1170 (10th Cir. 2000); Wyoming Farm Bureau Federation v. Babbitt, 199 F.3d 1224 (9th Cir. 1998), reversing, 987 F.Supp. 1349 (D. Wyo. 1997).

4. Robert B. Keiter, "Grizzlies, Wolves, and Law in the Greater Yellowstone Ecosystem: Wildlife Management Amidst Jurisdictional Complexity and Tension," *Wyoming Law Review* 22 (2022): 303, 340–53; see chap. 3 herein.

5. Arthur D. Middleton et al., "The Role of Private Lands in Conserving Yellowstone's Wildlife in the Twenty-First Century," *Wyoming Law Review* 22 (2022): 237; see chap. 4 herein.

6. David N. Cherney, "Securing the Free Movement of Wildlife: Lessons from the American West's Longest Land Mammal Migration," *Environmental Law* 41 (2011): 599; Matthew J. Kauffman, James E. Meacham, Hall Sawyer, Alethea Y. Steingisser, William J. Rudd, and Emiline Ostlind, *Wild Migrations: Atlas of Wyoming's Ungulates* (Corvallis: Oregon State University Press, 2018), 136–37; see chap. 4 herein.

7. Robert B. Keiter, "The Greater Yellowstone Ecosystem Revisited: Law, Science, and the Pursuit of Ecosystem Management in an Iconic Landscape," *University of Colorado Law Review* 91 (2020): 1, 92–96; see chap. 4 herein.

8. Kauffman et al., *Wild Migrations*, 138–43.

9. See Robert B. Keiter and Mark S. Boyce, eds., *The Greater Yellowstone Ecosystem: Redefining America's Wilderness Heritage* (New Haven, CT: Yale University Press, 1993); Richard West Sellars, *Preserving Nature in the National Parks: A History* (New Haven, CT: Yale University Press, 1997); Paul Schullery, *Searching for Yellowstone: Ecology and Wonder in the Last Wilderness* (New York: Houghton Mifflin, 1997); James A. Pritchard, *Preserving Yellowstone's Natural Conditions: Science and the Perception of Nature* (Lincoln: University of Nebraska Press, 1999); Michael J. Yochim, *Yellowstone and the Snowmobile: Locking Horns Over National Park Use* (Lawrence: University Press of Kansas, 2009).

10. Rick Reese, *Greater Yellowstone: The National Park and Adjacent Wildlands* (Helena: Montana Magazine, 1984), 64–100; Robert B. Keiter, "Taking Account of the Ecosystem on the Public Domain: Law and Ecology in the Greater Yellowstone Region," *University of Colorado Law Review* 60 (1989): 923, 975–82. See also Paul W. Hirt, *A Conspiracy of Optimism: Management of the National Forests Since World War Two* (Lincoln: University of Nebraska Press, 1994); David A. Clary, *Timber and the Forest Service* (Lawrence: University Press of Kansas, 1986); Bob Ekey, "The New World Agreement: A Call for Reform of the 1872 Mining Law," *Public Land and Resources Law Review* 18 (1997): 151; see chap. 5 herein.

11. See Justin Farrell, *The Battle for Yellowstone: Morality and the Sacred Roots of Environmental Conflict* (Princeton, NJ: Princeton University Press, 2015); Justin Farrell, *Billionaire Wilderness: The Ultra-Wealthy and the Remaking of the American West* (Princeton, NJ: Princeton University Press, 2020); Todd Wilkinson, "Unnatural Disaster: Will America's Most Iconic Wild Ecosystem Be Lost to a Tidal Wave of People?," *Mountain Journal*, February 14, 2019, https://mountainjournal.org/the-wildest-ecosystem-in-america-faces-death-by-too-many-people; see chap. 6 herein.

12. See Robert B. Keiter, "The Greater Yellowstone Ecosystem Revisited: Law, Science, and the Pursuit of Ecosystem Management in an Iconic Landscape," *University of Colorado Law Review* 91 (2020): 1.

13. Aubrey L. Haines, *The Yellowstone Story*, 2 vols. (Boulder: Colorado Associated University Press, 1977), 1:267–68, 2:94–97.

14. Haines, *The Yellowstone Story*, 2:322–23; Alfred Runte, *National Parks: The American Experience*, 4th ed. (Lincoln: University of Nebraska Press, 2010), 110–12.

15. Frank C. Craighead Jr., *Track of the Grizzly* (San Francisco: Sierra Club Books, 1982), 5, 239; Duncan T. Patten, "Defining the Greater Yellowstone Ecosystem," in *The Greater Yellowstone Ecosystem*, ed. Keiter and Boyce, 21–23; Rick Reese, *Greater Yellowstone: The National Park and Adjacent Wildlands*, 2nd ed. (Helena: Montana Magazine, 1991), 55–77.

16. For various definitions of ecosystem management, see Robert B. Keiter, *Keeping Faith with Nature: Ecosystems, Democracy, and America's Public Lands* (New Haven, CT: Yale University Press, 2003), 71–78.

17. On the concept of landscape conservation, see National Academies of Science, *A Review of the Landscape Conservation Cooperatives* (Washington, DC:

National Academies Press, 2016); Matthew McKinney, Lynn Scarlett, and Daniel Kemmis, *Large Landscape Conservation: A Strategic Framework for Policy and Action* (Cambridge, MA: Lincoln Institute of Land Policy, 2010); Robert F. Baldwin et al., "The Future of Landscape Conservation," *Bioscience* 68 (2018): 60–62. For background on the 30×30 initiative, see Eric Dinerstein et al., "A Global Deal for Nature: Guiding Principles, Milestones, and Targets," *Science Advances* 5 (2019): 1–18.

Chapter One

1. Duncan T. Patten, "Defining the Greater Yellowstone Ecosystem," in *The Greater Yellowstone Ecosystem: Redefining America's Wilderness Heritage*, ed. Robert B. Keiter and Mark S. Boyce (New Haven, CT: Yale University Press, 1993), 21–23; Rick Reese, *Greater Yellowstone: The National Park and Adjacent Wildlands*, 2nd ed. (Helena: Montana Geographic, 1991), 55–75; Greater Yellowstone Coordinating Committee, "About," accessed November 6, 2023, https://www.fedgycc.org/about (putting the Greater Yellowstone Area acreage at 22.7 million acres, with just over 15 million acres in federal ownership).

2. For an excellent collection of GYE-related maps and a general geographic, historical, and cultural overview of the GYE, see W. Andrew Marcus, James E. Meacham, Ann W. Rodman, Alethea Steingisser, and Justin T. Menke, *Atlas of Yellowstone*, 2nd ed., ed. Ross West (Oakland: University of California Press, 2022).

3. Paul Schullery, *Searching for Yellowstone: Ecology and Wonder in the Last Wilderness* (New York: Houghton Mifflin, 1997); James A. Pritchard, *Preserving Yellowstone's Natural Conditions: Science and the Perception of Nature* (Lincoln: University of Nebraska Press, 1999). See generally Joseph L. Sax, *Mountains Without Handrails: Reflections on the National Parks*, 2nd ed. (Ann Arbor: University of Michigan Press, 2018).

4. Robert B. Keiter, "The Greater Yellowstone Ecosystem Revisited: Law, Science, and the Pursuit of Ecosystem Management in an Iconic Landscape," *University of Colorado Law Review* 91 (2020): 96–136.

5. On the National Elk Refuge see Bruce L. Smith, *Where Elk Roam: Conservation and Biopolitics of Our National Elk Herd* (Guilford, CT: Lyon Press, 2012). On the Bureau of Land Management see James R. Skillen, *The Nation's Largest Landlord: The Bureau of Land Management in the American West* (Lawrence: University Press of Kansas, 2009); Don Aragon, "The Wind River Indian Tribes," *International Journal of Wilderness* 13 (2007): 14, 16.

6. On the Greater Yellowstone Coordinating Committee (GYCC), see Susan G. Clark, *Ensuring Greater Yellowstone's Future: Choices for Leaders and Citizens* (New Haven, CT: Yale University Press, 2008). On the GYCC's Vision process, see Greater Yellowstone Coordinating Committee, *Draft Vision for the Future: A Framework for Coordination in the Greater Yellowstone Area* (National Park Service, 1990); Bruce Goldstein, "Can Ecosystem Management Turn an Administrative Patchwork Into a Greater Yellowstone Ecosystem?," *Northwest Environmental Journal* 8 (1992): 285; Pamela Lichtman and Tim W. Clark, "Rethinking the 'Vision' Exercise in the Greater Yellowstone Ecosystem," *Society and Natural Resources* 7 (1994): 459; John

Freemuth and R. McGreggor Cawley, "Science, Expertise, and the Public: The Politics of Ecosystem Management in the Greater Yellowstone Ecosystem," *Landscape and Urban Planning* 40 (1998): 211.

7. William R. Travis, *Ranchland Dynamics in the Greater Yellowstone Ecosystem: A Report to Yellowstone Heritage* (Boulder, CO: Center of the American West, 2003); Patricia H. Gude, Andrew J. Hansen, Ray Rasker, and Bruce Maxwell, "Rates and Drivers of Rural Residential Development in the Greater Yellowstone," *Landscape and Urban Planning* 77 (2006): 131; Hannah Gosnell, Julia Hobson Haggerty, and William R. Travis, "Ranchland Ownership Change in the Greater Yellowstone Ecosystem, 1990–2001: Implications for Conservation," *Society and Natural Resources* 19, no. 8 (2006): 743.

8. On the role of the extractive resource industries in Wyoming, see Samuel Western, *Pushed Off the Mountain, Sold Down the River: Wyoming's Search for Its Soul* (Moose, WY: Homestead Publishing, 2002); Angus M. Thuermer Jr., "Legislators Want $50M to Sue Feds Over Environmental Laws," *WyoFile*, November 6, 2023, https://wyofile.com/legislators-want-50m-to-sue-feds-over-environmental-laws/. The quotations are from the author's interviews with Angus Thuermer Jr. and Mike Brennan, in Keiter, "Greater Yellowstone Ecosystem Revisited," 13.

9. For a detailed description and critical analysis of the town of Jackson, Wyoming, see Justin Farrell, *Billionaire Wilderness: The Ultra-Wealthy and the Remaking of the American West* (Princeton, NJ: Princeton University Press, 2020).

10. On the Grand Targhee Ski Resort expansion proposal, see US Forest Service, US Department of Agriculture, Caribou Targhee National Forest, *Grand Targhee Resort Master Development Plan Projects EIS, 2020*, 85 Fed. Reg. 52542; Julia Tellman, "County Named Cooperating Agency on Grand Targhee Expansion," *Teton Valley News*, December 12, 2020, https://www.tetonvalleynews.net/news/county-named-cooperating-agency-on-grand-targhee-expansion/article_00bd4a3f-e77f-586e-bbb7-3a0799c34dc3.html. The transformation of Pinedale, Wyoming, is captured in Alexandra Fuller, "Boomtown Blues: How Natural Gas Changed the Way of Life in Sublette County," *New Yorker*, February 5, 2007, https://www.newyorker.com/magazine/2007/02/05/boomtown-blues; see also Alexandra Fuller, *The Legend of Colton H. Bryant* (New York: Penguin Press, 2008).

11. On growth in Bozeman and Gallatin County see Todd Wilkinson, "Unnatural Disaster: Will America's Most Iconic Wild Ecosystem Be Lost to a Tidal Wave of People?," *Mountain Journal*, February 14, 2019, https://mountainjournal.org/the-wildest-ecosystem-in-america-faces-death-by-too-many-people. On the Yellowstone Club see Farrell, *Billionaire Wilderness*, 49–75.

12. Author's interviews with Bill Berg, Scott Christensen, and Louisa Willcox, in Keiter, "Greater Yellowstone Ecosystem Revisited," 16–17.

13. Justin Farrell, *The Battle for Yellowstone: Morality and the Sacred Roots of Environmental Conflict* (Princeton, NJ: Princeton University Press, 2015), 61. On the New West concept see Thomas M. Power and Richard Barrett, *Cowboy Economics: Pay and Prosperity in the New American West*, 2nd ed. (Washington, DC: Island Press, 2001); William Riebsame, ed., *Atlas of the New West: Portrait of a Changing Region* (New York: W. W. Norton, 1997).

14. Farrell, *Battle for Yellowstone*, 100–103 (identifying 183 environmental organizations in the region, with a combined annual budget of $150 million, 500 employees, and over 700 board members). See also "Northern Rockies Regional Office," Earthjustice, https://earthjustice.org/about/offices/northern-rockies.

15. Farrell, *Battle for Yellowstone*, 70–73; John D. Leshy. Robert L. Fischman, and Sarah A. Krakoff, *Federal Public Land and Resources Law*, 8th ed. (St. Paul, MN: Foundation Press, 2022), 850; see also chap. 6 herein.

16. Kekek Jason Stark, Autumn L. Bernhardt, Monte Mills, and Jason A. Robison, "Re-Indigenizing Yellowstone," *Wyoming Law Review* 22, no. 2 (2022): 397, 453–67; John D. Leshy, *Our Common Ground: A History of America's Public Lands* (New Haven, CT: Yale University Press, 2021), 567–74; see also chapters 2, 4, conclusion herein.

17. On early Park Service policy see Richard West Sellars, *Preserving Nature in the National Parks: A History* (New Haven, CT: Yale University Press, 1997), 69–90. On early Forest Service policy see Samuel P. Hays, *Conservation and the Gospel of Efficiency: The Progressive Conservation Movement, 1890–1920* (Cambridge, MA: Harvard University Press, 1959), 27–48. On the "balance of nature" theory, see Daniel B. Botkin, *Discordant Harmonies: A New Ecology for the Twenty-First Century* (New York: Oxford University Press, 1990).

18. See Frank B. Golley, *A History of the Ecosystem Concept in Ecology: More than the Sum of the Parts* (New Haven, CT: Yale University Press, 1993); Joel B. Hagen, *An Entangled Bank: The Origins of Ecosystem Ecology* (New Brunswick, NJ: Rutgers University Press, 1992); Donald Worster, *Nature's Economy: A History of Ecological Ideas* (New York: Cambridge University Press, 1985).

19. William D. Newmark, "Legal and Biotic Boundaries of Western North American National Parks: A Problem of Congruence," *Biological Conservation* 33 (1985): 197; see also Newmark, "Extinction of Mammal Populations in Western North American National Parks," *Conservation Biology* 9 (1985): 512. On conservation biology see Michael E. Soule, "What Is Conservation Biology?," *BioScience* 35, no. 11 (December 1985): 727. On biodiversity conservation strategy see Jocelyn L. Aycrigg et al., "Completing the System: Opportunities and Challenges for a National Habitat Conservation System," *BioScience* 66, no. 9 (2016): 774; Mark L. Shaffer and Bruce A. Stein, "Safeguarding Our Precious Heritage," in *Precious Heritage*, ed. Bruce A. Stein, Lynn S. Kutner, and Jonathan S. Adams (New York: Oxford University Press, 2000), 307–10; Robert B. Keiter, "Toward a National Conservation Network Act: Transforming Landscape Conservation on the Public Lands into Law," *Harvard Environmental Law Review* 42 (2018): 90–93.

20. A. Starker Leopold, S. A. Cain, C. M. Cottam, I. N. Gabrielson, and T. N. Kimball, "Wildlife Management in the National Parks" (March 4, 1963), in *America's National Park System: The Critical Documents*, ed. Lary M. Dilsaver (Lanham, MD: Rowman and Littlefield, 1994), 237; Robert B. Keiter, *To Conserve Unimpaired: The Evolution of the National Park Idea* (Washington, DC: Island Press, 2013), 148–52. See also chapter 2 herein.

21. National Forest Management Act, Pub. L. No. 94-588, 90 Stat. 2949 (1976) (codified at 16 U.S.C. § 1604); see Charles F. Wilkinson and H. Michael Anderson,

"Land and Resource Planning in the National Forests," *Oregon Law Review* 64 (1985): 1.

22. Louis S. Warren, ed., *American Environmental History* (Malden, MA: Blackwell, 2003), 244–97; Richard N. L. Andrews, *Managing the Environment, Managing Ourselves: A History of American Environmental Policy* (New Haven, CT: Yale University Press, 1998), 201–54; Samuel P. Hays, *Beauty, Health, and Permanence: Environmental Politics in the United States, 1955–1985* (New York: Cambridge University Press, 1987), 3–4, 26–34, 88, 111.

23. Rachel Carson, *Silent Spring* (Boston: Houghton Mifflin, 1962); Aldo Leopold, *A Sand County Almanac with Essays from Round River* (New York: Ballantine, 1966), 262.

24. Hays, *Beauty, Health, and Permanence*; Hal K. Rothman, *Devil's Bargain: Tourism in the Twentieth-Century American West* (Lawrence: University Press of Kansas, 1998).

25. Multiple Use–Sustained Yield Act of 1960, 16 U.S.C. §§ 528–31; Wilderness Act of 1964, 16 U.S.C. §§ 1131–36; Land and Water Conservation Act, 16 U.S.C. §§ 460l-4–460l-11; National Wildlife Refuge Administration Act, 16 U.S.C. §§ 668dd–668ee; Wild and Scenic Rivers Act of 1968, 16 U.S.C. §§ 1271–87; National Trails Act of 1968, 16 U.S.C. §§ 1241–49; Wallace Stegner "Wilderness Letter" (1960), reprinted in Wallace Stegner, *The Sound of Mountain Water* (Lincoln: University of Nebraska Press, 1985), 145–53.

26. National Environmental Policy Act, 42 U.S.C. §§ 4321–61; see Thomas v. Peterson, 753 F.2d 754 (9th Cir. 1985); Conner v. Burford, 848 F.2d 1441 (9th Cir. 1988).

27. Endangered Species Act, 16 U.S.C. §§ 1531–43; see Tennessee Valley Authority v. Hill, 437 U.S. 153 (1978); Babbitt v. Sweet Home Chapter, 515 US 687 (1995).

28. Clean Water Act, 33 U.S.C. § 1251 et seq.; Clean Air Act, 42 U.S.C. § 7401 et seq.

29. Comprehensive Environmental Response, Compensation, and Liability Act, 42 U.S.C. § 9601 et seq.; Resources Conservation and Recovery Act, 42 U.S.C. § 6901 et seq.

30. National Forest Management Act, 16 U.S.C. § 1604; Wilkinson and Anderson, "Land and Resource Planning," 40–46, 154–88, 290–96.

31. Federal Land Management and Policy Act, 43 U.S.C. §§ 1701–84.

32. 54 U.S.C. § 100101(b)(2); William J. Lockhart, "External Threats to Our National Parks: An Argument for Substantive Protection," *Stanford Environmental Law Review* 16 (1997): 3, 60–72; Robert B. Keiter, "National Park Protection: Putting the Organic Act to Work," in *Our Common Lands: Defending the National Parks*, ed. David J. Simon (Washington, DC: Island Press, 1988), 75, 76–78.

33. Alaska National Interest Lands Conservation Act, 16 U.S.C. §§ 3101–3223. In ANILCA Congress expressly sought to "preserve in their natural state extensive, unaltered . . . ecosystems." 16 U.S.C. § 3101(b); see also 16 U.S.C. § 3103(b) (urging that national park and other conservation unit boundaries adhere to "topographical or natural features").

34. On the Northwest spotted owl–timber controversy and resulting Northwest Forest Plan, see Kathie Durbin, *Tree Huggers: Victory, Defeat, and Renewal in the Northwest Ancient Forest Campaign* (Seattle: The Mountaineers, 1996); Steven Lewis Yaffee, *Wisdom of the Spotted Owl: Policy Lessons for a New Century* (Washington, DC: Island Press, 1994); Robert B. Keiter, *Keeping Faith with Nature: Ecosystems, Democracy, and America's Public Lands* (New Haven, CT: Yale University Press, 2003), 79–126.

35. Seattle Audubon Society v. Lyons, 871 F.Supp. 1291, 1311 (W.D. Wash. 1994) (emphasis in original).

36. On the Clinton administration and ecosystem management, see Keiter, *Keeping Faith with Nature*, 113–26; James R. Skillen, *Federal Ecosystem Management: Its Rise, Fall, and Afterlife* (Lawrence: University Press of Kansas, 2015), 150–82.

37. US Forest Service, *The Wilderness Story*, https://www.fs.usda.gov/managing-land/wilderness/wilderness-stories; General Accounting Office, *Federal Land Management: Status and Uses of Wilderness Study Areas* (Washington, DC: GAO, 1993), 19, 65, https://www.gao.gov/assets/rced-93-151.pdf.

38. On the Clinton administration's BLM-managed national monuments see Mark Squillace, "The Monumental Legacy of the Antiquities Act of 1906," *Georgia Law Review* 37 (2003): 473, 507–14. The National Landscape Conservation System Act is found at 16 U.S.C. §§ 7201–3.

39. National Wildlife Refuge System Improvement Act of 1997, Pub. L. No. 105-57, 111 Stat. 1255 (codified at 16 U.S.C. §§ 668dd, 668ee); National Parks Omnibus Management Act of 1998, Pub. L. No. 105-391, 112 Stat. 3495 (codified at 54 U.S.C. § 100701); Great American Outdoors Act, Pub. L. No. 116-152, 134 Stat. 682 (2020) (codified at 54 U.S.C. §§ 200401–200402).

40. Idaho Statutes §§ 55-2101 et seq. (2018); Montane Code Annotated §§ 76-6-201 et seq. (2018); Wyoming Statutes. §§ 34-1-201.

41. USFS Roadless Area Rule, 36 C.F.R. pt. 294 (2020); USFS Planning Regulations, 36 C.F.R. pt. 219 (2020).

42. US Forest Service, "Land Management Plan Direction for Old Growth Forest Conditions Across the National Forest System" 88 Fed. Reg. 88042 (Dec. 20, 2023).

43. National Park Service, *Management Policies 2006* (Washington, DC: National Park Service, US Department of the Interior, 2006), 13, 23, 24.

44. US Fish and Wildlife Service, "Policy on Maintaining the Biological Integrity, Diversity, and Environmental Health of the National Wildlife Refuge System," 66 Fed. Reg. 3810 (Jan. 16, 2001).

45. Bureau of Land Management, US Department of the Interior, "Conservation and Landscape Health," 88 Fed. Reg. 19583 (April 3, 2023); Jamie Pleune, "BLM's Conservation Rule and Conservation as a 'Use,'" *Environmental Law Reporter* 53 (2023): 10824.

46. Secretary of the Interior Order No. 3289 (September 14, 2009) (establishing the Landscape Conservation Cooperatives); Bureau of Land Management, *Greater Sage-Grouse*, accessed December 16, 2023, https://www.blm.gov/programs

/fish-and-wildlife/sage-grouse; Western Watersheds Project v. Zinke, 441 F.Supp.3d 1042 (D. Idaho 2020); Western Watersheds Project v. Bernhardt, 2021 WL 517035 (D. Idaho 2021); Bureau of Land Management, "Conservation and Landscape Health Rule," 89 Fed. Reg. 40308 (May 9, 2024), https://www.federalregister.gov/documents/2024/05/09/2024-08821/conservation-and-landscape-health.

47. On the Age of the Anthropocene, see Richard Monastersky, "Anthropocene: The Human Age," *Nature* 519 (2015): 144; Stanley C. Finney and Lucy E. Edwards, "The Anthropocene Epoch: Scientific Decision or Political Statement?," *GSA Today* 26 (2016): 4; Elizabeth Kolbert, "Enter the Anthropocene—Age of Man," *National Geographic*, March 2011.

48. Sellars, *Preserving Nature in the National Parks*, 204–66; Charles F. Wilkinson, *Crossing the Next Meridian: Land, Water, and the Future of the American West* (Washington, DC: Island Press, 1993), 114–74; Reese, *Greater Yellowstone*, 36–63.

49. Robert B. Keiter, "Taking Account of the Ecosystem on the Public Domain: Law and Ecology in the Greater Yellowstone Region," *University of Colorado Law Review* 60 (1989): 956–84; Clark, *Ensuring Greater Yellowstone's Future*, 123–26.

50. Keiter, "Greater Yellowstone Ecosystem Revisited," 96–136. On Yellowstone wolf restoration see Hank Fischer, *Wolf Wars: The Remarkable Inside Story of the Restoration of Wolves to Yellowstone* (Helena, MT: Falcon Press, 1995).

51. Keiter, "Greater Yellowstone Ecosystem Revisited," 49–69, 105–36; Char Miller, *Public Lands, Public Debates: A Century of Controversy* (Corvallis: Oregon State University Press, 2012), 162; Arthur D. Middleton et al., "Conserving Transboundary Wildlife Migrations: Recent Insights from the Greater Yellowstone Ecosystem," *Frontiers in Ecology and the Environment* 18, no. 2 (2020): 83.

Chapter Two

1. The Washburn quote is found in John Ise, *Our National Park Policy: A Critical History* (Baltimore: Johns Hopkins University Press, 1961), 15; the Hayden quote appears in Ferdinand V. Hayden, *Preliminary Report of the United States Geological Survey of Montana and Portions of Adjacent Territories* (Washington, DC: Government Printing Office, 1872), 4, 162. The Lane Letter on national park management was sent from Secretary of the Interior Franklin Lane to National Park Service Director Stephen Mather on May 13, 1918, and is found in Lary M. Dilsaver, ed., *America's National Park System: The Critical Documents* (Lanham, MD: Rowman and Littlefield, 1994), 48–52. Secretary Lane, however, did contemplate an educational role for the new park system as he endorsed using the parks for university-based studies and new museums.

2. Wright's national park resource management policy views are found in George M. Wright, Joseph S. Dixon, and Ben H. Thompson, "Fauna of the National Parks of the United States: A Preliminary Survey of Faunal Relations in National Parks" (Fauna Series no. 1, May 1932), in *America's National Park System*, ed. Dilsaver, 104–10. On park buffer zones, see George M. Wright and Ben H. Thompson, *Fauna of the National Parks of the United States: Wildlife Management in the National Parks* (Fauna Series no. 2, July 1934) (Washington, DC: Government Printing Office, 1935). See also Robert B. Keiter, *To Conserve Unimpaired: The Evolution of the*

National Park Idea (Washington, DC: Island Press, 2013), 177–78, 205–6; Richard West Sellars, *Preserving Nature in the National Parks: A History*, 2nd ed. (New Haven, CT: Yale University Press, 2008), 93–112.

3. Paul Schullery, *Searching for Yellowstone: Ecology and Wonder in the Last Wilderness* (New York: Houghton Mifflin, 1997), 167–73; Sellars, *Preserving Nature in the National Parks*, 195–201.

4. A. Starker Leopold, S. A. Cain, C. M. Cottam, I. N. Gabrielson, and T. N. Kimball, "Wildlife Management in the National Parks," in *America's National Park System*, ed. Dilsaver, 237–51; National Academy of Sciences, "A Report by the Advisory Committee to the National Park Service on Research," in Dilsaver, *America's National Park System*, 253–62. See also Sellars, *Preserving Nature in the National Parks*, 214–17; Keiter, *To Conserve Unimpaired*, 148–52, 178–81.

5. "Historic Tribes," Yellowstone National Park, https://www.nps.gov/yell/learn/historyculture/historic-tribes.htm.

6. Sam Kalen, "Rekindling Yellowstone's Early History," *Wyoming Law Review* 22 (2022): 217, 219–21.

7. For a detailed description of the early Native American presence in Yellowstone, see Mark David Spence, *Dispossessing the Wilderness: Indian Removal and the Making of the National Parks* (New York: Oxford University Press, 1999), 65–106. See also C. Adrian Heidenreich, "The Native Americans' Yellowstone," *Montana: The Magazine of Western History* 35, no. 4 (1985): 2–17, http://www.jstor.org/stable/4518923; Isaac Kantor, "Ethnic Cleansing and America's Creation of National Parks," *Public Land and Resources Law Review* 28 (2007): 41; Kekek Jason Stark, Autumn L. Bernhardt, Monte Mills, and Jason A. Robison, "Re-Indigenizing Yellowstone," *Wyoming Law Review* 22 (2022): 397, 453–61, 467–72.

8. James A. Pritchard, *Preserving Yellowstone's Natural Conditions: Science and the Perception of Nature* (Lincoln: University of Nebraska Press, 1999), 201–50; Schullery, *Searching for Yellowstone*, 219–31. On the "natural regulation" policy, see Mark S. Boyce, "Natural Regulation or the Control of Nature?" in *The Greater Yellowstone Ecosystem: Redefining America's Wilderness Heritage*, ed. Robert B. Keiter and Mark S. Boyce (New Haven, CT: Yale University Press), 183–208. For critics of the "natural regulation" policy, see Alston Chase, *Playing God in Yellowstone: The Destruction of America's First National Park* (New York: Atlantic Monthly Press, 1986); Stephen Budiansky, *Nature's Keepers: The New Science of Nature Management* (New York: Free Press, 1995); Frederic H. Wagner et al., *Wildlife Policies in the US National Parks* (Washington, DC: Island Press, 1995); Frederic H. Wagner, *Yellowstone's Destabilized Ecosystem: Elk Effects, Science, and Policy Conflict* (New York: Oxford University Press, 2006).

9. For more detailed discussion of bear, wolf, bison, and elk management in the GYE, see chaps. 3 and 4 herein. Additional discussion of wildfire management policy is found later in this chapter and in chapter 5 herein.

10. On the Redwood controversy and the Redwood Amendment, see William J. Lockhart, "External Threats to Our National Parks: An Argument for Substantive Protection," *Stanford Environmental Law Review* 16 (1997): 3, 60–72; Keiter, *To Conserve Unimpaired*, 207–9.

11. National Park Service, *State of the Parks, 1980: A Report to Congress*,

partially reprinted in *America's National Park System*, ed. Dilsaver, 405–8; Robert B. Keiter, "On Protecting the National Parks from the External Threat Dilemma," *Land and Water Law Review* 20 (1985): 355, 396–408.

12. The quoted provisions are found at National Park Service, *Management Policies 2006* (Washington, DC: National Park Service, US Department of the Interior, 2006), 36–37.

13. National Park Service Advisory Board Science Committee, *Revisiting Leopold: Resource Stewardship in the National Parks* (2012), 8, 11–18, https://www.nps.gov/calltoaction/pdf/leopoldreport_2012.pdf.

14. National Park Service, Director's Order #100: Resource Stewardship for the 21st Century (December 20, 2016), https://www.nps.gov/subjects/policy/upload/DO_100_12-20-2016.pdf.

15. Kurt Repanshek, "National Park Service Scuttles Director's Order Pertaining to Natural Resource Protection," *National Parks Traveler*, August 25, 2017, https://www.nationalparkstraveler.org/2017/08/national-park-service-scuttles-directors-order-pertaining-natural-resource-protection; Executive Order No. 14008, "Tackling the Climate Crisis at Home and Abroad" (January 27, 2017), 86 Fed. Reg. 7619 (Feb. 1, 2021); US Department of the Interior, US Department of Agriculture, US Department of Commerce, and Council on Environmental Quality, *Conserving and Restoring America the Beautiful* (2021); National Park Service, US Department of the Interior, *Climate Change Response Strategy: 2023 Update*.

16. For a history of national park wildfire policy, see Hal K. Rothman, *Blazing Heritage: A History of Wildland Fire in the National Parks* (New York: Oxford University Press, 2007); Bruce M. Kilgore, "Origin and History of Wildland Fire Use in the US National Park System," *George Wright Forum* 24, no. 3 (2007).

17. The 1988 Yellowstone fires are recounted in Micah Morrison, *Fire in Paradise: The Yellowstone Fires and the Politics of Environmentalism* (New York: HarperCollins, 1993); Rocky Barker, *Scorched Earth: How the Fires of Yellowstone Changed America* (Washington, DC: Island Press, 2005); see also David Carle, *Burning Questions: America's Fight with Nature's Fire* (Westport, CT: Praeger, 2002); Michael J. Yochim, *Protecting Yellowstone: Science and the Politics of National Park Management* (Albuquerque: University of New Mexico Press, 2013), 45–79..

18. US Department of the Interior and US Department of Agriculture, *Final Report of the Fire Management Policy Review Team* (Washington, DC, 1989); Norman L. Christensen et al., "Interpreting the Yellowstone Fires of 1988," *BioScience* 39, no. 10 (1989): 678.

19. Mary Ann Franke, *Yellowstone in the Afterglow: Lessons from the Fires* (Mammoth, WY: Yellowstone Center for Resources, 2000); Dennis H. Knight, "The Yellowstone Fire Controversy," in *The Greater Yellowstone Ecosystem: Redefining America's Wilderness Heritage*, ed. Robert B. Keiter and Mark C. Boyce (New Haven, CT: Yale University Press, 1991), 87–103; Christensen et al., "Interpreting the Yellowstone Fires of 1988."

20. National Park Service, *Management Policies* (Washington, DC: US Department of the Interior, 2006). 49; National Park Service, Yellowstone National Park, *Wildland Fire Management Plan Environmental Assessment and Finding of*

No Significant Impact (2012), 1–2, https://parkplanning.nps.gov/document.cfm?documentID=49667.

21. Leopold et al., "Wildlife Management in the National Parks," 240.

22. National Park Service, *Management Policies 2006*, 43–44, 47; 54 U.S.C. § 100752; WildEarth Guardians v. National Park Service, 703 F.3d 1178 (10th Cir. 2013); Davis v. Latschar, 202 F.3d 359 (DC Cir. 2000).

23. John M. Syslo, Christopher S. Guy, and Todd M. Koel, "Feeding Ecology of Native and Nonnative Salmonids During the Expansion of a Nonnative Apex Predator in Yellowstone Lake, Yellowstone National Park," *Transactions of the American Fisheries Society* 145 (2016); James R. Ruzycki, David Beauchamp, and Daniel L. Yule, "Effects of Introduced Lake Trout on Native Cutthroat Trout in Yellowstone Lake," *Ecological Applications* 13 (2003); Todd M. Koel, Patricia E. Bigelow, Philip D. Doepke, Brian D. Ertel, and D. L. Mahony, "Non-native Lake Trout Result in Yellowstone Cutthroat Trout Decline and Impacts to Bears and Anglers," *Fisheries* 20 (2005): 10–19.

24. The data recited in this paragraph is found in Todd M. Koel et al., "Non-native Lake Trout Induce Cascading Changes in the Yellowstone Lake," *Yellowstone Science* 25, no. 1 (2017): 42-50, https://www.nps.gov/articles/non-native-lake-trout-induce-cascading-changes-in-the-yellowstone-lake.htm; Todd M. Koel et al., "Yellowstone Lake Ecosystem Restoration: A Case Study for Invasive Fish Management," *Fishes* 5, no. 2 (2020): 18, https://doi.org/10.3390/fishes5020018. See also Todd Wilkinson, "It All Started with a Few Trout. Now Yellowstone's Iconic Birds Face 'Collapse,'" *National Geographic* (July 2018), https://www.nationalgeographic.com/environment/article/yellowstone-lake-trout-trumpeter-swan-avian-collapse-animals.

25. National Park Service, Yellowstone National Park, *Native Fish Conservation Plan Finding of No Significant Impact (FONSI)*, (2011), 2, https://parkplanning.nps.gov/document.cfm?parkID=111&projectID=30504&documentID=41145; National Park Service, Yellowstone National Park, *Native Fish Conservation Plan: Environmental Assessment* (2011), https://parkplanning.nps.gov/document.cfm?parkID=111&projectID=30504&documentID=37967.

26. Koel et al., "Yellowstone Lake Ecosystem Restoration"; Brett French, "Yellowstone Lake's Native Cutthroat Getting Bigger as Population Rebounds," *Billings Gazette*, May 22, 2023, https://billingsgazette.com/outdoors/yellowstone-lakes-native-cutthroat-trout-getting-bigger-as-population-rebounds/article_19ce1144-f8c6-11ed-93cc-63f592eeb1ad.html.

27. Mike Koshmrl, "Perpetual Netting: Can Yellowstone Win Its Lake Trout Fight," *WyoFile*, October 29, 2022, https://wyofile.com/perpetual-netting-can-yellowstone-win-its-lake-trout-fight/.

28. Koel et al., "Yellowstone Lake Ecosystem Restoration"; Hayley C. Glassic et al., "Yellowstone Cutthroat Trout Recovery in Yellowstone Lake: Complex Interactions among Invasive Species, Suppression, Disease, and Climate Change," *Fisheries* (October 16, 2023); Mike Koshmrl, "Climate Change, Lake Trout Could Impede Yellowstone Cutthroat Trout Recovery Indefinitely," *WyoFile*, February 6, 2024, https://wyofile.com/studies-climate-change-lake-trout-could-impede-yellowstone-lake-cutthroat-recovery-indefinitely/.

29. On the history and ecology of the Grand Teton mountain goats and bighorn sheep, see Blake Lowrey, Robert A. Garrott, Hollie Miyasaki, Gary Fralick, and Sarah Dewey, "Seasonal Resource Selection by Introduced Mountain Goats in the Southwest Greater Yellowstone Area," *Ecosphere* 8 (April 2017); Carson J. Butler et al., "Respiratory Pathogens and Their Association with Population Performance in Montana and Wyoming Bighorn Sheep Populations," *PLOS ONE* 13 (2018); Blake Lowrey, Robert A. Garrott, D. E. McWhirter, P. J. White, N. J. DeCesare, and S. T. Stewart, "Niche Similarities Among Introduced and Native Mountain Ungulates," *Ecological Applications* 28 (2018).

30. National Park Service, *Environmental Assessment: Mountain Goat Management Plan* (December 2018), 3. The disease threat quotation is from the EA; Hadley Hammer, "Don't Shred on Them: A Young Star Skier Speaks Up for Bighorns," *Mountain Journal*, November 11, 2021, https://mountainjournal.org/a-star-back country-skier-takes-a-stand-for-wild-bighorns-in-the-tetons.

31. Wyoming Game and Fish Commission, Press Release, "Game and Fish Commission Passes Resolution on Grand Teton Mountain Goat Removal," January 16, 2020, https://wgfd.wyo.gov/News/Game-and-Fish-Commission-Passes-Resolution-on-Gran.

32. National Park Service, *Finding of No Significant Impact*: *Mountain Goat Management Plan, Environmental Assessment*, September 27, 2019, 19; Grand Teton National Park, "Non-Native Mountain Goat Management Qualified Volunteer Program," last updated October 6, 2020, https://www.nps.gov/grte/getinvolved/mountain-goat-management-volunteer.htm; Mike Koshmrl, "Grand Teton Goat Killing Kerfuffle Reaches DC," *Jackson Hole News and Guide*, February 24, 2020, https://www.jhnewsandguide.com/jackson_hole_daily/local/grand-teton-goat-killing-kerfuffle-reaches-d-c/article_7231c5b0-d1f0-5f25-825e-2cf465917fc0.html; Maggie Mullen, "Grand Teton National Park Will Use Volunteer Culling to Remove Non-Native Goats," *Wyoming Public Media*, August 6, 2020, https://www.wyomingpublicmedia.org/post/grand-teton-national-park-will-use-volunteer-culling-remove-non-native-goats#stream/0; Elizabeth Gamillo, "To Protect Big Horn Sheep, Authorities Kill 58 Mountain Goats in Grand Teton National Park," *Smithsonian*, March 2, 2022, https://www.smithsonianmag.com/smart-news/to-protect-bighorn-sheep-authorities-kill-58-mountain-goats-in-grand-teton-national-park-180979662/.

33. National Park Service, *Management Policies 2006*, 11; National Rifle Association v. Potter, 628 F.Supp. 903, 909 (D.D.C. 1986); see also Fund for Animals v. Norton, 294 P.Supp.2d 92, 105 (D.D.C. 2003); Bicycle Trails Council of Marin v. Babbitt, 82 F.2d 1445 1453–54 (9th Cir. 1996).

34. Michael J. Yochim, *Yellowstone and the Snowmobile: Locking Horns over National Park Use* (Lawrence: University Press of Kansas, 2009), 13–86 (recounting the early history of snowmobiling in the park through the mid-1970s).

35. Executive Order No. 11644 (1972) (requiring the federal land management agencies to zone their lands for off-road vehicle travel); Executive Order No. 11989 (1977) (requiring federal land management agencies to ban off-road vehicle travel that causes "considerable adverse effects"); Michael J. Yochim, "Snow Machines

in the Gardens: The History of Snowmobiles in Glacier and Yellowstone National Parks," *Montana Western History Magazine* 53, no. 3 (Autumn 2003): 2–15; Yochim, *Yellowstone and the Snowmobile*, 124–29.

36. Yochim, *Yellowstone and the Snowmobile*, 87–149; National Park Service, "Record of Decision: Winter Use Plans for the Yellowstone and Grand Teton National Parks and John D. Rockefeller Jr. Memorial Parkway," 65 Fed. Reg. 80908, 80917 (Dec. 22, 2000).

37. National Park Service, "Winter Use Plan, Final Supplemental Environmental Impact Statement for the Yellowstone and Grand Teton National Parks and the John D. Rockefeller, Jr., Memorial Parkway, Wyoming and Montana," 68 Fed. Reg. 8618 (Feb. 24, 2003); National Park Service, "Winter Use Plan Final Rule for the Yellowstone and Grand Teton National Parks and John D. Rockefeller Jr. Memorial Parkway," 68 Fed. Reg. 69268 (Dec. 11, 2003); Fund for Animals v. Norton, 294 F.Supp.2d 92, 109, 116 (D.D.C. 2003). See Yochim, *Yellowstone and the Snowmobile*, 149–76.

38. International Snowmobile Manufacturers Association v. Norton, 304 F.Supp.2d 1278 (D. Wyo. 2004); National Park Service, "Special Regulations, Areas of the National Park System," 69 Fed. Reg. 65348 (Nov. 10, 2004), codified at 36 C.F.R. part 7. Wyoming business interests unsuccessfully challenged the 2004 interim rule in Wyoming Lodging and Restaurant Association v. US Dept. of the Interior, 298 F.Supp.2d 1197 (D. Wyo. 2005). The early snowmobile litigation is recounted in Joanna M. Hooper, "Blowing Snow: The National Park Service's Disregard for Science, Law, and Public Opinion in Regulating Snowmobiling in Yellowstone National Park," *Environmental Law Reporter* 34 (2004): 10975.

39. Greater Yellowstone Coalition v. Kempthorne, 577 F.Supp.2d 183 (D.D.C. 2008); Wyoming v. US Dept. of the Interior, 587 F.3d 1245, 1249–51 (10th Cir. 2009) (describing the Wyoming federal district court's 2008 decision and finding the case moot). The proposed congressional bills included National Parks Snowmobile Restrictions Act, H.R. 1465, 107th Cong., 1st Sess. (2001); Yellowstone Protection Act, S. 965, 108th Cong., 1st Sess. (2003); National Park Service Winter Access Act, S. 365, 107th Cong., 1st Sess. (2001). See also Yochim, *Yellowstone and the Snowmobile*, 149–76.

40. National Park Service, Yellowstone National Park, *Winter Use Plan / Supplemental Environmental Impact Statement* (2013), https://www.nps.gov/yell/learn/management/winter-use-archive.htm. On the Sylvan Pass controversy, Yochim, *Yellowstone and the Snowmobile*, 202–8; Kurt Repanshek, "Ruminating on the Unexploded Ordnance, Climate Change, and Maintaining Winter Access to Yellowstone National Park," *National Parks Traveler*, May 16, 2011, https://www.nationalparkstraveler.org/2011/05/ruminating-unexploded-ordnance-climate-change-and-maintaining-winter-access-yellowstone-national-park; Carl Alan Hamming, "Yellowstone National Park and the Winter Use Debate: Community Resilience and Tourism Impacts in the Gateway Community of West Yellowstone, MT" (master's thesis, Montana State University, April 2016), https://scholarworks.montana.edu/items/07fd404d-a53b-4e09-a780-d2c9091a7dd8; Brett French, "Yellowstone Plans Ceasing Winter Wildlife Surveillance, Concentrating on Sum-

mer," *Spokesman-Review*, October 25, 2021, https://www.spokesman.com/stories/2021/oct/25/yellowstone-plans-ceasing-winter-wildlife-surveill/.

41. The proposed revisions sought to dilute the statutory "impairment" standard by requiring park managers to consider visitor enjoyment along with resource conditions when assessing the effect of recreational activities. The revisions also equated visitor enjoyment with recreational use and reduced the protection for natural soundscapes. Sellars, *Preserving Nature in the National Parks*, 301–6; Keiter, *To Conserve Unimpaired*, 75–76.

42. Michael J. Yochim, "Kayaking Playground or Nature Preserve?: Whitewater Boating Conflicts in Yellowstone National Park," *Montana: The Magazine of Western History* (Spring 2005): 52; Mike Koshmrl, "Parks' Packrafting Bill Floats into Uncharted Waters," *Jackson Hole News and Guide*, October 14, 2015, https://www.jhnewsandguide.com/news/environmental/parks-packrafting-bill-floats-into-uncharted-waters/article_1bc504d8-5cd9-5489-9da3-4f807e530edd.html.

43. See Jackson Hole Pathway System Map, https://www.tetoncountywy.gov/DocumentCenter/View/4158/County-Pathway-Map-PDF (the Teton County Pathway System map showing the potential loop ride that includes the Moose-Wilson Road inside the park).

44. National Park Service, *Moose-Wilson Corridor Final Comprehensive Management Plan/Environmental Impact Statement* (2015).

45. Keiter, *To Conserve Unimpaired*, 41–52.

46. For national park historic and current visitation statistics, see "Welcome to Visitor Use Statistics," National Park Service, https://irma.nps.gov/STATS/. On the Yellowstone visitor survey, see Todd Wilkinson, "Cam Sholly's Agenda for Safeguarding Yellowstone," *Mountain Journal*, June 10, 2019, https://mountainjournal.org/yellowstone-is-confronting-many-major-threats.

47. River Runners for Wilderness v. Martin, 593 F.3d 1064 (9th Cir. 2010) (limiting the number of commercial and individual rafts in the Grand Canyon); Bicycle Trails Council of Marin v. Babbitt, 82 F.3d 1445 (9th Cir. 1996) (prohibiting mountain bikes on designated trails); Southern Utah Wilderness Alliance v. National Park Service, 387 F.Supp.2d 1178 (D. Utah 2005) (prohibiting off-road vehicle travel). See also National Park Service, *Management Policies 2006*, 10–11.

48. National Park Service, *2022 National Park Visitor Spending Effects: Economic Contributions to Local Communities, States, and the Nation* (2023), 24, 34, https://www.nps.gov/subjects/socialscience/vse.htm.

49. On the objections by Cody, Wyoming, to the Sylvan Pass winter closure and the Fishing Bridge campground closure, see Yochim, *Yellowstone and the Snowmobile*, 202–8; Schullery, *Searching for Yellowstone*, 187–90; Pritchard, *Preserving Yellowstone's Natural Conditions*, 262–65.

50. On the summer 2022 flooding event and aftermath, see Ron Wirtz, "Montana's Silent Yellowstone Flood: When Tourists Leave," *Federal Reserve Bank of Minneapolis*, July 15, 2022, https://www.minneapolisfed.org/article/2022/montanas-silent-yellowstone-flood-when-tourists-leave; Douglas Scott, "Despite Flooding and Closures Yellowstone's 2022 Visitation Numbers Are Surprising, at Least to Me," *Outdoor Society*, January 20, 2023, https://outdoor-society.com/despite-flooding-and-closures-yellowstones-2022-visitation-numbers-are-surprising/.

51. National Park Service, *Intermountain Region Infrastructure Fact Sheets* (2018), https://npgallery.nps.gov/GetAsset/18032801-9bde-4dd7-bfd8-59d237e36aa9/original.

52. Federal Lands Recreation Enhancement Act, 16 U.S.C. §§ 6801–14; Great American Outdoors Act, Pub. L. No. 116-152, 134 Stat. 682 (2020) (codified at 54 U.S.C. §§ 200401–200402).

53. Regarding drones in national parks, see National Park Service Director Policy Memorandum 14-05 (2014); 36 C.F.R. § 1.5. Regarding the debate over cell towers in national parks see Christopher Ketcham, "Wi-Fi in the Wilderness," *Sierra Magazine*, July/August 2020, https://www.sierraclub.org/sierra/2020-4-july-august/feature/wi-fi-wilderness; Kurt Repanshek, "Yellowstone National Park Working to Remove Unsightly Cell Towers," *National Parks Traveler*, June 2, 2021, https://www.nationalparkstraveler.org/2021/06/yellowstone-national-park-working-remove-unsightly-cell-towers; Jimmy Tobias, "Wiring the Wild," *High Country News* 52, no. 35 (March 2020): 26; Holly Doremus, "Foreword," in Joseph L. Sax, *Mountains Without Handrails: Reflections on the National Parks*, 2nd ed. (Ann Arbor: University of Michigan Press, 2018), xiv–xvi.

54. 16 U.S.C. § 673c (permitting elk hunting in the park subject to federal-state agreement); Defenders of Wildlife v. Everson, 984 F.3d 918 (10th Cir. 2020).

55. Ted Kerasote, "This Tiny Parcel in Paradise Could Be Devoured" (guest essay), *New York Times*, December 6, 2023, https://www.nytimes.com/2023/12/06/opinion/development-national-parks-wyoming.html.

56. Tyler Pratt, "Thousands of Wyomingites Reject 'Kelly Parcel' Auction," *Wyoming Public Radio*, December 1, 2023, https://www.wyomingpublicmedia.org/natural-resources-energy/2023-12-01/thousands-of-wyomingites-reject-proposed-kelly-parcel-auction; Mike Koshmrl, "Wyoming Legislature's Two Chambers OK 'Kelly Parcel' Sale to Feds for $100M," *WyoFile*, February 22, 2024, https://wyofile.com/kelly-parcel-sale-survives-midnight-house-run-but-with-new-baggage/; Angus M. Thuermer Jr., "After Decades of Political Maneuvering, Grand Teton Buys Wyoming's Kelly Parcel," *WyoFile*, December 30, 2024, https://wyofile.com/after-decades-of-political-maneuvering-grand-teton-buys-wyomings-kelly-parcel-today/.

Chapter Three

1. Thomas Mangelsen and Todd Wilkinson, *The Grizzlies of Pilgrim Creek* (New York: Rizzoli, 2015).

2. 16 U.S.C. § 1531(b).

3. A history of the grizzly bear and its relationship to western settlement can be found in Mangelsen and Wilkinson, *The Grizzlies of Pilgrim Creek*, and Thomas McNamee, *The Grizzly Bear* (New York: Knopf, 1984). The quotes from the Lewis and Clark Expedition are found in Mangelsen and Wilkinson, *The Grizzlies of Pilgrim Creek*, 17; McNamee, *The Grizzly Bear*, 38.

4. Alice Wondrak Biel, *Do (Not) Feed the Bears: The Fitful History of Wildlife and Tourists in Yellowstone* (Lawrence: University Press of Kansas, 2006), 7–62.

5. For a detailed account of the Craighead–National Park Service controversy, see Pritchard, *Preserving Yellowstone's Natural Conditions*, 237–53; Biel, *Do (Not)*

Feed the Bears, 86–112. See also Frank C. Craighead Jr., *Track of the Grizzly* (San Francisco: Sierra Club Books, 1982). For a comprehensive scientific account of Yellowstone's grizzly bears see John Craighead, Jay Sumner, John Alexander Mitchell, and Craighead Wildlife-Wildlands Institute, *The Grizzly Bears of Yellowstone: Their Ecology in the Yellowstone Ecosystem* (Washington, DC: Island Press, 1995).

6. Richard West Sellars, *Preserving Nature in the National Parks* (New Haven, CT: Yale University Press, 1997), 252; Pritchard, *Preserving Yellowstone's Natural Conditions*, 246.

7. Endangered Species Act of 1973, Pub. L. No. 93-205, 87 Stat. 888, codified at 16 U.S.C. §§ 1531-1544; 50 C.F.R. § 17.11 (2023).

8. Under the ESA, according to the court, the Park Service's decision to leave the developed area largely unchanged did not "jeopardize" or "take" any bears and there was no obligation to prepare a recovery plan for the grizzly bear. National Wildlife Federation v. National Park Service, 669 F.Supp.384 (D. Wyo. 1977); Paul Schullery, *Searching for Yellowstone: Ecology and Wonder in the Last Wilderness* (New York: Houghton Mifflin, 1997), 187–90; Pritchard, *Preserving Yellowstone's Natural Conditions*, 263–65; Michael J. Yochim, *Protecting Yellowstone: Science and the Politics of National Park Management* (Albuquerque: University of New Mexico Press, 2013), 13–44.

9. Congressional Research Service, Library of Congress, 99th Cong., *Greater Yellowstone Ecosystem: An Analysis of Data Submitted by Federal and State Agencies* (Washington, DC: Comm. Print, 1986).

10. On revision of the grizzly bear management guidelines and related zoning scheme, see US Fish and Wildlife Service et al., *Interagency Grizzly Bear Guidelines* (1986), https://npshistory.com/publications/wildlife/interagency-grizzly-bear-guidelines.pdf. On the Vision process, see Bruce Goldstein, "Can Ecosystem Management Turn an Administrative Patchwork Into a Greater Yellowstone Ecosystem?," *Northwest Environmental Journal* 8 (1992): 285; for further analysis of the Vision process, see chap. 1 herein.

11. 16 U.S.C. § 1533(f) (recovery plans); US Fish and Wildlife Service, *Grizzly Bear Recovery Plan* (1993), https://fwp.mt.gov/binaries/content/assets/fwp/conservation/wildlife-reports/bears/grizzly_bear_recovery_plan.pdf; Fund for Animals v. Babbitt, 903 F.Supp. 96 (D.D.C. 1995). The court concluded that the agency's Management Situation zoning scheme adequately protected bear habitat and that the FWS's plan to augment the Yellowstone grizzly population at five-year intervals with bears from elsewhere addressed genetic diversity concerns.

12. 16 U.S.C. § 1536 (consultation requirement); Thomas v. Peterson, 753 F.2d 754 (1985); Bennett v. Spear, 520 US 154 (1997); 16 U.S.C. § 1539 ("take" prohibition); Babbitt v. Sweet Home Chapter, 515 US 687 (1995).

13. Richard J. Knight, D. M. Mattson, and B. M. Blanchard, *Movement and Habitat Use of the Yellowstone Grizzly Bear: Interagency Grizzly Bear Study Team Report* (National Park Service, 1984); David J. Mattson, R. R. Knight, and B. M. Blanchard, "The Effects of Developments and Primary Roads on Grizzly Bear Habitat Use in Yellowstone National Park, Wyoming," *International Conference on Bear Research and Management* 7 (1987): 259; David J. Mattson and Matthew M. Reid, "Conservation of the Yellowstone Grizzly Bear," *Conservation Biology* 5, no. 3 (1991): 364.

14. See chapter 5 for further analysis of GYE national forest management policy and wildlife conservation.

15. See US Fish and Wildlife Service, "Reevaluation of the Record of Decision for the Final EIS and Selection of the Alternative for Grizzly Bear Recovery in the Bitterroot Ecosystem," 66 Fed. Reg. 33623 (June 22, 2001); Sarah van Wetering, "Bitterroot Grizzly Bear Reintroduction: Management by Citizen Committee?," in *Across the Great Divide: Explorations in Collaborative Conservation and the American West*, ed. Philip Brick, Donald Snow, and Sarah van de Wetering (Washington, DC: Island Press, 2001), 150; Daniel Kemmis, *This Sovereign Land: A New Vision for Governing the West* (Washington, DC: Island Press, 2001), 1–18.

16. US Forest Service, US Department of Agriculture, *Forest Plan Amendment for Grizzly Bear Habitat Conservation for the Greater Yellowstone Area National Forests: Record of Decision* (2006); Interagency Conservation Strategy Team, *Final Conservation Strategy for the Grizzly Bear in the Greater Yellowstone Area* (2007).

17. US Fish and Wildlife Service, "Endangered and Threatened Wildlife and Plants; Final Rule Designating the Greater Yellowstone Area Population of Grizzly Bears as a Distinct Population Segment; Removing the Yellowstone Distinct Population Segment of Grizzly Bears from the Federal List of Endangered and Threatened Wildlife; 90 Day Finding on a Petition to List as Endangered the Yellowstone Distinct Population Segment of Grizzly Bears," 72 Fed. Reg. 14866, 14926 (March 29, 2007).

18. 16 U.S.C. § 1533(a)(1).

19. Greater Yellowstone Coalition v. Servheen, 665 F.3d 1015 (9th Cir. 2011), reversing in part, 672 F.Supp.2d 1105 (D. Mont. 2009). The district court ruled that the FWS's delisting decision violated two statutory provisions, namely the "other manmade factors" and the "adequate regulatory mechanism" requirements. 16 U.S.C. §§ 1533(a)(1)(D), (E). On appeal, the Ninth Circuit rejected the district court's conclusion that the delisting decision ignored the "adequate regulatory mechanism" requirement, ruling that future legal protections for the bear did not have to be as stringent as those required under the ESA.

20. Cecily M. Costello et al., "Influence of Whitebark Pine Decline on Fall Habitat Use and Movements of Grizzly Bears in the Greater Yellowstone Ecosystem," *Ecology and Evolution* 4 (2014): 2004; Charles C. Schwartz, Jennifer Fortin-Noreus, Justin E. Teisberg, and Mark A. Haroldson, "Body and Diet Composition of Sympatric Black and Grizzly Bears in the Greater Yellowstone Ecosystem," *Journal of Wildlife Management* 78 (2014): 68; Kerry A. Gunter et al., "Dietary Breadth of Grizzly Bears in the Greater Yellowstone Ecosystem," *Ursus* 25, no. 1 (2014): 60; Jennifer K. Fortin, Jasmine V. Ware, Heiko T. Jansen, Charles C. Schwartz, and Charles T. Robbins, "Temporal Niche Switching by Grizzly Bears but Not American Black Bears in Yellowstone National Park," *Journal of Mammalogy* 94, no. 4 (2013): 833.

21. Mike Koshmrl, "States, Land Managers Spar Over Grizzlies," *Jackson Hole News and Guide*, September 14, 2016, https://www.jhnewsandguide.com/news/environmental/states-land-managers-spar-over-grizzlies/article_374cdc1f-cb53-5265-bfa9-26577ad6a302.html; Kurt Repanshek, "USFWS Proposal to Delist Greater Yellowstone Ecosystem Grizzly Bears Controversial," *National Parks*

Traveler, November 6, 2016, https://www.nationalparkstraveler.org/2016/10/usfws-proposal-delist-greater-yellowstone-ecosystem-grizzly-bears-controversial; Dan Wenk, former Yellowstone superintendent, interview with author.

22. US Fish and Wildlife Service, "Endangered and Threatened Wildlife and Plants; Removing the Greater Yellowstone Ecosystem Population of Grizzly Bears from the Federal List of Endangered and Threatened Wildlife," 82 Fed. Reg. 30502 (June 30, 2017). The ESA provision enabling the FWS to "relist" a delisted species is found at 16 U.S.C. § 1533(g), 82 Fed. Reg. at 30628. See also US Fish and Wildlife Service, Grizzly Bear Recovery Office, *Grizzly Bear Recovery Plan Supplement: Revised Demographic Recovery Criteria for the Yellowstone Ecosystem* (2017), https://www.fws.gov/node/68795 (establishing a Demographic Monitoring Area that surrounds the Primary Conservation Area, in which grizzly bear numbers would be monitored to assess population numbers and trends, recognizing that the bears were notably expanding beyond the PCA). Matthew Brown, "States Divvy Up Yellowstone Area Grizzly Hunt," *Billings Gazette*, January 4, 2016, https://billingsgazette.com/article_2a95a10e-dd25-566f-af9a-51ecaa25d755.html; Karin Brulliard, "Grizzly Bear Trophy Hunt in Yellowstone Area Approved by Wyoming," *Washington Post*, May 23, 2018, https://www.washingtonpost.com/news/animalia/wp/2018/05/23/wyoming-may-legalize-trophy-hunting-of-grizzly-bears-for-the-first-time-in-40-years/; Mike Koshmrl, "Chamber: Griz Hunt May Be Tourism Killer," *Jackson Hole News and Guide*, June 8, 2016, https://www.jhnewsandguide.com/news/environmental/chamber-griz-hunt-may-be-tourism-killer/article_9aa9e601-0a15-57a3-b068-6535c61f9c09.html. Author's interviews with Doug McWhorter, biologist, Wyoming Fish and Game Department; Tricia O'Connor, Bridger Teton National Forest supervisor; Mary Gibson Scott, former Grand Teton National Park superintendent.

23. Crow Indian Tribe v. United States, 965 F.3d 662 (9th Cir. 2020), affirming as modified, 343 F.Supp.3d 999 (D. Mont. 2018). According to the district court: "The Service does not have unbridled discretion to draw boundaries around every potentially healthy population of a listed species without considering how that boundary will affect the members of the [remnant populations] on either side of it." Regarding the FWS's decision to drop its previous bear translocation commitment, the court observed that the possibility of relisting the Yellowstone bears did not justify delisting them in the absence of "concrete, enforceable mechanisms in place to ensure [their] long-term genetic health."

24. State of Wyoming, Office of the Governor, "Petition: To Establish the Greater Yellowstone Ecosystem (GYE) Grizzly Bear (*Ursus acrtos horribilis*) Distinct Population Segment (DPS) and Remove the GYE Grizzly Bear DPS from the Federal List of Endangered and Threatened Species" (January 10, 2022); Angus M. Thuermer Jr., "Wyo Approves Tri-State Grizzly Hunting Pact in Delisting Push," *WyoFile*, December 2, 2021, https://wyofile.com/wyo-approves-tri-state-grizzly-hunting-pact-in-delisting-push/; Angus M. Thuermer Jr., "Wyo Asks Feds to Delist Yellowstone Ecosystem Grizzlies," *WyoFile*, January 11, 2022, https://wyofile.com/wyo-asks-feds-to-delist-yellowstone-ecosystem-grizzlies/; Mike Koshmrl, "Grizzly Hunt in Wyoming Could Target Upwards of 39 Bears," *WyoFile*, March 20, 2023, https://wyofile.com/grizzly-hunt-in-wyoming-could-target-upwards-of-39-bears/.

25. Center for Biological Diversity v. Haaland, 603 F.Supp.3d 1094 (D. Wyo. 2022), affirmed in part and reversed in part, Western Watersheds Project v. Haaland, 69 F.4th 689 (10th Cir. 2023); Angus M. Thuermer Jr., "Grizzly Conflicts Central to New Upper Green River Grazing Debate," *WyoFile*, July 6, 2021, https://www.wyofile.com/grizzly-conflicts-central-to-new-upper-green-river-grazing-debate/.

26. Brulliard, "Grizzly Bear Trophy Hunt in Yellowstone Area Approved by Wyoming,"; Kurt Repanshek, "Groups Challenge Law Allowing Wyoming to Stage Grizzly Bear Hunts," *National Parks Traveler*, February 21, 2019, https://www.nationalparkstraveler.org/2019/02/groups-challenge-law-allowing-wyoming-stage-grizzly-bear-hunts; Mark Davis, "Game and Fish Commission Decides Against Grizzly Hunting Season," *Casper Star Tribune*, April 30, 2019, https://trib.com/news/state-and-regional/game-and-fish-commission-decides-against-grizzly-hunting-season/article_88771802-0a69-5a38-9c1c-f7230123c4e5.html. The Upper Green River grazing–grizzly bear issue is explored more fully in chapter 6.

27. Grizzly Bear Management Act of 2018, H.R. 6877, 115th Cong. (2018); Grizzly Bear State Management Act of 2021, S. 997, 117th Cong. (2021).

28. US Fish and Wildlife Service, *Grizzly Bear in the Lower-48 States: 5-Year Status Review—Summary and Evaluation* (March 2021), https://ecosphere-documents-production-public.s3.amazonaws.com/sams/public_docs/species_nonpublish/942.pdf; US Fish and Wildlife Service, *Grizzly Bear Recovery Office, Special Status Assessment for the Grizzly Bear in the Lower-48 States: A Biological Report* (January 2021), https://www.fws.gov/node/70376.

29. Blair Miller, "Feds to Consider Delisting Grizzly Bears near Yellowstone, in Northwest Montana," *Daily Montanan*, February 3, 2023, https://dailymontanan.com/2023/02/03/feds-to-consider-delisting-grizzly-bears-near-yellowstone-in-northwest-montana/.

30. Montana Code Annotated § 87-6-106(4); op-ed, "Prominent Scientists Push Back Against Delisting Grizzly Bears," *Mountain Journal*, January 13, 2022, https://mountainjournal.org/prominent-scientists-say-removing-grizzly-bears-from-federal-protection-in-west-is-bad-idea; Blair Miller, "Fish and Wildlife Commission Adopts New Administrative Rules on Grizzly, Wolf Management," *Daily Montanan*, December 14, 2023, https://dailymontanan.com/2023/12/14/fish-and-wildlife-commission-adopts-new-administrative-rules-on-grizzly-wolf-management/#:~:text=The%20commission%20in%20the%20past,grizzly%20bear%20on%20public%20land.

31. Julia Barton, "If Grizzlies Delisted, Here's What Montana Plans to Do," *Mountain Journal*, April 3, 2024, https://mountainjournal.org/is-montana-grizzly-bear-management-plan-really-grizzly-hunting-plan.

32. Interagency Grizzly Bear Committee, "A Reassessment of Chao 2 Estimates for Population Monitoring of Grizzly Bears in the Greater Yellowstone Ecosystem" (April 6, 2021), https://www.usgs.gov/publications/a-reassessment-chao2-estimates-population-monitoring-grizzly-bears-greater-yellowstone; Mike Koshmrl, "Yellowstone Grizzly Population Jumps with Count Revision," *Jackson Hole Daily*, April 11, 2021, https://trib.com/outdoors/yellowstone-grizzly-population-jumps-with-count-revision/article_5d1e9041-e9c7-5101-94aa

-a8c66ce92fd0.html; Koshmrl, "Will Yellowstone's Grizzly Bears Remain Forever Isolated?," *WyoFile*, June 22, 2023, https://wyofile.com/will-yellowstones-grizzly-bears-remain-forever-isolated/; Koshmrl, "Wyoming Sues Over Feds Tardiness on Grizzly Delisting Decision," *WyoFile*, June 1, 2023, https://wyofile.com/wyoming-sues-over-feds-tardiness-on-grizzly-delisting-decision/; Koshmrl, "Grizzly Hunt in Wyoming Could Target Upwards of 39 Bears," *WyoFile*, March 20, 2023, https://wyofile.com/grizzly-hunt-in-wyoming-could-target-upwards-of-39-bears/; Rob Chaney, "Montana Moves Grizzly Bears to Yellowstone as Legal Fights Swirl," *Billings Gazette*, August 5, 2024, https://billingsgazette.com/news/state-and-regional/article_81bc4f50-d3d0-5fdf-b26c-bbae66e73ece.html.

33. Should the FWS delist the Yellowstone grizzly bear population, the ESA will still play a role in the GYE, since the agency decided in late 2023, after years of legal wrangling, to list the wolverine as a threatened species, and it has a presence in the GYE. Although the wolverine's habitat is typically in snow-covered, high-elevation areas, the FWS's listing decision will likely require the GYE federal agencies to consult with the FWS during planning processes and on project proposals to avoid any possible jeopardy to the animal. Climate change and habitat fragmentation represent the biggest threat to wolverines. Not surprisingly, the GYE states have challenged the FWS's wolverine listing decision. Angus M. Thuermer Jr., "Wolverines to Be Protected by Endangered Species Act in Lower 48, Wyoming," *WyoFile*, November 30, 2023, https://wyofile.com/wolverines-to-be-protected-by-endangered-species-act-in-wyoming-lower-48/; Amanda Eggert, "Montana Signals Intent to Sue Over Federal Wolverine Protections," *Montana Free Press*, January 26, 2024, https://montanafreepress.org/2024/01/26/montana-signals-intent-to-sue-over-federal-wolverine-protections/.

34. Bernard DeVoto, ed., *The Journals of Lewis and Clark* (Boston: Houghton Mifflin, 1953), 11, 105, 423.

35. John James Audubon, *Audubon and His Journals: Missouri River Journals* (New York: Charles Scribner's Sons, 1897), 1:491.

36. Francis Parkman, *The California and Oregon Trail: Being Sketches of Prairie and Rocky Mountain Life, 1849* (New York T.Y. Crowell, 1901), https://www.loc.gov/item/01016630/.

37. The Roosevelt quote is found in Rick McIntyre, ed., *War Against the Wolf: America's Campaign to Exterminate the Wolf* (Stillwater, MN: Voyageurs Press, 1995), 108. On wolf bounties, see 75–132; Hank Fischer, *Wolf Wars: The Remarkable Inside Story of the Restoration of Wolves in Yellowstone* (Helena, MT: Falcon Press, 1995), 17–18.

38. For historical accounts of wolves in America, see Bruce Hampton, *The Great American Wolf* (New York: Henry Holt, 1997); Barry Lopez, *Of Wolves and Men* (New York: Scribner's, 1978).

39. The Park Service's evolving attitude toward the wolf is explained in Schullery, *Searching for Yellowstone*, 160–61; Sellars, *Preserving Nature in the National Parks*, 119–23. The Aldo Leopold wolf restoration proposal is from Aldo Leopold, "Review of the Wolves of North America," *Journal of Forestry* 42, no. 12 (1944): 928–29; see also Leopold, "Thinking Like a Mountain," in *A Sand Country Almanac*

(New York: Oxford University Press, 1949), 137–41. The Leopold Report–related quote is from Interior Secretary Stewart Udall's 1964 Memo to the Director of the National Park Service on Management of the National Park System, in *America's National Park System*, ed. Dilsaver, 273. For the predator control report, see A. Starker Leopold, S. A. Cain, C. M. Cottam, I. M. Gabrielson, and T. L. Kimball, "Predator and Rodent Control in the United States," reprinted in *Transactions of the Twenty-Ninth North American Wildlife Conference* (1964), 27, 35.

40. The Endangered Species Act, Pub. L. No. 93-205, 87 Stat. 884 (1973), codified at 16 U.S.C. §§ 1531–43. The wolf is listed at 50 C.F.R. 17.11 (2021); Section § 10(j) is codified in the Endangered Species Act at 16 U.S.C. § 1539(j). See also US Fish and Wildlife Service, *Northern Rocky Mountain Wolf Recovery Plan* (1987), https://www.pinedaleonline.com/wolf/pdf/NRMWolfRecoveryPlan.pdf.

41. For an excellent summary of attitudes toward the wolf restoration, see Justin Farrell, *The Battle for Yellowstone: Morality and the Sacred Roots of Environmental Conflict* (Princeton, NJ: Princeton University Press, 2015), 168–216.

42. US Fish and Wildlife Service, *The Reintroduction of Gray Wolves to Yellowstone National Park and Central Idaho Final Environmental Impact Statement* (1994), https://fwp.mt.gov/binaries/content/assets/fwp/conservation/wildlife-reports/wolf/eis_1994.pdf; US Fish and Wildlife Service, *Record of Decision, The Reintroduction of Gray Wolves to Yellowstone National Park and Central Idaho* (1994), https://npshistory.com/publications/yell/gray-wolf-reintro-eis-1994.pdf. The political maneuvering behind the wolf reintroduction is recounted in Fischer, *Wolf Wars*; Thomas McNamee, *The Return of the Wolf to Yellowstone* (New York: Henry Holt, 1997).

43. Fischer, *Wolf Wars*, 155–57, 161–63; Wyoming Farm Bureau v. Babbitt, 987 F.Supp. 1349 (D. Wyo. 1997), reversed, 199 F.3d 1224 (10th Cir. 2000) (holding that the "outside the current range" prohibition on the use of section 10(j) applied only when a population of wolves was present, not just a few solitary wolves); United States v. McKittrick, 142 F.3d 1170 (9th Cir. 1998) (interpreting section 10(j)'s "outside of the current range" language similarly).

44. William Ripple and Robert Beschta, "Restoring Yellowstone's Willows with Wolves," *Biological Conservation* 138 (2007): 514; Robert L. Beschta, "Cottonwood, Elk, and Wolves in the Lamar Valley of Yellowstone National Park," *Ecological Applications* 13 (2003): 1295. Recent studies take a more nuanced view of the ecological impact that wolves have had within the park, including on the elk population. See chaps. 5, 11, and 12 in P. J. White et al., eds., *Yellowstone Wildlife in Transition* (Cambridge, MA: Harvard University Press, 2013), 69–93, 179–94, 195–207; N. Thompson Hobbs, Danielle B. Johnston, Kristin N. Marshall, Evan C. Wolf, and David J. Cooper, "Does Restoring Apex Predators to Food Webs Restore Ecosystems? Large Carnivores in Yellowstone as a Model System," *Ecological Monographs* 94, no. 2 (January 30, 2024), https://esajournals.onlinelibrary.wiley.com/doi/10.1002/ecm.1598. See generally Douglas W. Smith, Daniel R. Stahler, and Daniel R. MacNulty, eds., *Yellowstone Wolves: Science and Discovery in the World's First National Park* (Chicago: University of Chicago Press, 2020).

45. On the wolf population numbers, see Douglas W. Smith, Kerry M. Murphy,

and Debra S. Guernsey, *Yellowstone Wolf Project: Annual Report 2000*, at 1, https://www.nps.gov/yell/learn/nature/upload/Wolfrpt00.pdf. On the projected economic benefits attached to wolf reintroduction, see John Duffield, "An Economic Analysis of Wolf Recovery in Yellowstone: Park Visitor Attitudes and Values," in National Park Service et al., *Wolves for Yellowstone? A Report to the United States Congress* (Washington, DC, 1990).

46. US Fish and Wildlife Service, "Endangered and Threatened Wildlife and Plants; Final Rule Designating the Northern Rocky Mountain Population of Gray Wolves as a Distinct Population Segment and Removing This Distinct Population Segment from the Federal List of Endangered and Threatened Wildlife," 73 Fed. Reg. 10514 (Feb. 27, 2008). Between the initial proposed delisting and publication of the final rule, Wyoming revised its statutes and modified its wolf management plan to satisfy the FWS that adequate regulatory mechanisms were in place to safeguard the state's wolf population.

47. Defenders of Wildlife v. Safari Club International, 565 F.Supp.2d 1160, 1163 (D. Mont. 2008). According to the court, the number of wolves on the landscape as well as documented dispersals from each subpopulation was not evidence that genetic material had been exchanged. Under long-standing administrative law doctrine, federal agencies must provide "a reasoned analysis" when changing course by revising a rule or established policy. Motor Vehicle Manufacturers Association v. State Farm Mutual Automobile Insurance Co., 463 US 29, 42 (1983).

48. Defenders of Wildlife v. Salazar, 729 F.Supp.2d 1207, 1228 (D. Mont. 2010). As a legal matter, the court based its ruling both on the statutory language and on the agency's failure to provide a reason for its changed position.

49. Defenders of Wildlife v. Salazar, 776 F.Supp.2d 1178, 1183 (D. Mont. 2011).

50. The rider is found in Pub. L. No. 112-10 § 1713, 125 Stat. 38 (2011). For the ensuing litigation, see Alliance for the Wild Rockies v. Salazar, 672 F.3d 1170 (9th Cir. 2012), affirming, 800 F.Supp.2d 1123 (D. Mont. 2011). For the Molloy quote, see 800 F.Supp.2d at 1126.

51. Wyoming v. US Dept. of the Interior, 2010 WL 4814950, at *6 (D. Wyo. Nov. 18, 2010). The Wyoming wolf management law is found at Wyoming Statutes. Annotated § 23-1-304(a) (2012).

52. US Fish and Wildlife Service, "Removal of the Gray Wolf in Wyoming from the Federal List of Endangered and Threatened Wildlife, and Removal of the Wyoming Wolf Population's Status as an Experimental Population," 77 Fed. Reg. 55530 (Sept. 10, 2012). The litigation is found at Defenders of Wildlife v. Zinke, 849 F.3d 1077 (D.C. Cir. 2017), affirming in part and reversing in part, Defenders of Wildlife v. Jewell, 68 F.Supp.2d 193 (D.D.C. 2014). The DC Circuit Court of Appeals concluded that the state was committed to sustaining more than the required minimum wolf population level and that the Wyoming plan adequately provided for genetic connectivity between the GYE wolves and their more northern counterparts. The FWS's final decision removing Wyoming's wolves from federal ESA protection is found at "Endangered and Threatened Wildlife and Plants; Reinstatement of Removal of Federal Protections for Gray Wolves in Wyoming," 82 Fed. Reg. 20284 (May 1, 2017).

53. Douglas W. Smith, P. J. White, Daniel R. Stahler, Adrian Wydeven, and David C. Hallac, "Managing Wolves in the Yellowstone Area: Balancing Goals Across Jurisdictional Boundaries," *Wildlife Society Bulletin* 40 (2016): 436; Jim Robbins, "A Famous Alpha Wolf's Daughter, Spitfire, Is Killed by a Hunter," *New York Times*, November 30, 2018, https://www.nytimes.com/2018/11/30/science/wolf-spitfire-killed.html. See also Nate Blakeslee, *American Wolf: A True Story of Survival and Obsession in the American West* (New York: Crown, 2017).

54. Robert B. Keiter, "Grizzlies, Wolves, and Law in the Greater Yellowstone Ecosystem: Wildlife Management Amidst Jurisdictional Complexity and Tension," *Wyoming Law Review* 22 (2022): 303, 338–40; Ellis Juhlin, "Hunting Bills Seek to Address Growing Elk Populations in Montana," *Montana Public Radio*, March 30, 2023, https://www.mtpr.org/montana-news/2023-03-30/hunting-bills-seek-to-address-growing-elk-populations-in-montana; see also Angus M. Thuermer Jr., "'Decade of the Elk' for Hunters as Herds Top Goals by 32%," *WyoFile*, September 15, 2020, https://wyofile.com/decade-of-the-elk-for-hunters-as-herds-top-goals-by-32/.

55. Mont. H.B. 224 (2021), amending Mont. Code § 87-1-901; Mont. H.B. 225 (2021), amending Mont. Code § 87-1-304; Mont. S.B. 267, codified at Mont. Code § 87-6-214(1)(d); Mont. S.B. 314, codified at Mont. Code § 87-1-901.

56. Joshua Partlow, "'Unprecedented Killing': The Deadliest Season for Yellowstone's Wolves," *Washington Post*, March 4, 2022, https://www.washingtonpost.com/climate-environment/2022/03/04/yellowstone-wolves-hunting/.

57. Idaho S. 1211 (2021), codified at Idaho Statutes § 22-5304, § 22-5306, § 36-201, § 36-401, § 36-1107.

58. Keiter, "Grizzlies, Wolves, and Law," 340–44.

59. US Fish and Wildlife Service, "Endangered and Threatened Wildlife and Plants; Finding for the Gray Wolf in the Northern Rocky Mountains and the Western United States," 89 Fed. Reg. 8391 (Feb. 7, 2024); Amanda Eggert, "Federal Government Denies Petitions to Restore Protections for Gray Wolves," *Montana Free Press*, February 5, 2024, https://montanafreepress.org/2024/02/05/federal-government-denies-petitions-to-restore-protections-for-gray-wolves/.

60. Amanda Eggert, "Environmental Groups Sue USFWS over Its Decision Not to Restore Protections for Northern Rockies Wolves," *Montana Free Press*, April 8, 2024, https://montanafreepress.org/2024/04/08/environmental-groups-sue-usfws-over-wolf-management/; Flathead-Lolo-Bitterroot Citizen Task Force v. Montana, 2023 WL 8064884 (D. Mont. 2023), affirmed in part, vacated in part, and remanded, 98 F.4th 1180 (9th Cir. 2024); Isabel Hicks, "Court Decision Restricts Wolf Trapping and Snaring Season in Western Montana," *Bozeman Daily Chronicle*, November 22, 2023, https://www.bozemandailychronicle.com/news/agriculture/court-decision-restricts-wolf-trapping-and-snaring-season-in-western-montana/article_107a4fa6-8960-11ee-b589-5bf898c961ac.html; Blair Miller, "9th Circuit Upholds Length-of-Season Restrictions for Montana Wolf Trapping, Snaring," *Daily Montanan*, April 24, 2024, https://dailymontanan.com/2024/04/24/9th-circuit-upholds-length-of-season-restrictions-for-montana-wolf-trapping-snaring/; Dac Collins, "Federal Judge Restricts Wolf Trapping in Idaho to 'Pro-

tect' Grizzly Bears," *Outdoor Life*, March 21, 2024, https://www.outdoorlife.com/conservation/idaho-wolf-trapping-restrictions-grizzly-bears/.

61. Montana Fish, Wildlife, and Parks, *Montana Gray Wolf Conservation and Management Plan* (draft, 2023); https://fwp.mt.gov/binaries/content/assets/fwp/aboutfwp/public-comments/draft-wolf-plan/wmp2023_.pdf; MTFP Staff, "What's at Stake with a New Wolf Plan," *Montana Free Press*, October 30, 2023, https://montanafreepress.org/2023/10/30/whats-at-stake-with-a-new-wolf-plan/#; Blair Miller, "Montana Releases Draft of First Wolf Management Plan Update in 20 Years," *Daily Montanan*, October 26, 2023, https://dailymontanan.com/2023/10/26/montana-releases-draft-of-first-wolf-management-plan-update-in-20-years/.

62. Mike Koshmrl, "Fury Over Wyoming Wolf Torture Allegations Sparks Demands for Steeper Penalties, Reform," *WyoFile*, April 8, 2024, https://wyofile.com/fury-over-wyoming-wolf-torture-allegations-sparks-demands-for-steeper-penalties-reform/; Amanda Eggert, "Groups Cite Wyoming Wolf Incident as They Plan Lawsuit for Renewed Endangered Species Status," *Montana Free Press*, April 29, 2024, https://montanafreepress.org/2024/04/29/wyoming-wolf-incident-spurs-lawsuit/; Mike Koshmrl, "The Right to Snowmobile over Wildlife Could Soon Be Explicitly Protected in Wyoming," *WyoFile*, September 4, 2024, https://wyofile.com/the-right-to-snowmobile-over-wildlife-could-soon-be-explicitly-protected-in-wyoming/.

63. Western governors are on record supporting a more robust state role in administering the ESA. Western Governors' Association, Policy Resolution 2024-03, *Species Conservation and the Endangered Species Act* (November 8, 2023), https://westgov.org/resolutions/article/policy-resolution-2024-03-species-conservation-and-the-endangered-species-act. In contrast, national polling data suggests widespread support for continued federal management of wolves. Center for Biological Diversity, "Poll: Majority of Americans Oppose Trump Plan to End Wolf Protections," May 28, 2019, https://biologicaldiversity.org/w/news/press-releases/majority-of-americans-oppose-trump-plan-to-end-wolf-protections-2019-05-28/.

64. Farrell, *Battle for Yellowstone*, 168–216. For example, Professor Justin Farrell found that opposition to wolves in the Yellowstone area relates to "rugged American individualism" as reflected in "long-standing commitments to anti-federalism, private property rights, and defining the reintroduced wolf as an ecological outsider . . . [as well as] commitments pertain[ing] to human dominionism, the belief that humans sit atop a (sometimes God-ordained) natural hierarchy, thus giving humanity certain rights and duties vis-à-vis nonhuman animals" (195). See also Martin A. Nie, *Beyond Wolves: The Politics of Wolf Recovery and Management* (Minneapolis: University of Minnesota Press, 2003), 32–53; Barry Lopez, *Of Wolves and Men* (New York: Scribner's, 1978), 138–39.

Chapter Four

1. Bernard DeVoto, ed., *The Journals of Lewis and Clark* (Boston: Houghton Mifflin, 1953), 125.

2. George Catlin, *Letters and Notes on the Manners, Customs and Condition of the*

North American Indians (1844), reprint, with an introduction by Marjorie Halpin, 2 vols. (New York: Dover, 1973), 1:260–62; see also Mark David Spence, *Dispossessing the Wilderness: Indian Removal and the Making of the National Parks* (New York: Oxford University Press, 1999), 9–10.

3. Dayton Duncan, *Out West: An American Journey* (New York: Viking, 1987), 200–201; Yellowstone Park Act, 17 Stat. 32 (1872), codified at 16 U.S.C. §§ 21–22.

4. Richard West Sellars, *Preserving Nature in the National Parks: A History* (New Haven, CT: Yale University Press, 1997), 75–76.

5. Sellars, *Preserving Nature in the National Parks*, 75–76, 116–18, 156–58; James A. Pritchard, *Preserving Yellowstone's Natural Conditions: Science and the Perception of Nature* (Lincoln: University of Nebraska Press, 1999), 174–84.

6. National Academy of Sciences, *Revisiting Brucellosis in the Greater Yellowstone Area* (Washington, DC: The National Academies Press, 2017), 10, https://doi.org/10.17226/24750; Robert B. Keiter and Peter H. Froelicher, "Bison, Brucellosis, and Law in the Greater Yellowstone Ecosystem," *Land and Water Law Review* 28 (1993): 22–32.

7. National Academy of Sciences, *Revisiting Brucellosis*, 2, 8, 10–11, 48–51.

8. Keiter and Froelicher, "Bison, Brucellosis, and Law," 45–48.

9. State of Montana v. United States, No. CV-95-6-H-CCL (D. Mont. 1995) (Complaint and Settlement Agreement); Intertribal Bison Co-op. v. Babbitt, 25 F. Supp.2d 1135 (D. Mont. 1998), affirmed, Greater Yellowstone Coalition v. Babbitt, 175 F.3d 1149 (9th Cir. 1999). The *Montana v. United States* litigation is further described in Robert B. Keiter, "Greater Yellowstone's Bison: Unraveling of an Early American Wildlife Conservation Achievement," *Journal of Wildlife Management* 61 (1997): 8. See also Fund for Animals, Inc. v. Lujan, 794 F.Supp. 1015 (D. Mont. 1991), affirmed, 962 F.2d 1391 (9th Cir. 1992); Greater Yellowstone Coal. v. Babbitt 952 F.Supp. 1435 (D. Mont. 1998), affirmed, 108 F.3d 1385 (9th Cir. 1997).

10. National Park Service et al., *Record of Decision for Final Environmental Impact Statement and Bison Management Plan for the State of Montana and Yellowstone National Park* (2000), https://www.nps.gov/yell/learn/management/upload/yellbisonrod.pdf.

11. National Park Service, *Record of Decision, Remote Vaccination Program to Reduce the Prevalence of Brucellosis in Yellowstone Bison* (2014), https://parkplanning.nps.gov/document.cfm?parkID=111&documentID=58229.

12. 9 C.F.R. § 78.40 (2021); "Brucellosis Class Free States and Certified Brucellosis-Free Herds; Revisions to Testing and Certification Requirements," 75 Fed. Reg. 81,090 (Dec. 27, 2010) (to be codified at 9 C.F.R. pt. 78); Mont. Admin. Rules § 32.2.43 (2021).

13. Yellowstone National Park, National Park Service, Bison Management, https://www.nps.gov/yell/learn/management/bison-management.htm#:~:text=Currently%2C%20the%20park's%20bison%20population,start%20their%20own%20bison%20herds.

14. Montana Fish, Wildlife, and Parks, *Bison Translocation, Bison Quarantine Phase IV Environmental Assessment Decision Notice* (2010); Robert B. Keiter, "The Greater Yellowstone Ecosystem Revisited: Law, Science, and the Pursuit of Ecosys-

tem Management in an Iconic Landscape," *University of Colorado Law Review* 91 (2020): 75–87.

15. Associated Press, "Despite Protests, 88 Bison from Yellowstone National Park Will Be Moved to Mogul Ted Turner's Ranch," *Los Angeles Times*, February 2, 2010, https://www.latimes.com/archives/blogs/la-unleashed/story/2010-02-02/despite-protests-88-bison-from-yellowstone-national-park-will-be-moved-to-mogul-ted-turners-ranch.

16. Cally Carswell, "Latest: Bison Transferred to Fort Peck Indian Reservation," *High Country News*, November 24, 2014, https://www.hcn.org/issues/46.20/latest-bison-transferred-to-fort-peck-indian-reservation; Citizens for Balanced Use v. Maurier, 303 P.3d 794 (Mont. 2013); National Park Service, "Bison Bellows: The 'Buffalo' Treaty," https://www.nps.gov/articles/bison-bellows-1-21-16.htm#:~:text=This%20treaty%2C%20often%20referred%20to,prairie%20ecosystems%20and%20their%20culture.

17. National Park Service, *The Use of Quarantine to Identify Brucellosis-Free Yellowstone Bison for Relocation Elsewhere, Environmental Assessment* (January 14, 2016), https://parkplanning.nps.gov/document.cfm?documentID=70262.

18. National Academy of Sciences, *Revisiting Brucellosis*.

19. See Cold Mountain v. Garber, 375 F.3d 884 (9th Cir. 2004); W. Watersheds Project v. Salazar, 766 F.Supp.2d 1095 (D. Mont. 2011), aff'd, 494 Fed. Appx 740 (9th Cir. 2012); All. for the Wild Rockies v. US Dept. of Agriculture, 938 F.Supp.2d 1034 (D. Mont. 2013, aff'd, 772 F.3d 592 (9th Cir. 2014). But see Cottonwood Environmental Law Center v. Bernhardt, Order, No. CV 18–12-BU-SHE (D. Mont., December 10, 2020) (confirming preparation of a supplemental environmental analysis to the IBMP due to new Blackfeet hunting plans, but denying the request to temporarily enjoin bison hunting outside Yellowstone or to vacate the IBMP).

20. National Park Service, US Department of Interior, *Yellowstone National Park Bison Management Plan Final Environmental Impact Statement* (June 2024), 51–52, https://parkplanning.nps.gov/document.cfm?parkID=111&projectID=94496&documentID=137582; see also Buffalo Field Campaign, *Yellowstone Buffalo Slaughter Totals*, https://www.buffalofieldcampaign.org/yellowstone-buffalo-slaughter-totals; Keiter, "Greater Yellowstone Ecosystem Revisited," 77.

21. Yellowstone Forever, "Bison Conservation and Transfer Program," https://www.yellowstone.org/bison-conservation/.

22. US Forest Service, US Department of Agriculture, *Record of Decision, Custer Gallatin National Forest Land Management Plan* (2022), 10, 26–27, https://www.fs.usda.gov/Internet/FSE_DOCUMENTS/fseprd1008522.pdf; National Park Service, "Notice of Intent to Prepare an Environmental Impact Statement for a Bison Management Plan for Yellowstone National Park, Idaho, Montana, Wyoming," 87 Fed. Reg. 4653 (Jan. 28, 2022).

23. Montana Department of Fish, Wildlife, and Parks, *Final Programmatic Environmental Impact Statement, Bison Conservation and Management in Montana* (2020), https://fwp.mt.gov/binaries/content/assets/fwp/conservation/wildlife-reports/bison/final-programmatic-bison-eis-december-2019-1.pdf.

24. Complaint, United Property Owners of Montana, Inc. v. Montana Dept. of Fish, Wildlife, and Parks, Mont. 10th Dist., Fergus County (filed March 9, 2020).

25. Governor's Office, "Governor Gianforte Announces Settlement Agreement Ending FWP Bison Plan," Montana.gov., April 20, 2021, https://news.mt.gov/Governors-Office/governor-gianforte-announces-settlement-agreement-ending-fwp-bison-plan; Amanda Eggert, "FWP: New Bison Herds on Hold," *Montana Free Press*, April 21, 2021, https://montanafreepress.org/2021/04/21/fwp-new-bison-herds-on-hold/.

26. Clarifying Definition of Bison, Montana H.B. 318, Leg. 67th Sess. (Mont. 2021); Require County Approval to Move Bison, Montana H.B. 302, Leg., 67th Sess. (Mont. 2021); Jim Bailey, "'Bad' Bison Bills in Montana Set Back Conservation of America's Official Mammal," *Mountain Journal*, March 29, 2021, https://mountainjournal.org/montana-bills-for-bison-represent-major-blow-to-conservation.

27. National Park Service, US Department of the Interior, *Yellowstone National Park Bison Management Plan, Final Environmental Impact Statement* (June 2024), 1–8, https://parkplanning.nps.gov/document.cfm?parkID=111&projectID=94496&documentID=137582; US National Park Service, Yellowstone National Park, *Record of Decision, Bison Management Plan* (2024), https://parkplanning.nps.gov/document.cfm?documentID=138363.

28. Montana governor Greg Gianforte, letter to Yellowstone Superintendent Cam Sholly, October 10, 2023; Montana governor Greg Gianforte, letter to Interior Secretary Deb Haaland, July 1, 2024.

29. Buffalo Field Campaign v. Williams, 579 F.Supp.3d 186 (D.D.C. 2022); Buffalo Field Campaign v. Zinke, 289 F.Supp.3d 103 (D.D.C. 2018); US Fish and Wildlife Service, "Endangered and Threatened Wildlife and Plants; 90-Day Finding for Three Petitions to List the Yellowstone Bison," 87 Fed. Reg. 34228 (June 6, 2022).

30. Montana Fish, Wildlife, and Parks, Elk Management in Areas with Brucellosis: 2021 Work Plan (October 2020), https://fwp.mt.gov/binaries/content/assets/fwp/conservation/final-brucellosis-2021-work-plan.pdf.

31. National Academies of Science, *Revisiting Brucellosis*, 5. Nonetheless, the report also urges continued separation between bison and cattle to avoid disease transmission.

32. DeVoto, ed., *The Journals of Lewis and Clark*, 28; Paul Schullery, *Searching for Yellowstone: Ecology and Wonder in the Last Wilderness* (New York: Houghton Mifflin, 1997), 42–50; Yellowstone Park Act, 17 Stat. 32 (1872), codified at 16 U.S.C. § 22.

33. Schullery, *Searching for Yellowstone*, 77, 121.

34. The debate over early wildlife abundance in Yellowstone is described and analyzed in Schullery, *Searching for Yellowstone*, 41–50. See also Frederic H. Wagner, *Yellowstone's Destabilized Ecosystem: Elk Effects, Science and Policy Conflict* (New York: Oxford University Press, 2006), 15–47.

35. Pritchard, *Preserving Yellowstone's Natural Conditions*, 137–38; Schullery, *Searching for Yellowstone*, 148–54.

36. Sellars, *Preserving Nature in the National Parks*, 246–49.

37. The principal works addressing Yellowstone's Northern Range and elk herd are Douglas Houston, *The Northern Yellowstone Elk: Ecology and Management* (New York: Macmillan, 1982); Alston Chase, *Playing God in Yellowstone: The Destruction*

of America's First National Park (New York: Atlantic Monthly Press, 1986); Wagner, *Yellowstone's Destabilized Ecosystem*. See also Yellowstone National Park, *Yellowstone's Northern Range: Complexity and Change in a Wildland Ecosystem* (1997), http://npshistory.com/publications/yell/northern_range.pdf; Schullery, *Searching for Yellowstone*, 235–36.

38. Luke E. Painter et al., "Aspen Recruitment in the Yellowstone Region Linked to Reduced Herbivory After Large Carnivore Restoration," *Ecosphere* 8 (2018): 9; William Ripple and Robert Beschta, "Restoring Yellowstone's Willows with Wolves," *Biological Conservation* 138 (2007): 514; Daniel Fortin et al., "Wolves Influence Elk Movements: Behavior Shapes a Trophic Cascade in Yellowstone National Park," *Ecology* 86 (2005): 1320; Robert L. Beschta, "Cottonwood, Elk, and Wolves in the Lamar Valley of Yellowstone National Park," *Ecological Applications* 13 (2003): 1395. But in *Yellowstone's Wildlife in Transition*, ed. P. J. White, Robert A. Garrett, and Glenn E. Plumb (Cambridge, MA: Harvard University Press, 2013), see P. J. White and Robert A. Garrott, "Predation: Wolf Restoration and the Transition of Yellowstone Elk," 69, and N. Thompson Hobbs and David J. Cooper, "Have Wolves Restored Riparian Willows in Northern Yellowstone," 179.

39. See chap. 3 herein; Robert B. Keiter, "Grizzlies Wolves, and Law in the Greater Yellowstone Ecosystem: Wildlife Management Amidst Jurisdictional Complexity and Tension," *Wyoming Law Review* 22 (2022): 303, 340–44.

40. Isabel Hicks, "Montana Livestock Officials Talk Brucellosis, Wildlife at Three Forks Meeting," *Bozeman Daily Chronicle*, May 17, 2023, https://www.bozemandailychronicle.com/news/agriculture/montana-livestock-officials-talk-brucellosis-wildlife-at-three-forks-meeting/article_b33d7346-f4da-11ed-9c15-43d4e5ae4f61.html; Amanda Eggert, "Chronic Wasting Disease Found in Yellowstone National Park," *WyoFile*, November 20, 2023, https://wyofile.com/chronic-wasting-disease-found-in-yellowstone-national-park/.

41. Mark S. Boyce, *The Jackson Elk Herd: Intensive Wildlife Management in North America* (New York: Cambridge University Press, 1989), 1–12; see also Bruce L. Smith, *Where Elk Roam: Conservation and Biopolitics of Our National Elk Herd* (Guilford, CT: Lyons Press, 2012).

42. Kelly M. Profitt, Neil Anderson, Paul Lukacs, Margaret M. Riordan, Justin A. Gude, and Julee Shamhart, "Effects of Elk Density on Elk Aggregation Patterns and Exposure to Brucellosis," *Journal of Wildlife Management* 79, no. 3 (2015): 373; Brant A. Schumaker, Dannele E. Peck, and Mandy E. Kauffman, "Brucellosis in the Greater Yellowstone Area: Disease Management at the Wildlife Livestock Interface," *Human Wildlife Interactions* 6, no. 1 (2012): 48.

43. The court decisions, in the order noted, are Parker Land & Cattle Co. v. United States, 796 F.Supp. 477 (D. Wyo. 1992); Parker Land & Cattle Co. v. Wyoming Game & Fish Commission, 845 P.2d 1040 (Wyo. 1993); Fund for Animals v. Clark, 27 F.Supp.2d 8 (D.D.C. 1998); Wyoming v. United States, 279 F.3d 1214 (10th Cir. 2002).

44. US Fish and Wildlife Service and National Park Service, *Record of Decision, National Elk Refuge, Grand Teton National Park, Final Bison and Elk Management Plan and Environmental Impact Statement* (2007).

45. Defenders of Wildlife v. Salazar, 651 F.3d 112, 117-118 (D.C. Cir. 2011).

46. Mike Koshmrl, "Elk Feeding Tweaked to Battle Brucellosis," *Jackson Hole News and Guide*, March 23, 2016, https://www.jhnewsandguide.com/news/environmental/elk-feeding-tweaked-to-battle-brucellosis/article_a82ddc30-dc11-5c31-be73-e99015402558.html; Aaron M. Foley, Paul C. Cross, David A Christianson, Brandon M. Scurlock, and Scott Creely, "Influences of Supplemental Feeding on Winter Elk Calf-Cow Ratios in the Southern Greater Yellowstone Ecosystem," *Journal of Wildlife Management* 79 (2015): 887; Kari Boroff, Mandy Kauffman, Dannele Peck, Eric Maichak, Brandon Scurlock, and Brant Schumaker, "Risk Assessment and Management of Brucellosis in the Southern Greater Yellowstone Area (II): Cost-benefit Analysis of Reducing Elk Brucellosis Prevalence," *Preventive Veterinary Medicine* 134 (2016): 39; Tom Hallberg, "Brucellosis Discovered in Teton County Herd," *Jackson Hole News & Guide*, November 17, 2018, https://www.jhnewsandguide.com/news/environmental/brucellosis-discovered-in-teton-county-herd/article_8d806ea8-2b21-576b-a8cf-a943cc9cec71.html.

47. Fish and Wildlife Service, "Intent to Prepare an Updated Bison and Elk Management Plan for the National Elk Refuge in Wyoming; Environmental Impact Statement," 88 Fed. Reg. 50168 (Aug. 1, 2023); Mike Koshmrl, "Nat'l Refuge Tries to Wean Its Elk Off Feed, and Fails So Far," *WyoFile*, January 25, 2022, https://wyofile.com/natl-refuge-tries-to-wean-its-elk-off-feed-and-fails-so-far/.

48. Greater Yellowstone Coalition v. Kimball, No. 06-CV-37, 2007 WL 9709798 (D. Wyo. Aug. 24, 2007), affirmed sub nom., Greater Yellowstone Coalition v. Tidwell, 572 F.3d 1115 (10th Cir. 2009).

49. Western Watersheds Project v. Christiansen, 348 F.Supp.3d 1204 (D. Wyo. 2018); US Forest Service, *Dell Creek and Forest Park Elk Feedgrounds: Long Term Special Use Permits Draft Environmental Impact Statement* (November 2023), https://www.fs.usda.gov/project/?project=60949; Mike Koshmrl, "Future of Bridger-Teton Elk Feedgrounds Uncertain," *WyoFile*, December 14, 2023, https://wyofile.com/future-of-bridger-teton-elk-feedgrounds-uncertain/.

50. Elizabeth S. Williams, Michael W. Miller, Terry J. Kreeger, Richard H. Kahn, and E. Tom Thorne, "Chronic Wasting Disease of Deer and Elk: A Review with Recommendations for Management," *Journal of Wildlife Management* 66 (2002): 551; Smith, *Where Elk Roam*, 102–15; Todd Wilkinson, "The Coming Plague: Chronic Wasting Disease, Cousin to Mad Cow, Is Bearing Down on Yellowstone National Park and America's Most Famous Elk Herd," *Mountain Journal*, January 25, 2017 (quoting Dr. Mary Wood, Wyoming Game and Fish Department chief wildlife veterinarian); Eggert, "Chronic Wasting Disease Found in Yellowstone National Park."

51. Smith, *Where Elk Roam*, 112–13; Eric J. Maichak et al., "Effects of Management, Behavior, and Scavenging on Risk of Brucellosis Transmission in Elk of Western Wyoming," *Journal of Wildlife Diseases* 45, no. 2 (2009): 398; Wilkinson, "The Coming Plague" (quoting the Center for Disease Control on the potential for transmission from infected animals or soil to humans).

52. Wyoming Game and Fish Department, *Wyoming Chronic Wasting Disease Management Plan* (2020), https://wgfd.wyo.gov/sites/default/files/content/PDF/Get%20Involved/CWD/WGFD_DRAFTCWDManagementPlan_030320.pdf

https://wgfd.wyo.gov/WGFD/media/content/PDF/Get%20Involved/CWD/Final-WGFD-CWD-Management-Plan-7-2020-with-appendices.pdf; Wyoming Statutes Annotated § 23-1-305 (2021); Angus M. Thuermer Jr., "Legislature Strips Game and Fish of Elk Feedground Closure Power," *WyoFile*, March 30, 2021, https://wyofile.com/legislature-strips-game-and-fish-of-elk-feedground-closure-power/.

53. Wyoming Game and Fish Department, *Draft Wyoming Elk Feedgrounds Management Plan* (2023), 8 (copy available from the author); Brett French, "Wyoming Elk Feedground Draft Management Plan Out for Public Comment," *Spokesman-Review*, August 2, 2023, https://www.spokesman.com/stories/2023/aug/01/wyoming-elk-feedground-draft-management-plan-out-f/.

54. Wyoming Game and Fish Department, *Wyoming Elk Feedgrounds Management Plan* (2024), 8, 50–77; Jonathan D. Cook et al., "Evaluating Management Alternatives for Wyoming Elk Feedgrounds in Consideration of Chronic Wasting Disease" (Department of Interior, US Geological Survey, 2023), https://pubs.usgs.gov/of/2023/1015/ofr20231015.pdf; Mike Koshmrl, "Fewer Elk. Sicker Elk. That's What the Experts Expect if Wyoming Keeps on Feeding," *WyoFile*, December 14, 2023, https://wyofile.com/fewer-elk-sicker-elk-thats-what-the-experts-expect-if-wyoming-keeps-on-feeding/; John Carter, "Wyoming Game and Fish Should Vote No on Elk Feedground Plan," *WyoFile*, March 11, 2024, https://wyofile.com/wyoming-game-and-fish-should-vote-no-on-elk-feedground-plan/.

55. Maxine Speier, "Montana Wildlife Officials Ask Wyoming to Stop Feeding Elk," *Montana Public Radio*, December 14, 2017, https://www.mtpr.org/montana-news/2017-12-14/montana-wildlife-officials-ask-wyoming-to-stop-feeding-elk; Chronic Wasting Disease Research and Management Act, S. 4111, 117th Cong. (2022); Staff Report, "CWD Research and Management Act Approved by Congress," *Outdoor News*, December 24, 2022, https://www.outdoornews.com/2022/12/24/cwd-research-and-management-act-approved-by-congress/.

56. Matthew J. Kauffman, James E. Meacham, Hall Sawyer, Alethea Y. Steingisser, William J. Rudd, and Emiline Ostlind, *Wildlife Migrations: Atlas of Wyoming's Ungulates* (Corvallis: Oregon State University Press, 2018), 126–45.

57. Arthur Middleton, "Elk in the Greater Yellowstone Ecosystem," in Kauffman et al., *Wildlife Migrations*, 127.

58. Joel Berger, "The Last Mile: How to Sustain Long Distance Migration in Mammals," *Conservation Biology* 18 (2004): 320, 323; Kauffman et al., *Wildlife Migrations*. See also Rob Ament, Renee Callahan, Laramie Maxwell, Grace Stonesipher, Elizabeth Fairbank, and Abigail Breuer, *Wildlife Connectivity: Opportunities for State Legislation* (Bozeman, MT: Center for Large Landscape Conservation, 2019), https://largelandscapes.org/wp-content/uploads/2019/03/Wildlife_Connectivity_Opportunities_for_State-Legislation_2019.pdf; US Public Lands and Rivers Conservation, *How to Conserve Wildlife Migrations in the American West* (PEW, October 2022), https://www.pewtrusts.org/en/research-and-analysis/reports/2022/10/how-to-conserve-wildlife-migrations-in-the-american-west.

59. Angus M. Thuermer Jr., "Game and Fish Proposes New Migration Corridor Protections," *WyoFile*, March 5, 2019, https://wyofile.com/game-and-fish-proposes-new-migration-corridor-protections/; Angus M. Thuermer Jr., "Gov Wants

Lawmakers to Back Off from Wildlife Migration Bill," *WyoFile*, October 22, 2019, https://wyofile.com/gov-wants-lawmakers-to-back-off-from-wildlife-migration-bill/; Governor Mark Gordon, State of Wyoming, Wyoming Mule Deer and Antelope Migration Corridor Protection, Executive Order No. 2020-1 (2020); Montana Department of Fish, Wildlife, and Parks, Strategy for Wildlife Movement and Migration (2020), https://fwp.mt.gov/conservation/strategy-for-wildlife-movement-and-migration; Idaho Department of Fish and Game, Idaho Action Plan (V3.0) for Implementing the Department of the Interior Secretarial Order 3362: "Improving Habitat Quality in Western Big-Game Winter Range and Migration Corridors" (September 10, 2020).

60. Western Governors' Association, Policy Resolution 07-01, *Protecting Wildlife Migration Corridors and Crucial Wildlife Habitat in the West* (2007); Western Governors' Association, Policy Resolution 2019-08, *Wildlife Migration Corridors and Habitat* (2019), https://www.trcp.org/wp-content/uploads/2019/06/WGA_Wildlife_Migration_Corridors_and_Habitat-Res-6-12-19.pdf; US Forest Service, NFMA Planning Rule, 36 C.F.R. §§ 219.9(a)(1), 219.10(a)(1) (referencing "connectivity"); Secretary of the Interior, Order No. 3362 (February 9, 2019); US Geological Service, *Ungulate Migrations in the Western United States*, vols. 1 and 2 (2020/2022); "USDA Commits to Big Game Conservation Partnership with the State of Wyoming: Initial Investments Forthcoming," Natural Resources Conservation Service, US Department of Agriculture (May 20, 2022), https://www.nrcs.usda.gov/wps/portal/nrcs/detail/national/newsroom/releases/?cid=NRCSEPRD1926821; Wildlife Corridors Conservation Act, H.R. 6448, 114th Cong. (2016), H.R. 2795, 166th Cong. (2019); "Colorado Joins Wave of States Protecting Wildlife Corridors," Center for Large Landscape Conservation, 2021, https://largelandscapes.org/news/state-corridors-legislation/.

61. For a description of the process leading to the Path of the Pronghorn wildlife corridor, see David N. Cherney, "Securing the Free Movement of Wildlife: Lessons from the American West's Longest Land Mammal Migration," *Environmental Law* 41 (2011): 599.

62. Kauffman et al., *Wildlife Migrations*, 136–37; Gordon, Wyoming Mule Deer and Antelope Migration Corridor Protection, Executive Order No. 2020-1.

63. Mike Koshmrl, "Gas Field May Cut Off Jackson's Pronghorn," *Jackson Hole News and Guide*, June 27, 2018, https://www.jhnewsandguide.com/news/environmental/gas-field-may-cut-off-jackson-s-pronghorn/article_2981c8de-541f-59f3-8f90-48fe810e0290.html.

64. Upper Green River Alliance v. US Bureau of Land Management, 598 F.Supp. 3d 1303, 1315 (D. Wyo. 2022), affirmed sub nom. Western Watersheds Project v. US Bureau of Land Management, 76 F.4th 1286 (10th Cir. 2023); Mike Koshmrl, "Wyoming Sides with Industry, Oks 'Path of Pronghorn Lease As-Is," *WyoFile*, October 6, 2023, https://wyofile.com/wyoming-sides-with-industry-oks-path-of-the-pronghorn-lease-as-is/; Mike Koshmrl, "Path of the Pronghorn at 'High Risk' of Being Lost, New Analysis Finds," *WyoFile*, November 3, 2023, https://wyofile.com/path-of-the-pronghorn-at-high-risk-of-being-lost-new-analysis-finds/.

65. Wyoming Game and Fish Department, *Dry Piney Wildlife Crossing Project*

Complete (October 4, 2023), https://wgfd.wyo.gov/News/Dry-Piney-wildlife-crossing-project-complete.

66. Ben Goldfarb, "When Wildlife Safety Turns Into Fierce Political Debate," *High Country News*, January 1, 2020, https://www.hcn.org/issues/52.1/wildlife-when-wildlife-safety-turns-into-fierce-political-debate.

Chapter Five

1. Rick Reese, *Greater Yellowstone: The National Park and Adjacent Wildlands* (Helena, MT: Montana Geographic, 1984), 76–78; Congressional Research Service, Library of Congress, 99th Cong., *Greater Yellowstone Ecosystem: An Analysis of Data Submitted by Federal and State Agencies* (Washington, DC: Comm. Print, 1986), 71.

2. US Forest Service, US Department of Agriculture, *Targhee National Forest Land Management Plan* (1985), 328; Dennis Glick et al., *An Environmental Profile of the Greater Yellowstone Ecosystem* (Greater Yellowstone Coalition, 1991), 115.

3. Glick et al., *An Environmental Profile*, 115.

4. Richard J. Knight, D. J. Mattson, and B. M. Blanchard, *Movements and Habitat Use of the Yellowstone Grizzly Bear* (Missoula, MT: Interagency Grizzly Bear Study Team Report, 1984); David J. Mattson, "Human Impacts on Bear Habitat Use," *International Conference on Bear Research and Management* 8 (1990): 33; David J. Mattson and Matthew M. Reid, "Conservation of the Yellowstone Grizzly Bear," *Conservation Biology* 5, no. 3 (1991): 364; see also Resources Limited v. Robertson, 35 F.3d 1300 (9th Cir. 1994); Swan View Coalition, Inc. v. Turner, 824 F.Supp. 923 (D. Mont. 1992).

5. US Forest Service, *Record of Decision, Final Environmental Impact Statement for the Revised Forest Plan, Targhee National Forest* (1997), 9–11.

6. Intermountain Forest Industry Ass'n v. Lyng, 683 F.Supp. 1330 (D. Wyo. 1988); Mountain States Legal Fdn. v. Robertson, No. C87–0274-B (D. Wyo. October 19, 1987). See Robert B. Keiter, "Taking Account of the Ecosystem on the Public Domain: Law and Ecology in the Greater Yellowstone Region," *University of Colorado Law Review* 60 (1989): 923, 973–74.

7. Gallatin Range Consolidation and Protection Act of 1993, Pub. L. No. 103-91, 107 Stat. 987; Gallatin Land Consolidation Act of 1998, Pub. L. No. 105-267, 111 Stat. 2371; George Draffin and Janine Blaeloch, "The Gallatin Land Exchanges," in *Commons or Commodity: The Dilemma of Federal Land Exchanges* (2000); US Forest Service, *Custer Gallatin National Forest Land Management Plan* (2022), 158–82, https://www.fs.usda.gov/Internet/FSE_DOCUMENTS/fseprd1008515.pdf.

8. US Forest Service, *Custer Gallatin National Forest Land Management Plan*, 76–77; US Forest Service, *Record of Decision, Custer Gallatin National Forest Land Management Plan* (2022), 12, 33, https://www.fs.usda.gov/Internet/FSE_DOCUMENTS/fseprd1008522.pdf.

9. US Forest Service, *Beaverhead-Deerlodge National Forest Plan* (2009); US Forest Service, *Record of Decision for the Final Environmental Impact Statement and Revised Land and Resource Management Plan: Beaverhead-Deerlodge National*

Forest (2009), 18–21, https://www.fs.usda.gov/Internet/FSE_DOCUMENTS/stelprdb5052838.pdf.

10. Forest Jobs and Recreation Act of 2009, S. 1470, 111th Cong. (2009); S. 268, 112th Cong. (2011); S. 37, 113th Cong. (2014); Forest Jobs and Recreation, S. Rep. No. 113-165 (2014). See Ray Ring, "Taking Control of the Machine," *High Country News*, July 14, 2009, https://www.hcn.org/issues/41.12/taking-control-of-the-machine; Ted Fellman, "Collaboration and the Beaverhead-Deerlodge Partnership: The Good, the Bad, and the Ugly," *Public Land and Resources Law Review* 30 (2009): 79.

11. See Martin Nie and Michael Fiebig, "Managing the National Forests Through Place-Based Legislation," *Ecology Law Quarterly* 37, no. 1 (2010): 1.

12. US Forest Service, *Shoshone National Forest Land Management Plan* (2015), 117, https://www.fs.usda.gov/Internet/FSE_DOCUMENTS/stelprd3842886.pdf; US Forest Service, *Bridger- Teton National Forest Land and Resource Management Plan* (2015), 434, https://www.fs.usda.gov/Internet/FSE_DOCUMENTS/stelprd3840286.pdf; Steven W. Hays, Lucas Townsend, Thale Dillon, Todd A. Morgan, and John D. Shaw, *Montana's Forest Products Industry and Timber Harvest, 2018* (Ft. Collins, CO: Rocky Mountain Research Station, 2021), 22, https://www.fs.usda.gov/rm/pubs_series/rmrs/rb/rmrs_rb035.pdf; Kate C. Marcillie, Thale Dillon, Lucas P. Townsend, Todd A. Morgan, and John D. Shaw, *Wyoming's Forest Products Industry and Timber Harvest, 2018* (Ft. Collins, CO: Rocky Mountain Research Station, 2021), https://www.fs.usda.gov/rm/pubs_series/rmrs/rb/rmrs_rb033.pdf; Eric A. Simmons, Steven W. Hayes, Todd A. Morgan, Charles E. Keegan III, and Chris Witt, *Idaho's Forest Products Industry and Timber Harvest 2011 with Trends through 2013* (Ft. Collins, CO: Rocky Mountain Research Station, 2014), 19, https://www.fs.usda.gov/rm/pubs_series/rmrs/rb/rmrs_rb019.pdf.

13. On the worsening wildfire problem on public lands, see Stephen J. Pyne, *The Pyrocene: How We Created an Age of Fire, and What Happens Next* (Berkeley: University of California Press, 2021); Edward Struzik, *Firestorm: How Wildfire Will Shape Our Future* (Washington, DC: Island Press, 2017). See also Healthy Forests Restoration Act,16 U.S.C. §§ 6501–6517; Collaborative Forest Stewardship Act, 16 U.S.C. §§ 6591, 7303; Congressional Research Service, *Forest Service Appropriations: Ten-Year Data and Trends FY2011–FY2020* (2020), https://crsreports.congress.gov/product/pdf/R/R46557/2; US Forest Service, *Confronting the Wildfire Crisis: A Strategy for Protecting Communities and Improving Resilience in America's Forests* (2022), https://www.fs.usda.gov/sites/default/files/Confronting-Wildfire-Crisis.pdf.

14. Julie Cart, "Hundreds Flee, Homes Threatened as Winds Intensify Wyoming Blaze," *Los Angeles Times*, July 26, 2001, https://www.latimes.com/archives/la-xpm-2001-jul-26-mn-26690-story.html; Star Tribune Staff, "A Fire Burning Western Wyoming Has Now Destroyed 55 Homes," *Casper Star Tribune*, September 27, 2018, https://trib.com/news/state-and-regional/a-fire-burning-in-western-wyoming-has-now-destroyed-55-homes/article_e7d0d587-69a3-5874-a3fb-58229f4a2a75.html; Helena Dore, "One Year Later, Bridger Foothills Fire Stirs Feelings of Loss, Appreciation," *Bozeman Daily Chronicle*, September 3, 2021,

https://www.bozemandailychronicle.com/news/county/one-year-later-bridger-foothills-fire-stirs-feelings-of-loss-appreciation/article_47c21952-4bd7-55cf-9b0c-692ccc347d20.html; US Forest Service, *Custer Gallatin National Forest Land Management Plan* (2022), 233 ("approximately 809,759 acres" have burned since 1980).

15. US Forest Service, *Shoshone National Forest Plan* (2015), 13, https://www.fs.usda.gov/Internet/FSE_DOCUMENTS/stelprd3842886.pdf; US Forest Service, *Custer Gallatin Land Management Plan—Sharing the Decision* (2022), 12, https://www.fs.usda.gov/Internet/FSE_DOCUMENTS/fseprd991254.pdf.

16. See Native Ecosystems Council v. Marten, 2020 WL 3064496 (D. Mont., 2020); Native Ecosystems Council v. Erickson, 804 F.3d App'x 651 (9th Cir. 2020); Alliance for the Wild Rockies v. Marten, 464 F.Supp.3d 1169 (D. Mont. 2020); Cottonwood Environmental Law Center v. Marten, 2022 WL 1439127 (9th Cir. 2022).

17. US Forest Service, *South Plateau Landscape Area Treatment Project, Final Environmental Assessment* (2022), https://www.fs.usda.gov/project/?project=57353; Isabel Hicks, "Forest Service Sued Over South Plateau Logging Project," *Billings Gazette*, September 25, 2023, https://billingsgazette.com/news/state-regional/logging-west-yellowstone-forest-service-lawsuit/article_d89a086b-4118-5f54-9a6d-5269afaeca6a.html; US Forest Service, *Draft Environmental Impact Statement: Teton to Snake Fuels Management Project* (2015), i; Angus M. Thuermer Jr., "Wildfire Worries, Wilderness, Collide Above Wilson," *WyoFile*, August 18, 2015, https://wyofile.com/wildfire-worries-wilderness-collide-above-wilson/; Mike Koshmrl, "'Teton-to-Snake' Plan Finally Goes Into Action," *Jackson Hole News and Guide*, September 27, 2017, https://www.jhnewsandguide.com/news/environmental/teton-to-snake-plan-finally-goes-in-action/article_128fb4e4-b327-52d7-9a07-1af84d7f64a7.html.

18. Reese, *Greater Yellowstone*, 66–72; Keiter, "Taking Account of the Ecosystem," 975–76; Samuel Western, *Pushed Off the Mountain, Sold Down the River* (Moose, WY: Homestead Publishing, 2002), 71–74.

19. Sierra Club v. Peterson 717 F.2d 1409 (D.C. Cir. 1983) (Bridger-Teton National Forest); Conner v. Burford, 848 F.2d 1441 (9th Cir. 1988) (Gallatin National Forest); Thomas v. Peterson, 753 F.2d 754 (9th Cir. 1985) (Endangered Species Act consultation). In *Park County Resource Council v. US Dept. of Agriculture*, 817 F.2d 609 (10th Cir. 1987), a case involving Wyoming's Shoshone National Forest, the Tenth Circuit ruled that in-depth environmental analysis could be deferred to the drilling permit stage, because that was when the Forest Service would have site-specific information about potential impacts.

20. 30 U.S.C. §§ 181–241; United States ex rel. McLennan v. Wilbur, 284 US 414 (1931).

21. Pub. L. No. 100-203, subtitle B, 101 Stat. 1330 (codified at 30 U.S.C. § 226(g)–(h) (1987)).

22. Rocky Mountain Oil and Gas Association v. US Forest Service, 12 Fed. App'x 498 (9th Cir. 2001), cert. denied, 534 US 1018 (2001); see also Joseph L. Sax and Robert B. Keiter, "The Realities of Regional Resource Management: Glacier National Park and Its Neighbors Revisited," *Ecology Law Quarterly* 33 (2006): 233, 275–80.

23. Energy Policy Act of 2005, Pub. L. No. 109-58, 119 Stat. 594 (codified at 42 U.S.C. § 15801).

24. WildEarth Guardians v. Zinke, 368 F.Supp.3d 41 (D.D.C. 2019); US Environmental Protection Agency, "National Environmental Policy Act Guidance on Consideration of Greenhouse Gas Emissions and Climate Change," 88 Fed. Reg. 1196, 1197 (Jan. 9, 2023) ("When conducting climate change analyses in NEPA reviews, agencies should consider: (1) the potential effects of a proposed action on climate change, including by assessing both GHG emissions and reductions from the proposed action; and (2) the effects of climate change on a proposed action and its environmental impacts.").

25. Bureau of Land Management, US Department of the Interior, *Record of Decision for the Pinedale Anticline Oil and Gas Exploration and Development Project Environmental Impact Statement* (2000).

26. Bureau of Land Management, *Record of Decision for the Bureau of Land Management, Final Supplemental Environmental Impact Statement for the Pinedale the Anticline Oil and Gas Exploration and Development Project* (2008); Theodore Roosevelt Conservation Partnership v. Salazar, 661 F.3d 66 (D.C. Cir. 2011), affirming, 744 F.Supp.2d 151 (D.D.C. 2010).

27. Hall Sawyer, Jon P. Beckmann, Renee G. Seidler, and Joel Berger, "Long-Term Effects of Energy Development on Winter Distribution and Residency of Pronghorn in the Greater Yellowstone Ecosystem," *Conservation Science and Practice* (September 2019): 8 (noting a 47% decline in the Sublette pronghorn herd unit population from 2005 to 2017); Wyoming Game and Fish Department, Wyoming Game and Fish Department Ungulate Migration Corridor Strategy, 2 (revised January 28, 2019) ("Mule deer have declined by about 40% in the past twenty years."); David R. Edmunds, Cameron L. Aldridge, Michael S. O'Donnell, and Adrian P. Monroe, "Greater Sage-Grouse Population Trends Across Wyoming," *Journal of Wildlife Management* (May 2017): 2 (noting Wyoming's greater sage-grouse "populations [have] declined 29–76% during 2007–2013, resulting in populations that represented approximately 4–25% of the size of the same populations in the 1970s and 1980s"); Kirk Johnson, "In Pinedale, Wyo., Residents Adjust to Air Pollution," *New York Times*, March 9, 2011, https://www.nytimes.com/2011/03/10/us/10smog.html; Dustin Bleizeffer, "Winter Ozone Spikes near Pinedale Prompt Health Advisories," *WyoFile*, March 21, 2023, https://wyofile.com/winter-ozone-spikes-near-pinedale-prompt-health-advisories/.

28. Alexandra Fuller, "Boomtown Blues: How Natural Gas Changed the Way of Life in Sublette County," *New Yorker*, February 5, 2007; Mike Koshmrl, "'Path of the Pronghorn' Protections Delayed as Development Proceeds," *WyoFile*, December 17, 2022, https://wyofile.com/path-of-the-pronghorn-protections-delayed-as-development-proceeds/.

29. Bureau of Land Management, *Normally Pressurized Land Natural Gas Development Project, Record of Decision* (2018).

30. Upper Green River Alliance v. Bureau of Land Management, 598 F.Supp.3d 1303 (D. Wyo. 2022), affirmed sub nom., Western Watersheds Project v. US Bureau of Land Management 76 F.4th 1286 (10th Cir. 2023).

31. US Forest Service, *Final Environmental Impact Statement on Oil and Gas Leasing in the Bridger-Teton National Forest Management Areas* (2003); Jeff Gearino, "Bridger-Teton Areas Off-Limits for Oil and Gas," *Casper Star Tribune*, March 8, 2003, https://trib.com/news/bridger-teton-areas-off-limits-for-oil-and-gas/article_d3c3dcfd-b3cd-5f0d-860b-19792424fce3.html; US Department of Agriculture and US Department of Interior, *Lewis and Clark National Forest Oil and Gas Leasing, Final Environmental Impact Statement Record of Decision* (1997).

32. Florence R. Shepard and Susan L. Marsh, *Saving Wyoming's Hoback: The Grassroots Movement that Stopped Natural Gas Development* (Salt Lake City: University of Utah Press, 2016).

33. US Department of Agriculture, *Record of Decision, Oil and Gas Leasing on Portions of the Wyoming Range* (2017).

34. US Forest Service, *Record of Decision, Proposed Oil and Gas Leasing, Shoshone National Forest* (1995); US Forest Service, *Final Oil and Gas Leasing Environmental Impact Statement: Shoshone National Forest* (1992); Wyoming Outdoor Council v. US Forest Service, 165 F.3d 43 (D.C. Cir. 1999); see also Wyoming Outdoor Council v. Bosworth, 284 F.Supp.2d 81 (D.D.C. 2003); Wyoming Outdoor Council v. Dombeck, 148 F.Supp.2d 1 (D.D.C. 2001).

35. US Forest Service, *Record of Decision for the Land Management Plan Revision, Shoshone National Forest* (2015); US Forest Service, *Land Management Plan, 2015 Revision, Shoshone National Forest* (2015), 82–83.

36. US Forest Service, *Final Environmental Impact Statement for the Caribou National Forest Revised Plan*, Vol. 1 (2003), 1–19, 3–138.

37. US Forest Service, "Record of Decision for Oil and Gas Leasing on Lands Administered by the Targhee National Forest," 65 Fed. Reg. 51285 (Aug. 23, 2000).

38. US Forest Service, *Final Environmental Impact Statement for the Custer Gallatin National Forest Land and Resource Management Plan*, vol. 2 (2005), 166–67.

39. US Forest Service, *Final Environmental Impact Statement for the Beaverhead-Deerlodge National Forest Land and Resource Management Plan* (2009), 135; US Forest Service, *Decision Notice and Finding of No Significant Impact for the Tendoy Project* (2021), 6.

40. Matt Kirby and Steve Kandell, "Oil and Gas Drilling Is Getting Dangerously Close to Our National Parks," *The Hill*, September 14, 2021, https://thehill.com/opinion/energy-environment/572243-oil-and-gas-drilling-is-getting-dangerously-close-to-our-national/.

41. General Mining Law of 1872, 30 U.S.C. §§ 21–42; Belk v. Meagher, 104 US 279 (1881); United States v. Coleman, 390 US 599 (1968); Gordon Morris Bakken, *The Mining Law of 1872: Past, Politics, and Prospects* (Albuquerque: University of New Mexico Press, 2008); John D. Leshy, *The Mining Law: A Study in Perpetual Motion* (Washington, DC: Resources for the Future, 1987); Todd Wilkinson, *Ripple Effects: How to Save Yellowstone and America's Most Iconic Wildlife Ecosystem* (Deadwood, OR: Wyatt-McKenzie Publishing, 2022), 28–74.

42. Marc Humphries, "New World Gold Mine and Yellowstone National Park," in *American National Parks: Current Issues and Developments*, ed. Rony Mateo (Hauppauge, NY: Nova Science Publishers, 2004), 51–56; Michael J. Yochim, *Protecting*

Yellowstone: Science and the Politics of National Park Management (Albuquerque: University of New Mexico Press, 2013), 80–119; Bob Ekey, "The New World Agreement: A Call for Reform of the 1872 Mining Law," *Public Land and Resources Law Review* 18 (1997): 151; US Forest Service and Montana Department of State Lands, *Preliminary Draft Environmental Impact Statement, Crown Butte Mines, Inc., New World Project* (1995), 20.

43. William J. Lockhart, "External Threats to Our National Parks: An Argument for Substantive Protection," *Stanford Environmental Law Journal* 16 (1997): 3, 14–35; Peter Dykstra, "Defining the Mother Lode: Yellowstone National Park v. the New World Mine," *Ecology Law Quarterly* 24 (1997): 299, 302–4.

44. Ekey, "The New World Agreement," 154–56; Department of the Interior, "Notice of Proposed Withdrawal, Montana," 60 Fed. Reg. 45732 (Sept. 1, 1995); Department of the Interior, "Amendments to Proposed Withdrawal, Montana," 61 Fed. Reg. 49450 (Sept. 20, 1996); Beartooth Alliance v. Crown Butte Mines, 904 F.Supp. 1168 (D. Mont. 1995).

45. Ekey, "The New World Agreement," 159–62; Kurt Repanshek, "Land Deal Closes the Book on the New World Mine Proposal on Yellowstone National Park's Doorstep," *National Parks Traveler*, June 15, 2010, https://www.national parkstraveler.org/2010/06/land-deal-closes-book-new-world-mine-proposed-yellowstone-national-parks-doorstep6045.

46. US Department of the Interior, Public Land Order No. 7875, Department of the Interior, "Emigrant Crevice Mineral Withdrawal, Montana," 83 Fed. Reg. 51701 (Oct. 12, 2018); US Department of the Interior, "Notice of Application for Withdrawal and Notification of Public Meeting, Montana," 81 Fed. Reg. 83867 (Nov. 22, 2016); US Department of Agriculture, *Emigrant Crevice Mineral Withdrawal, Draft Environmental Assessment* (2018); Yellowstone Gateway Protection Act, Pub. L. No. 116-9 § 1204, 133 Stat. 580 (2018).

47. Michael Wright, "State Approves Exploratory Drilling Near Yellowstone National Park," *Bozeman Daily Chronicle*, July 26, 2017, https://www.bozeman dailychronicle.com/news/environment/state-approves-exploratory-drilling-near-yellowstone-national-park/article_b49c3eff-8ae5-5ea7-8ead-dde0e1610d76.html; Park County Environmental Council v. Montana Dept. of Environmental Quality, 477 P.3d 288 (2020); Amanda Eggert, "State Supreme Court Blocks Proposed Gold Mine Near Yellowstone," *Montana Free Press*, December 10, 2020, https://montanafreepress.org/2020/12/10/state-supreme-court-blocks-proposed-gold-mine-near-yellowstone/.

48. Montana Code Annotated § 75-1-102 (2022); Michael Wright, "Exec Vows to Open Controversial Mine on Border of Yellowstone," *Billings Gazette*, April 4, 2019, https://billingsgazette.com/outdoors/exec-vows-to-open-controversial-mine-on-border-of-yellowstone/article_34ec8143-ce53-5273-8fd0-a0c4b2b521aa.html; Brett French, "Gold Mine Near Yellowstone Purchased to Avoid Development," *Spokesman-Review*, October 5, 2023, https://www.spokesman.com/stories/2023/oct/05/gold-mine-near-yellowstone-purchased-to-avoid-deve/.

49. US Forest Service, Caribou-Targhee National Forest, *Decision Notice and Finding of No Significant Impact for the Kilgore Gold Exploration Project* (2021);

Idaho Conservation League v. US Forest Service, 429 F.Supp.3d 719 (D. Idaho 2019); Idaho Conservation League v. US Forest Service, 2020 WL 2115436 (D. Idaho 2020); Idaho Conservation League v. US Forest Service, 2023 WL 5000514 (D. Idaho 2023).

50. William H. Lee, *A History of Phosphate Mining in Southeastern Idaho* (Washington, DC: US Geological Survey, 2001).

51. US Government Accountability Office, *Phosphate Mining: Oversight Has Strengthened, but Financial Assurances and Oversight Still Need Improvement*, GAO-12-505 (Washington, DC, May 2012), https://www.gao.gov/assets/gao-12-505.pdf; Greater Yellowstone Coalition v. Lewis, 628 F.2d 1143 (9th Cir 2010, affirming, 641 F.Supp.2d 1120 (D. Idaho 2009); Center for Biological Diversity v. US Bureau of Land Management, 2023 3796675 (D. Idaho 2023); Bureau of Land Management, "Notice of Intent to Prepare an Environmental Impact Statement for the Caldwell Canyon Revised Mine and Reclamation Plan, Caribou County, Idaho," 88 Fed. Reg. 81427 (Nov. 22, 2023), https://www.federalregister.gov/documents/2023/11/22/2023-25756/notice-of-intent-to-prepare-an-environmental-impact-statement-for-the-caldwell-canyon-revised-mine.

52. Ray Levy-Uyeda, "Can a Mining Corporation Ever Truly Be a Good Neighbor?," *The Guardian*, September 2, 2020, https://www.theguardian.com/environment/2020/sep/02/mining-corporation-montana-good-neighbour-agreement; Sarah M. Zuzulock and James R. Kuipers, "The Good Neighbor Agreement: A Proactive Approach to Water Management through Community Enforcement of Site-Specific Standards," *Greener Management International* 53 (Spring 2006): 73–88.

53. US Forest Service, Stillwater Mining Company, *East Boulder Mine Amendment 004 Final Environmental Impact Statement for Pre-Decisional Administrative Review* (November 2023), https://www.fs.usda.gov/project/?project=61385; Amanda Eggert, "Forest Service Forwards Plan to Keep East Boulder Mine Operating," *Montana Free Press*, December 1, 2023, https://montanafreepress.org/2023/12/01/forest-service-forwards-plan-to-keep-east-boulder-mine-operating/; Brett French, "Conservation Groups, Landowners Unite Against East Boulder Mine Tailings Dams," *Billings Gazette*, February 28, 2023, https://billingsgazette.com/news/state-and-regional/conservation-groups-landowners-unite-against-east-boulder-mine-tailings-dams/article_c3b63984-b7b2-11ed-8a0c-c325ac7fd477.html.

54. 16 U.S.C. §1131(c). In addition, the Wilderness Act provides that wilderness areas "may also contain ecological, geological or other features of scientific, education, scenic, or historical value."

55. Wilderness Act of 1964, 16 U.S.C. §§ 1131(c), 1132(b), 1133; John D. Leshy, *Our Common Ground: A History of America's Public Lands* (New Haven, CT: Yale University Press, 2021), 463–70; Peter S. White, L. Yung, D. N. Cole, and R. J. Hobbs, "Conservation at Large Scales: Systems of Protected Areas and Protected Areas in the Matrix," in *Beyond Naturalness: Rethinking Park and Wilderness Stewardship in an Era of Rapid Change*, ed. David N. Cole and Laurie Yung (Washington, DC: Island Press, 2010), 197; Reed F. Noss et al., "Core Areas: Where Nature Reigns," in *Continental Conservation: Scientific Foundations of Regional Reserve Networks*, ed. Michael E. Soule and John Terborgh (Washington, DC: Island Press, 1999), 99.

56. Wilkinson, *Ripple Effects*, 39–60; White et al., "Conservation at Large Scales," 209–12.

57. US Forest Service, *Shoshone National Forest Land Management Plan* (2015), 14–17, 123; US Forest Service, *Record of Decision, Custer Gallatin National Forest Land Management Plan* (Jan. 2022), 20–23, https://www.fs.usda.gov/Internet/FSE_DOCUMENTS/fseprd1008522.pdf; US Forest Service, *Revised Forest Plan for the Caribou National Forest* (2003), 2–13; US Forest Service, *Record of Decision, Final Environmental Impact Statement (FEIS) for the Revised Forest Plan, Targhee National Forest Plan* (1997), 18.

58. US Forest Service, *Record of Decision for the Final Environmental Impact Statement and Revised Land and Resource Management Plan: Beaverhead-Deerlodge National Forest* (2009), 18–21; Forest Jobs and Recreation Act of 2009, S. 1470, 111th Cong. (2009); Fellman, "Collaboration and the Beaverhead-Deerlodge Partnership," 79; Nie and Fiebig, "Managing the National Forests Through Place-Based Legislation," 1, 23–31.

59. Pub. L. No. 95-150, 78 Stat. 890 (95th Cong., 1977).

60. Montana Wilderness Association v. McAllister, 666 F.3d 549 (9th Cir. 2011), affirming, 658 F.Supp.2d 1249 (D. Mont. 2009); Matthew Koehler, "Groups Object to 'Undemocratic' Gallatin Community Collaborative Process," *Smokey Wire*, March 31, 2016, https://forestpolicypub.com/2016/04/01/groups-object-to-undemocratic-gallatin-community-collaborative-process/.

61. US Forest Service, *Proposed Action—Revised Forest Plan, Custer Gallatin National Forest* (2018), https://www.fs.usda.gov/Internet/FSE_DOCUMENTS/fseprd567788.pdf; Gazette Staff, "Bozeman Group Proposes Forest Consider More Wilderness, Wildlife Management Areas," *Billings Gazette*, January 31, 2018, https://billingsgazette.com/outdoors/bozeman-group-proposes-forest-consider-more-wilderness-wildlife-management-areas/article_4e190ba6-354c-5e46-a76a-7cd703657d8f.html; US Forest Service, *Record of Decision, Custer Gallatin National Forest Land Management Plan* (2022), https://www.fs.usda.gov/Internet/FSE_DOCUMENTS/fseprd1008522.pdf.

62. Todd Wilkinson, "'Unbroken Wilderness': Some Call the Porcupine and Buffalo Horn 'Holy Land,'" *Mountain Journal*, May 14, 2020, https://mountainjournal.org/some-call-these-mountains-in-montana-holy-land-for-their-wildlife; Amanda Eggert, "The Custer Gallatin National Forest's New Guiding Doc," *Montana Free Press*, February 2, 2022, https://montanafreepress.org/2022/02/04/new-custer-gallatin-forest-plan/.

63. US Forest Service, *Custer Gallatin Forest Plan Record of Decision*, 21.

64. David Tucker, "Wilderness Proposal in Gallatin, Madison Ranges Sparks Debate," *Mountain Journal*, August 19, 2024, https://mountainjournal.org/wilderness-proposal-in-gallatins-madisons-sparks-debate; Nancy Watters and Rick Johnson, "Commentary: Our Best Chance for Protecting the Gallatin and Madison Ranges," *Daily Montanan*, September 23, 2024, https://dailymontanan.com/2024/09/23/our-best-chance-for-protecting-the-gallatin-and-madisonranges/#:~:text=The%20bill%20will%20designate%20124%2C000,Wildlife%20and%20Recreation%20Management%20Areas.

65. Wyoming County Commissioners Association, "Wyoming Public Lands Ini-

tiative: Principles and Guidelines," (November 2015), https://wcca.wygisc.org/wpli/hub/index.html; Rebecca Worby, "Can Wyoming Learn from Utah's Public Land Mistakes?," *High Country News* 49, no. 1 (June 12, 2017): 5, https://www.hcn.org/issues/49.10/wyoming-confronts-wilderness-study-limbo-public-lands-initiative.

66. Restoring Local Input and Access to Public Lands Act, H.R. 6939, 114th Cong. (2018); Wyoming Public Lands Initiative Act of 2021, S. 1750, 117th Cong. (2021); Angus Thuermer Jr., "The Wyoming Public Lands Initiative Risks Collapse," *High Country News*, March 1, 2018, https://www.hcn.org/articles/wilderness-tensions-mount-over-wilderness-study-areas-in-wyoming.

67. US Forest Service, *Record of Decision Final Environmental Impact Statement (FEIS) for the Revised Forest Plan, Targhee National Forest* (1997), 18; US Forest Service, *Revised Forest Plan for the Caribou National Forest* (2013), 2–13; Heather Randall, "National Monument for Island Park," *Teton Valley News*, September 12, 2014, https://www.tetonvalleynews.net/news/national-monument-for-island-park/article_549f0064-3abe-11e4-88cc-1b980a4723c5.html.

68. John D. Leshy, "Legal Wilderness: Its Past and Some Speculations on Its Future," *Environmental Law* 44 (2014): 549, 599; Nie and Fiebig, "Managing the National Forests Through Place-Based Legislation," 36–38.

69. US Forest Service, *Forest Service Roadless Area Conservation Final Environmental Impact Statement* (2000), 3–395; US Forest Service, "Special Areas; Roadless Area Conservation," 66 Fed. Reg. 3243, 3245 (Jan. 12, 2001) (codified at 36 C.F.R. § 294).

70. US Forest Service, *Roadless Area Conservation FEIS*, vol. 2, *Maps of Inventoried Roadless Areas*, https://www.fs.usda.gov/Internet/FSE_DOCUMENTS/stelprdb5057902.pdf.

71. McKinley J. Talty, Kelly Mott Lacroix, Gregory H. Aplet, and R. Travis Belote, "Conservation Value of National Forest Roadless Areas," *Conservation Science and Practice* 2, no. 11 (2020), https://www.researchgate.net/publication/347102999_Conservation_value_of_national_forest_roadless_areas.

72. Wyoming v. US Dept. of Agriculture, 661 F.3d 1230 (10th Cir. 2011), reversing, 570 F.Supp.2d 1309 (D. Wyo. 2008); Kootenai Tribe of Idaho v. Veneman, 313 F.3d 1094 (9th Cir. 2002), reversing 2001 WL 1141275 (D. Idaho, 2001).

73. 36 C.F.R. 294; Special Areas; Roadless Area Conservation; Applicability to the National Forests in Idaho, 73 Fed. Reg. 61,456 (Oct. 16, 2008); Jayne v. Sherman, 706 F.3d 994 (9th Cir. 2013).

74. Robert B. Keiter, "The Emerging Law of Outdoor Recreation on the Public Lands," *Environmental Law* 51 (2021): 89, 110–13.

75. US Forest Service, *Custer Gallatin National Forest Record of Decision*, 11–12. Notably, the Stillwater mining complex in the Custer Gallatin outperforms recreation in terms of local economic output ($93 million versus $85 million), but lags in the number of local jobs (1,250 versus 2,900).

76. Katherine A. Zeller, "Experimental Recreationist Noise Alters Behavior and Space Use of Wildlife," *Current Science* 34, no. 1 (2024): 2297, https://www.sciencedirect.com/science/article/abs/pii/S0960982224006730; Todd Wilkinson, "Study: Noise from Recreationists Causes Wildlife to Flee for Cover," *Yellowstonian*, June 6

2024; https://yellowstonian.org/noise-from-recreationists-causes-wildlife-flee/; Keiter, "The Emerging Law of Outdoor Recreation," 135.

77. Greater Yellowstone Coalition v. Timchak, 2006 WL 3386731 (D. Idaho 2006); Cory Hatch, "Court Rejects Permit for Helicopter Skiing," *Jackson Hole News and Guide*, November 22, 2006, https://www.jhnewsandguide.com/news/environmental/court-rejects-permit-for-helicopter-skiing/article_4c986520-285f-542d-b818-441610718f1f.html.

78. Montana Wilderness Association v. McAllister, 666 F.3d 549, 558 (9th Cir. 2011), affirming, 658 F.Supp.2d 1249 (D. Mont. 2009).

79. James T. Sylvester, *Montana Recreational Off-Highway Vehicles: Fuel-Use and Spending Patterns 2013* (Missoula: University of Montana, Bureau of Business and Economic Research, 2014), 2, https://www.bber.umt.edu/pubs/survey/MontanaOHVStudy2013.pdf.

80. Executive Order No. 11644, 37 Fed. Reg. 2877 (Feb. 9, 1972); Executive Order No. 11989, 42 Fed. Reg. 26959 (May 24, 1977).

81. WildEarth Guardians v. Montana Snowmobile Association, 790 F.3d 920 (9th Cir. 2015).

82. Keiter, "The Emerging Law of Outdoor Recreation," 125–27.

83. US Forest Service, *East Crazy Inspiration Divide Land Exchange, Draft Decision Notice and Finding of No Significant Impact* (September 2023), https://usfs-public.app.box.com/v/PinyonPublic/file/1318433898010; Amanda Eggert, "Forest Service Tentatively Approves Crazy Mountain Land Swap," *Montana Free Press*, September 27, 2023, https://montanafreepress.org/2023/09/27/forest-service-tentatively-approves-crazy-mountain-land-swap/; David Tucker, "Am I Taking Crazy Pills," *Mountain Journal*, March 30, 2024, https://mountainjournal.org/does-east-crazy-inspiration-divideland-exchange-need-more-scrutiny. National forest land exchanges are governed by the Federal Public Land Policy and Management Act, 43 U.S.C. §§ 1716, 1717 and related statues.

84. Ray Ring, "Armies of Skiers Are Coming to Yellowstone," *High Country News*, March 31, 1997, https://www.hcn.org/issues/101/3131; see chap. 6 for more on private land development in Teton Valley, Idaho.

85. Grand Teton Resort, *2018 Master Development Plan*, https://img1.wsimg.com/blobby/go/14a06f42-6140-4d0f-b9ee-e7b3e655ebcd/downloads/2018%20MDP.pdf?ver=1638226592298; Mike Koshmrl, "National Forest Mulls Keeping Teton Ski Resort Out of Sheep Range," *WyoFile*, January 13, 2022, https://wyofile.com/national-forest-mulls-keeping-teton-ski-resort-out-of-sheep-range/; Billy Arnold, "In Managing South Bowl Access, Grand Targhee Faces Backcountry Backlash," *Jackson Hole News and Guide*, December 21, 2022, https://www.jhnewsandguide.com/news/environmental/in-managing-south-bowl-access-grand-targhee-faces-backcountry-backlash/article_c35e9d63-475a-5d8b-907e-fe6e4448778c.html; Billy Arnold, "Idaho Officials Take Targhee to Task Over Development," *Jackson Hole News and Guide*, January 25, 2023, https://www.jhnewsandguide.com/news/environmental/idaho-officials-take-targhee-to-task-over-development/article_7b1aa517-8c98-5503-b9ce-de452d030f30.html.

86. Billy Arnold, "Jackson Hole Mountain Resort Wants Rock Springs, Green

River Canyons," *Jackson Hole News and Guide*, February 23, 2023, https://www.jhnewsandguide.com/news/environmental/jackson-hole-mountain-resort-wants-rock-springs-green-river-canyons/article_85c8237a-49b4-5e56-ae45-003fc261a5bc.html.

87. Ray Ring, "Big Sky, Big Mess in Montana," *High Country News*, March 31, 1997, https://www.hcn.org/issues/101/3127.

88. Boyne Resorts, "Vision for the Future of Big Sky Resort," Big Sky Montana, 2024, https://bigskyresort.com/2025; Ring, "Big Sky, Big Mess in Montana"; Wilkinson, "Unbroken Wilderness."; Todd Wilkinson, "The Spillover Effects of Big Sky's Ravenous Appetite for More," *Yellowstonian*, June 3, 2024, https://yellowstonian.org/spillover-effects-of-big-skys-ravenous-appetite/.

Chapter Six

1. Albert Harting and Dennis Glick, *Sustaining Greater Yellowstone: A Blueprint for the Future* (Bozeman, MT: Greater Yellowstone Coalition, 1994), 92–108; Arthur Middleton et al., "The Role of Private Lands in Conserving Yellowstone's Wildlife in the Twenty-First Century," *Wyoming Law Review* 22 (2022): 237.

2. Laura C. Gigliotti et al., "Wildlife Migrations Highlight Importance of Both Private Lands and Protected Areas in the Greater Yellowstone Ecosystem," *Biological Conservation* 275 (2022): 109752; Reed Noss, George Wuerthner, Ken Vance-Borland, and Carlos Carroll, *A Biological Conservation Assessment for the Greater Yellowstone Ecosystem* (Corvallis, OR: Conservation Science, 2001); Charles C. Schwartz, Patricia H. Gude, Lisa Landenburger, Mark A. Haroldson, and Shannon Podruzny, "Impacts of Rural Development on Yellowstone Wildlife; Linking Grizzly Bear *Ursus Arctos* Demographics with Projected Residential Growth," *Wildlife Biology* 18 (2012): 246; Eric K. Cole et al., "Changing Migratory Patterns in the Jackson Elk Herd," *Wildlife Management* 79 (2015): 887; US Geological Survey, ScienceBase-Catalogue, "Heart of the Rockies Initiative Conservation Atlas," last modified September 24, 2019, https://www.sciencebase.gov/catalog/item/5ba93ecae4b08583a5ca0990.

3. Andrew J. Hansen and Linda Phillips, "Trends in Vital Signs for Greater Yellowstone: Application of a Wildland Health Index," *Ecosphere* 9 (2018): 1, 11, 17–18; Patricia H. Gude, Andrew J. Hansen, Ray Rasker, and Bruce Maxwell, "Rates and Drivers of Rural Residential Development in the Greater Yellowstone," *Landscape and Urban Planning* 77 (2006): 131, 146.

4. Ray Rasker, Patricia H. Gude, and Mark Delorey, "The Effect of Protected Federal Lands on Economic Prosperity in the Non-Metropolitan West," *Journal of Regional Analysis and Policy* 43, no. 2 (2013): 110; Claire Cella, "An Economic Crossroads: Technology Has Broken the Barrier of Geography that Shielded Mountain Towns from Growth," *Mountain Outlaw* (2018), https://www.mtoutlaw.com/an-economic-crossroads/; Todd Wilkinson, *Ripple Effects: How to Save Yellowstone and America's Most Iconic Wildlife Ecosystem* (Deadwood, OR: Wyatt-MacKenzie Publishing, 2022), 159–88.

5. Hansen and Phillips, "Trends in Vital Signs for Greater Yellowstone"; Wilkin-

son, *Ripple Effects*, 177, 180; Gude et al., "Rates and Drivers," 146; Andre J. Hansen et al., "Ecological Causes and Consequences of Demographic Change in the New West," *BioScience* 52 (2002): 151.

6. Jackson/Teton County Affordable Housing, *2022 Jackson and Teton County Annual Housing Supply Plan*, 2, https://www.tetoncountywy.gov/DocumentCenter/View/22105/April-2022-Draft-HSP?bidId=; Mike Koshmrl, "Jackson Hole's Wealth Is Displacing Star Valley's Neediest," *WyoFile*, April 28, 2022; Amanda Eggert, "A Road Runs Through It," *Montana Free Press*, January 19, 2021, https://montanafreepress.org/2022/01/19/highway-191-wildlife-crossings/; Rob Sisson, "Waiting for Elk to Disappear from 'The Last Hundred Acres,'" *Mountain Journal*, February 23, 2021; Todd Wilkinson, "Is High Flying Bozeman, Montana Losing the Nature of Its Place?," *Mountain Journal*, November 24, 2020, https://mountainjournal.org/will-human-population-growth-destroy-the-american-serengeti. See also chapter 5 herein.

7. Angus M. Thuermer Jr., "Counties Grapple with Zoning, Liberties, as More Seek Wyo Hideouts," *WyoFile*, December 21, 2021, https://wyofile.com/counties-grapple-with-zoning-liberties-as-more-seek-wyo-hideouts/.

8. Caitlin Tan, "Sublette County Says 'No' to Further Expansion of Multi-Billionaire's Resort in Bondurant," *Wyoming Public Radio*, March 9, 2023, https://www.wyomingpublicmedia.org/natural-resources-energy/2023-03-09/sublette-county-says-no-to-further-expansion-of-multi-billionaires-resort-in-bondurant; Mike Koshmrl, "Bondurant Billionaire Buys 'Elite Enclave' up Greys River, Eyes Land Swap," *WyoFile*, February 24, 2023, https://wyofile.com/bondurant-billionaire-buys-elite-enclave-up-greys-river-eyes-land-swap/; Mike Koshmrl, "Spooked by Price Tag, Billionaire Cans Under-Construction Luxury Resort in Bondurant Area," *WyoFile*, July 9, 2024, https://wyofile.com/spooked-by-price-tag-billionaire-cans-under-construction-luxury-resort-in-bondurant-area/; Mike Koshmrl, "Development, Wildlife Collide Along Iconic Wyo Migration Paths," *WyoFile*, October 13, 2022, https://wyofile.com/development-wildlife-collide-along-iconic-wyo-migration-paths/.

9. William R. Travis, *Ranchland Dynamics in the Greater Yellowstone Ecosystem: A Report to Yellowstone Heritage* (Boulder, CO: Center of the American West, 2003); Hannah Gosnell, Julia Hobson Haggerty, and William R. Travis, "Ranchland Ownership Change in the Greater Yellowstone Ecosystem, 1990–2001: Implications for Conservation," *Society and Natural Resources* 19, no. 8 (2006): 743.

10. Kathleen Epstein, Julia H. Haggerty, and Hannah Gosnell, "With, Not for, Money: Ranch Management Trajectories of the Super-Rich in Greater Yellowstone," *Annals of the American Association of Geographers* 112, no. 2 (2022): 432–48; Todd Wilkinson, *Last Stand: Ted Turner's Quest to Save a Troubled Planet* (Lanham, MD: Lyons Press, 2013), 26–28, 45, 154–57.

11. US Forest Service, *2006 Forest Plan Amendment for Grizzly Bear Habitat Conservation for the Greater Yellowstone Area National Forest ROD* (2006), https://www.fs.usda.gov/Internet/FSE_DOCUMENTS/stelprdb5187774.pdf; John D. Leshy and Molly S. McCusic, "Where's the Beef: Facilitating Voluntary Retirements of Federal Land from Livestock Grazing," *New York University Environmental Law Review* 17 (2008): 368, 370.

12. US Geological Survey et al., *Yellowstone Grizzly Bear Investigations 2020: Annual Report of the Interagency Grizzly Bear Study Team* (2021), 103, https://www.usgs.gov/centers/norock/science/igbst-annual-reports; US Forest Service, US Department of Agriculture, *Shoshone National Forest Plan Final Environmental Impact Statement* (2015), 416; Andrew Pils and James Wilder, *Risk Analysis of Disease Transmission between Domestic Sheep and Goats and Rocky Mountain Bighorn Sheep* (Washington, DC: US Forest Service, US Department of Agriculture, 2018), https://www.fs.usda.gov/Internet/FSE_DOCUMENTS/fseprd592628.pdf.

13. National Academy of Sciences, *Revisiting Brucellosis in the Greater Yellowstone Area* (2020), 2, 10–11, 48–51; Robert B. Keiter, "The Greater Yellowstone Ecosystem Revisited: Law, Science, and the Pursuit of Ecosystem Management in an Iconic Landscape," *University of Colorado Law Review* 91 (2020): 142–43. See also Shawn Regan, Temple Stoellinger, and Jonathan Wood, "Opening the Range: Reforms to Allow Markets for Voluntary Conservation on Federal Grazing Lands," *Utah Law Review* (2023): 197.

14. Center for Biological Diversity v. Haaland, 603 F.Supp.3d 1094 (D. Wyo. 2022), reversed in part, 69 F.4th 689 (10th Cir. 2023); Mike Koshmrl, "Grizzlies, Environmentalists Win Toothless Victory in Court of Appeals," *WyoFile*, June 7, 2023, https://wyofile.com/grizzlies-environmentalists-win-toothless-victory-in-court-of-appeals/; US Forest Service, *Decision Notice, Finding of No Significant Impact, East Paradise Range Allotment Management Plan* (2021), https://www.fs.usda.gov/project/?project=41485; Amanda Eggert, "Conservationists Allege Yellowstone-Area Grazing Plan Threatens Grizzly Recovery," *Montana Free Press*, September 12, 2022, https://montanafreepress.org/2022/09/12/conservationists-allege-yellowstone-area-grazing-plan-threatens-grizzly-recovery/.

15. 16 U.S.C. § 1538(a)(1)(B) (2022); Babbitt v. Sweet Home Chapter, 515 US 687 (1995); Christy v. Hodel, 557 F.2d 1324 (9th Cir. 1988); United States v. McKittrick, 142 F.3d 1170 (9th Cir. 1998); 33 U.S.C. 1344 (2022); Greater Yellowstone Coalition v. Flowers, 359 F.3d 1257 (10th Cir. 2004); Robert B. Keiter, "The Old Faithful Protection Act: Congress, National Park Ecosystems, and Private Property Rights," *Public Land Law Review* 14 (1993): 5; Montana Code Annotated § 85-20-401 (1993).

16. Land and Water Conservation Act, 54 U.S.C. § 200301–200310 (2022); Great American Outdoors Act, 54 U.S.C. § 200401–200402 (2022); Healthy Forests Restoration Act, 16 U.S.C. 6513, 6518 (2022); Robert B. Keiter, "The Law of Fire Reshaping Public Land Policy in an Era of Ecology and Litigation," *Environmental. Law* 36 (2006): 301, 344–60.

17. Middleton et al., "The Role of Private Lands," 288–93 (describing federal conservation funding programs); Robert Bonnie, "The Importance of Working Lands to Yellowstone in the 21st Century," keynote address at the University of Wyoming's 150th anniversary of the Yellowstone Symposium, Cody, WY, May 20, 2022, https://westernlandowners.org/the-importance-of-private-working-lands-to-yellowstone-in-the-twenty-first-century/; Memorandum of Understanding between the US Department of Agriculture and State of Wyoming, *A Partnership to Preserve Big Game Habitat in Wyoming* (October 17, 2022), https://www.usda.gov

/sites/default/files/documents/mou-vilsack-gordon.pdf; Farm Service Agency, US Department of Agriculture, "USDA Commits to Big Game Conservation Partnership with the State of Wyoming: Initial Investments Forthcoming," news release, May 20, 2022, https://www.fsa.usda.gov/news-room/news-releases/2022/usda-commits-to-big-game-conservation-partnership-with-the-state-of-wyoming-initial-investments-forthcoming.

18. Mike Koshmrl, "Study Reveals Yellowstone Elk Reliance on Unprotected Private Land," *WyoFile*, November 29, 2022, https://wyofile.com/study-reveals-yellowstone-elk-reliance-on-unprotected-private-land/; Keiter, "Greater Yellowstone Ecosystem Revisited," 146.

19. Wyoming Statutes Annotated §§ 18-5-201, 202, 301, §§ 15-1-502, 505(1), 602 (2022); Montana Code Annotated §§ 76-2-101, 201, 301, 76-3-501 (2022); Idaho Code §§ 67–6503, 6504, 6511, 6513 (2022); Craig L. Shafer, "Land Use Planning: A Potential Force for Retaining Habitat Connectivity in Greater Yellowstone Ecosystem and Beyond," *Global Ecology and Conservation* 3 (2015): 256.

20. Montana Code Annotated § 76-1-601(2)(4) (2022); Michelle Bryan Mudd, DarAnn Dunning, and Melissa Hayes, "The Role of Fish and Wildlife Evidence in Local Land Use Regulation," *Public Land and Resources Law Review* 30 (2009): 107; Jackson/Teton County Comprehensive Plan, *Jackson and Teton County Communities* ES-2 (2012); Board of County Commissioners v. Crow, 65 P.3d 720 (Wyo. 2003); Greater Yellowstone Coalition v. Board of County Commissioners, 25 P.3d 168 (Mont. 2001); Keiter, "Greater Yellowstone Ecosystem Revisited," 147 n. 640.

21. David Tucker, "Mapping Our Values," *Mountain Journal*, October 20, 2023, https://mountainjournal.org/mapping-our-values-gallatin-county-sensitive-lands-plan.

22. *Gallatin Valley Sensitive Lands Protection Plan* (Adoption Draft, November 2023), https://gallatinvalleyplan.bozeman.net/.

23. Tucker, "Mapping Our Values."

24. Rees Consulting, *Western Greater Yellowstone Area Housing Needs Assessment* (2014); Darby Minow Smith, "Montana Has a Habitat Problem," *Sierra*, March 14, 2023, https://www.sierraclub.org/sierra/1-spring/feature/montana-has-habitat-problem; Nick Bowlin, "Bozeman's Boom Depends on Immigrants but Struggles to Support Them," *High Country News*, May 2024, 27–33, https://www.hcn.org/issues/56-5/bozemans-boom-depends-on-immigrants-but-struggles-to-support-them/.

25. Wyoming Statutes Annotated § 18-5-401 et seq. (2022); Jen Kocher, "Teton County's Affordable Housing Crisis Is Case Study for State," *Cowboy State Daily*, August 28, 2022, https://cowboystatedaily.com/2022/08/26/teton-countys-affordable-housing-crisis-is-case-study-for-state/; Hannah Merzbach, "Will Wyoming Erode Local Control Over Teton County's Housing Policy," *Jackson Hole Community Radio*, November 6, 2023, https://891khol.org/will-wyoming-erode-local-control-over-teton-countys-housing-policy/; Clair McFarland, "Supreme Court Could Force Jackson to Rethink $31,000 in Gov't Fees for Home Building," *Cowboy State Daily*, January 12, 2024, https://cowboystatedaily.com/2024/01/13/supreme-court-could-force-jackson-to-rethink-31-000-in-fees-for-home-building/; author's interview with Luther Propst, Teton County Commissioner.

26. Author's interviews with Anna Trentadue, Valley Advocates for Responsible Development; Michael Whitfield, Teton County, Idaho, Commissioner; Cathy Rinaldi, Greater Yellowstone Coalition Idaho Director and former Teton County, Idaho, Commissioner.

27. Teton County, Idaho, *Comprehensive Plan: A Vision and Framework, 2012–2030* (2012), 1–9, https://portal.laserfiche.com/Portal/DocView.aspx?id=142889&repo=r-e076e25b.

28. Nancy Keates, "Why Driggs Could Become the Next Jackson Hole," *Wall Street Journal*, September 21, 2023.

29. Connor Shea," High Noon Development Gets Reality Check from County P&Z," *Teton Valley News*, April 15, 2023, https://www.tetonvalleynews.net/businesses/high-noon-development-gets-reality-check-from-county-p-z/article_3cbcab00-dce4-11ed-be5b-7fbfa42de1f1.html; Jeanne Boner, "Jackson Hole Dinner Tees Up Dude Ranch," *Jackson Hole News and Guide*, April 11, 2023.

30. Hevenn Vanh, "Park County Voters Mobilize to Preserve County's Growth Policy They Say Is Threatened by Referendum 1," *KBZK*, March 6, 2024, https://www.kbzk.com/news/local-news/park-county-voters-mobilize-to-preserve-countys-growth-policy-they-say-is-threatened-by-referendum-1; Sean Batura, "County Growth Policy Repeal Fails, Unofficial Election Results Indicate," *Livingston Enterprise*, June 5, 2024, https://www.livingstonenterprise.com/news/county-growth-policy-repeal-fails-unofficial-election-results-indicate/article_deab53be-2331-11ef-bad0-3fb826b55f40.html; Keiter, "Greater Yellowstone Ecosystem Revisited," 147.

31. These estimates of GYE conservation easements and acreage derive from several sources: Keiter, "Greater Yellowstone Ecosystem Revisited," 148 n. 641 (assuming growth in these 2020 figures); additional research by my research assistant who puts the figure at 1.3 million acres (available from author); W. Andrew Marcus, James E. Meacham, Ann W. Rodman, Alethea Steingisser, and Justin T. Menke, *Atlas of Yellowstone*, 2nd ed., ed. Ross West (Oakland: University of California Press, 2022), 16–17 (maps depicting GYE conservation easements); US Forest Service and Ruckelshaus Institute, *Private Lands Conservation Toolkit and Training for Wyoming Land Managers* (2011), 18–19; Nancy A. McLaughlin, "Internal Revenue Code Section 170(h): National Perpetuity Standards for Federally Subsidized Conservation Easements," *Real Property, Trusts and Estates Law Journal* 45 (2011): 473.

32. Idaho Code § 55–2101 (2022); Montana Code Annotated § 76-6-206 (2022); Wyoming Statutes Annotated § 34-1-201 (2022); Jesse J. Richardson and Amanda C. Bernard, "Zoning for Conservation Easements," *Law and Contemporary Problems* 7 (2011): 83; Four B Properties, LLC v. The Nature Conservancy, 458 P.3d 832 (Wyo. 2020); see also Bullock v. Fox, 435 P.3d 1187 (Mont. 2019); Scott v. Lee and Donna Metcalf Charitable Trust, 358 P. 3d 879 (2015).

33. Nancy A. McLaughlin, "Rethinking the Perpetual Nature of Conservation Easements," *Harvard Environmental Law Review* 29 (2005): 421; Hicks v. Dowd, 157 P.3d 914 (Wyo. 2007).

34. Angus M. Thuermer Jr., "Ranch Owner Builds in the Path of Pronghorn Migration," *High Country News*, January 10, 2017, https://www.hcn.org/articles/ranch-owner-builds-in-the-path-of-pronghorn-migration.

35. Angus M. Thuermer Jr., "Cabin Removed from Path of the Pronghorn," *WyoFile*, July 18, 2017, https://wyofile.com/cabin-removed-path-pronghorn/.

36. Farm Service Agency, "Grassland CRP," https://www.fsa.usda.gov/programs-and-services/conservation-programs/crp-grasslands/index; Barb Gorges, "Habitat Leasing to Provide New Tool for Wyoming Conservation," *Wyoming Tribune Eagle*, February 3, 2023, https://www.wyomingnews.com/features/outdoors/habitat-leasing-to-provide-new-tool-for-wyoming-conservation/article;_151362f2-a0ed-11ed-a6d4-c750dc346cad.html; Helena Dore, "Agreement Sets Part of Paradise Valley Ranch Aside for Elk Winter Range," *Bozeman Chronicle*, December 8, 2021, https://www.bozemandailychronicle.com/news/environment/agreement-sets-part-of-paradise-valley-ranch-aside-for-elk-winter-range/article_fae0ca5b-4c73-5dc7-a610-f0520a30e8d0.html; Property and Environment Research Center, *Elk Occupancy Agreements*, Property and Environment Research Center (PERC), 2024, https://www.perc.org/elk-occupancy-agreements/.

37. Amanda Eggert, "Forever No More? Bill Seeks to Restrict State's Access to Perpetual Conservation Easements," *Montana Free Press*, February 22, 2023, https://montanafreepress.org/2023/02/22/legislature-conservation-easement-change/; Tom Kuglin, "Bill Eliminating FWP Conservation Easements Sees Pushback," *Helena Independent Record*, February 22, 2023, https://helenair.com/news/state-and-regional/govt-and-politics/bill-eliminating-perpetual-fwp-conservation-easements-sees-pushback/article_b5a93d21-d09c-5ca8-a71b-0b219a5d566a.html. "How Weed Revenue Went Down to the Wire—Again," *Montana Free Press*, May 8, 2023, https://montanafreepress.org/2023/05/08/how-weed-revenue-went-down-to-the-wire-again/; Blair Miller, "Montana Lawmakers Override Governor's Veto of Appropriations Bill Tied to Marijuana Revenue Measure," *Daily Montanan*, June 23, 2023, https://www.ktvq.com/news/68th-session/montana-lawmakers-override-governors-veto-of-appropriations-bill-tied-to-marijuana-revenue-measure.

38. Sarah Jane Keller, "Carnivores Not Condos: Ranches Provide Key Wildlife Passages Between Two Ecosystems," *Western Confluence* (2016), 11, https://westernconfluence.org/carnivores-not-condos/; Matthew McKinney, Lynn Scarlett, and Daniel Kemmis, *Large Landscape Conservation: A Strategic Framework for Policy and Action* (Cambridge, MA: Lincoln Institute of Land Policy, 2010); Sean Finn et al., *Recommended Practices for Landscape Conservation Design* (September 2018), 19–22, https://lccnetwork.org/sites/default/files/Resources/LCD-Recommended-Practices-v1-092818.pdf; Robert F. Baldwin et al., "The Future of Landscape Conservation," *BioScience* 68 (2018): 60.

39. High Divide Collaborative, https://highdivide.org/.

40. Headwaters Economics, *High Divide Region—Summary of Recreation Economy* (2014), https://headwaterseconomics.org/wp-content/uploads/High_Divide_Outdoor_Rec_Economy.pdf; Headwater Economics, *Home Construction in the High Divide* (2015), https://headwaterseconomics.org/economic-development/local-studies/high-divide/; Bray J. Beltran, *The High Divide Collaborative: Landscape-Scale Conservation Through Community-Based Collaboration* (Missoula, MT: Heart of the Rockies Initiative, 2017), https://alliancerally.org/wp-content/uploads/2017/05

/Rally2017_B18_Slides.pdf; Mountain Journal, "Ties Uniting People, Communities, and Nature," *Mountain Journal*, November 6, 2020, https://mountainjournal.org/high-divide-is-lens-for-thinking-about-struggles-of-rural-west.

41. For more detailed information about the High Divide Collaborative, see https://highdivide.org/,

42. Bob Freimark, "Bridging the Divide: Yellowstone to Yukon," Wilburforce Foundation, n.d., https://wilburforce.org/story/bridging-the-divide/; Rose A. Graves, Matthew A. Williamson, R. Travis Belote, and Jodi S. Brandt, "Quantifying the Contribution of Conservation Easements to Large-Landscape Conservation," *Biological Conservation* 232 (2019): 83; Abigail H. Sage, Vicken Hillis, Rose A. Graves, Morey Burnham, and Neil H. Carter, "Paths of Coexistence: Spatially Predicting Acceptance of Grizzly Bears Along Key Movement Corridors," *Biological Conservation* 266, no. 8 (2022): 109468; Karen Schumacher, "Agenda 21/2030: Collaborative Threats," *Redoubt News*, October 24, 2019, https://redoubtnews.com/2019/10/agenda-21-2030-collaborative-threats/; Karen Schumacher, "Y2Y: The Truth Is Out—Where Do We Go from Here?," *Redoubt News*, June 12, 2017, https://redoubtnews.com/2017/06/y2y-truth-where-from-here/.

43. Matthew J. Kauffman, James E. Meacham, Hall Sawyer, Alethea Y. Steingisser, William J. Rudd, and Emiline Ostlind, *Wild Migrations: Atlas of Wyoming's Ungulates* (Corvallis: Oregon State University Press, 2018), 142–43.

Chapter Seven

1. Samuel P. Hays, *Beauty, Health, and Permanence: Environmental Politics in the United States, 1955–1985* (New York: Cambridge University Press, 1987); Robert B. Keiter, *Keeping Faith with Nature: Ecosystems, Democracy, and America's Public Lands* (New Haven, CT: Yale University Press, 2003), 48–78.

2. Daniel B. Botkin, *Discordant Harmonies: A New Ecology for the Twenty-First Century* (New York: Oxford University Press, 1990); Jodi A. Hilty, Charles C. Chester, and Molly S. Cross, eds., *Climate and Conservation: Landscape and Seascape Science, Planning, and Action* (Washington, DC: Island Press, 2012); Keiter, *Keeping Faith with Nature*, 48–78, 186–95, 258–72.

3. Richard West Sellars, *Preserving Nature in the National Parks: A History* (New Haven, CT: Yale University Press, 1997), 214–17, 246–66; Robert B. Keiter, *To Conserve Unimpaired: The Evolution of the National Park Idea* (Washington, DC: Island Press, 2013), 148–52, 179–81.

4. See chap. 2 herein; Robert B. Keiter, "The Greater Yellowstone Ecosystem Revisited: Law, Science, and the Pursuit of Ecosystem Management in an Iconic Landscape," *University of Colorado Law Review* 91 (2020): 1, 35–48.

5. See chaps. 3 and 4 herein.

6. See chap. 3 herein; Robert B. Keiter, "Grizzlies, Wolves, and Law in the Greater Yellowstone Ecosystem: Wildlife Management Amidst Jurisdiction Complexity and Tension," *Wyoming Law Review* 22 (2022): 303, 304–53.

7. See chap. 4 herein; Keiter, "Greater Yellowstone Ecosystem Revisited," 70–80; Arthur Middleton et al., "The Role of Private Lands in Conserving Yellow-

stone's Wildlife in the Twenty-First Century," *Wyoming Law Review* 22 (2022): 237, 239–42, 254–60.

8. See chap. 5 herein; Keiter, "Greater Yellowstone Ecosystem Revisited," 21–26, 96–137, 140–43.

9. See chap. 5 herein; Keiter, "Greater Yellowstone Ecosystem Revisited," 124–37; US Forest Service, US Department of Agriculture, National Forest System Land Management Planning, 36 C.F.R., pt. 219 (2023); 77 Fed. Reg. 21162 (April 9, 2012); Todd Wilkinson, *Ripple Effects: How to Save Yellowstone and America's Most Iconic Wildlife Ecosystem* (Deadwood, OR: Wyatt McKenzie Publishers, 2022), 189–214.

10. Wilkinson, *Ripple Effects*, 159–88; Middleton et al., "Role of Private Lands," 240–60.

11. See chap. 6 herein; Middleton et al., "Role of Private Lands," 273–301; Keiter, "Greater Yellowstone Ecosystem Revisited," 137–54.

12. Justin Farrell, *Billionaire Wilderness: The Ultra-Wealthy and the Remaking of the American West* (Princeton, NJ: Princeton University Press, 2020), 247–50; Nick Bowlin, "Bozeman's Boom Depends on Immigrants but Struggles to Support Them," *High Country News*, May 1, 2024, https://www.hcn.org/issues/56-5/bozemans-boom-depends-on-immigrants-but-struggles-to-support-them/.

13. Justin Farrell, *The Battle for Yellowstone: Morality and the Sacred Roots of Environmental Conflict* (Princeton, NJ: Princeton University Press, 2015), 60–65; Paul Schullery, *Searching for Yellowstone: Ecology and Wonder in the Last Wilderness* (New York: Houghton Mifflin, 1997), 193–216.

14. See chap. 5 herein; Keiter, "Greater Yellowstone Ecosystem Revisited," 96–124, 155–56; Farrell, *Battle for Yellowstone*, 217–57.

15. See chap. 6 herein; Keiter, "Greater Yellowstone Ecosystem Revisited," 124–37.

16. See chaps. 3, 4, 6 herein; Keiter, "Greater Yellowstone Ecosystem Revisited," 48–96, 146–54; Middleton et al., "Role of Private Lands," 274–96.

17. Schullery, *Searching for Yellowstone*, 205; Keiter, "Greater Yellowstone Ecosystem Revisited," 155–60; see also Susan G. Clark, *Ensuring Greater Yellowstone's Future: Choices for Leaders and Citizens* (New Haven, CT: Yale University Press, 2008), 128–38 (reviewing the GYCC subcommittees).

18. These interstate differences over GYE issues are discussed in chapters 4, 5, and 6 herein; Keiter, "Grizzlies, Wolves, and Law," 331–32, 340–44.

19. Farrell, *Battle for Yellowstone*, 60–65; Schullery, *Searching for Yellowstone*, 193–216; Keiter, "Greater Yellowstone Ecosystem Revisited," 154–60, 175–77.

20. See chap. 1 herein; Farrell, *Battle for Yellowstone*, 100–105; Clark, *Ensuring Greater Yellowstone's Future*, 33–36.

21. For an overview of the role science plays in ecological conservation with references see Keiter, "Greater Yellowstone Ecosystem Revisited," 31–35; for examples of the role of science in the GYE see Robert B. Keiter and Mark S. Boyce, eds., *The Greater Yellowstone Ecosystem: Redefining America's Wilderness Heritage* (New Haven, CT: Yale University Press, 1991), parts 2, 3, and 4; Sellars, *Preserving Nature* (recounting the history of science in the National Park Service); Greater

Yellowstone Coalition v. Servheen, 665 F.3d 1015 (9th Cir. 2011); Crow Indian Tribe v. United States of America, 965 F.3d 662 (9th Cir. 2020).

22. James R. Skillen, *Federal Ecosystem Management: Its Rise, Fall, and Afterlife* (Lawrence: University Press of Kansas, 2015), 76–80, 142–60; Robert L. Fischman and Vicky J. Meretsky, "Managing Biological Integrity, Diversity, and Environmental Health in the National Wildlife Refuges: An Introduction to the Symposium," *Natural Resources Journal* 44 (2004): 931, 932–33.

23. See chap. 1 herein; Farrell, *Battle for Yellowstone*, 66–118; Keiter, "Greater Yellowstone Ecosystem Revisited," 13–21.

24. Keiter, "Greater Yellowstone Ecosystem Revisited," 160–69; Skillen, *Federal Ecosystem Management*, 65–76.

25. Endangered Species Act, 16 U.S.C. §§ 1531–43; see chaps. 3 and 5 herein; Keiter, "Grizzlies, Wolves, and Law," 320–332; Keiter, "Greater Yellowstone Ecosystem Revisited," 49–60, 161.

26. 16 U.S.C. § 1539(j); see chap. 3 herein; Keiter, "Grizzlies, Wolves, and Law," 332–44; Keiter, "Greater Yellowstone Ecosystem Revisited," 61–70, 161–62.

27. National Environmental Policy Act, 42 U.S.C. §§ 4321–61; National Parks Organic Act, 54 U.S.C. § 100101; Wilderness Act, 16 U.S.C. § 1133; National Wildlife Refuge Administration Act, 16 U.S.C. § 668dd(a)(4)(B); see chaps. 2, 4, and 5 herein; Keiter, "Greater Yellowstone Ecosystem Revisited," 152–64.

28. Clark, *Ensuring Greater Yellowstone's Future*, 40–42; Keiter, "Greater Yellowstone Ecosystem Revisited," 27–28, 173; 16 U.S.C. § 1604((a) (NFMA); 42 U.S.C. § 4332(D)(iv) (NEPA); National Park Service, *Management Policies 2006* (Washington, DC: National Park Service, US Department of the Interior, 2006), 13–14; 16 U.S.C. § 666dd(a)(4)(M) (FWS, NWRAA); 43 U.S.C. § 1712(c)(9) (BLM, FLPMA).

29. Martin Nie et al., "Fish and Wildlife Management on Federal Land: Debunking State Supremacy," *Environmental Law* 47 (2017): 797, 808–14; Montana Environmental Policy Act, Montana Code § 75-1-101 et seq.; see chap. 6 herein.

30. See chaps. 4, 5, and 6 herein for discussion of these cases.

31. US Constitution, art. IV, sec. 3, cl. 2 (Property Clause); Kleppe v. New Mexico, 426 US 529 (1976); see chaps. 3, 5 herein for discussion of the noted congressional legislation.

32. See chap. 5 herein.

33. See chaps. 2, 3, 4, and 5 herein.

34. See chaps. 3, 4, and 5 herein.

35. See chap. 4 herein.

36. Steven Hostetler et al., *Greater Yellowstone Climate Assessment: Past, Present, and Future Climate Change in Greater Yellowstone Watersheds* (Bozeman: Montana State University Institute on Ecosystems, 2021); US Fish and Wildlife Service, "Endangered and Threatened Wildlife and Plants; Threatened Species Status with Section 4(d) Rule for Whitebark Pine (*Pinus albicaulis*)," 87 Fed. Reg. 76882 (Dec. 15, 2022); Greater Yellowstone Coalition v. Servheen, 665 F.3d 1015 (9th Cir. 2011). See also Intergovernmental Panel on Climate Change, *Climate Change 2023: Synthesis Report for Policy Makers* (Geneva, Switzerland, 2023).

37. Hostetler et al., *Greater Yellowstone Climate Assessment*; Wilkinson, *Ripple*

Effects, 223–51; Cathy Whitlock et al., *Montana Climate Assessment* (Bozeman and Missoula: Montana State University and University of Montana, Montana Institute on Ecosystems, 2017), https://montanaclimate.org/.

38. Hostetler et al., *Greater Yellowstone Climate Assessment*, XIX, 201–2. See also John D. Leshy, "Federal Lands in the Twenty-First Century," *Natural Resources Journal* 50 (2010): 111, 122–31.

39. Leshy, "Federal Lands in the Twenty-First Century," 127–30; Alejandro E. Camacho, "Assisted Migration: Redefining Nature and Natural Resources Law Under Climate Change," *Yale Journal of Regulation* 27 (2010): 171; Keiter, "Greater Yellowstone Ecosystem Revisited," 169–75; Robert B. Keiter, "Wildlife Conservation, Climate Change, and Ecosystem Management," in *The Laws of Nature: Reflections on the Evolution of Ecosystem Management Law and Policy*, ed. Kalyani Robbins (Akron, OH: University of Akron Press, 2013), 235–60.

40. See chap. 6 herein.

41. See chap. 6 herein; Middleton et al., "Role of Private Lands," 282–301.

42. Charles F. Wilkinson, *Crossing the Next Meridian: Land, Water, and the Future of the West* (Washington, DC: Island Press, 1992), 3–27; Robert B. Keiter, "The Emerging Law of Outdoor Recreation on the Public Lands," *Environmental Law* 51 (2021): 89, 104–10; Jim Robbins, "Pandemic Crowds Bring 'Rivergeddon' to Montana Rivers," *New York Times*, November 23, 2020, https://www.nytimes.com/2020/11/23/us/pandemic-montana-wilderness-rush.html; Wilkinson, *Ripple Effects*, 189–214; Keiter, "Greater Yellowstone Ecosystem Revisited," 21, 124–37; see chap. 5 herein.

43. See chaps. 2, 5 herein; Keiter, "Greater Yellowstone Ecosystem Revisited," 134–35; Keiter, "Emerging Law of Outdoor Recreation," 145–53.

44. See chaps. 2 and 5 herein; Keiter, "Greater Yellowstone Ecosystem Revisited," 124–36; Keiter, "Emerging Law of Outdoor Recreation," 153–58.

45. See chap. 3 herein; Keiter, "Grizzlies, Wolves, and Law," 320–52; Todd Wilkinson, "This 'Bearish' Economy Is One Most States Would Love to Have," *Mountain Journal*, July 14, 2023, https://mountainjournal.org/here-is-a-bearest-economy-most-american-states-would-die-to-have; Todd Wilkinson, "How Greater Yellowstone Grizzly Bears Could Be Delisted and Remain Protected," *Mountain Journal*, July 18, 2023, https://mountainjournal.org/should-grizzlies-be-delisted-and-then-protected-like-bald-eagles; Tribal Heritage and Grizzly Bear Protection Act, H.R. 2531 (116th Congress, 2019).

46. See chaps. 3 and 6 herein; Keiter, "Grizzlies, Wolves, and Law," 320–52; Keiter, "Greater Yellowstone Ecosystem Revisited," 140–43, 159–75; Daniel R. Stahler, Bridgett M. vonHoldt, Elizabeth Heppenheimer, and Robert K. Wayne, "Yellowstone Wolves at the Frontiers of Genetic Research," in *Yellowstone Wolves: Science and Discovery in the World's First National Park*, ed. Douglas W. Smith, Daniel R. Stahler, and Daniel R. MacNulty (Chicago: University of Chicago Press, 2020), 99–102; Property and Environment Research Center (PERC), Grizzly Conflict Reduction Grazing Agreement (2023), https://www.perc.org/innovation-lab/grizzly-conflict-reduction-grazing-agreement/; Isabel Hicks, "Partnership to Reduce Grizzly Conflicts in the Gravelly Range Shows Early Promise," *Bozeman Daily Chron-*

icle, August 16, 2023, https://www.bozemandailychronicle.com/news/environment/partnership-to-reduce-grizzly-conflicts-in-gravelly-range-shows-early-promise/article_7261c2b8-3c52-11ee-8b4c-8b9a466b0434.html.

47. See chap. 4 herein; National Academy of Sciences, *Revisiting Brucellosis in the Greater Yellowstone Area* (Washington DC: National Academies Press, 2020), 2, 10–11, 48–51; Jonathan D. Cook et al., *Evaluating Management Alternatives for Wyoming Elk Feedgrounds in Consideration of Chronic Wasting Disease* (Department of Interior, US Geological Survey, 2023); Mike Koshmrl, "Fewer Elk, Sicker Elk, That's What the Experts Expect If Wyoming Keeps on Feeding," *WyoFile*, December 14, 202), https://wyofile.com/fewer-elk-sicker-elk-thats-what-the-experts-expect-if-wyoming-keeps-on-feeding/; Wilkinson, *Ripple Effects*, 81–124; Keiter, "Greater Yellowstone Ecosystem Revisited," 70–86.

48. See chaps. 3, 4, and 6 herein.

49. Yellowstone to Yukon Conservation Initiative, https://y2y.net/; Douglas Chadwick, *Yellowstone to Yukon: Destinations Series* (Washington, DC: National Geographic Society, 2000); Northern Rockies Ecosystem Protection Act, S. 1531 (1118th Cong., 2023); George Wuerthner, "Northern Rockies Ecosystem Protection Act: Conservation on the Large Scale Needed," *Rewilding Earth*, May 11, 2021, https://rewilding.org/northern-rockies-ecosystem-protection-act-conservation-on-the-large-scale-needed/.

50. Crown Managers Partnership, *We Manage Lands Together*, https://www.crownmanagers.org/; Tony Prato and Daniel B. Fagre, "Sustainable Management of the Crown of the Continent Ecosystem," *George Wright Forum* 27 (2010): 77–93; Robert B. Keiter, "Toward a National Conservation Network Act: Transforming Landscape Conservation on the Public Lands Into Law," *Harvard Environmental Law Review* 42 (2018): 51, 114–16, 117–19.

51. US Department of Agriculture and US Department of Interior, *Record of Decision for Amendments to Forest Service and Bureau of Land Management Planning Documents within the Range of the Northern Spotted Owl* (1994); US Forest Service, *Sierra Nevada Forest Plan Amendment: Final Supplemental Environmental Impact Statement* (2004), https://www.fs.usda.gov/Internet/FSE_DOCUMENTS/fsbdev3_046095.pdf; US Forest Service, Pacific Southwest Region, *Sierra Nevada Forest Plan Amendment: Final Supplemental Environmental Impact Statement* (2013), https://lccn.loc.gov/2013487363. See also Skillen, *Federal Ecosystem Management*, 183–221; Keiter, *Keeping Faith with Nature*, 162–69.

52. Desert Managers Group, https://www.denix.osd.mil/dmg/; US Fish and Wildlife Service, "San Luis Valley National Wildlife Refuge Complex, CO; Availability of Record of Decision for the Final Comprehensive Conservation Plan and Final Environmental Impact Statement," 80 Fed. Reg. 76998 (Dec. 11, 2015); Dan L. Flores, *American Serengeti: The Last Big Animals of the Great Plains* (Lawrence: University Press of Kansas, 2016); see also Keiter, "Toward a National Conservation Network Act," 114–25.

53. Keiter, "Toward a National Conservation Network Act," 91–110; Keiter, "Greater Yellowstone Ecosystem Revisited," 21–27. See also chapter 1 herein.

54. Columbia River Gorge National Scenic Area Act, 16 U.S.C. § 544; Tahoe

Regional Planning Authority Act, Pub. L. No. 91-148, 83 Stat. 360 (1969); Gary Spradling, "Regional Government for Lake Tahoe," *Hastings Law Journal* 22 (1971): 705.

55. Keiter, "Toward a National Conservation Network Act," 127–36.

Conclusion

1. See chap. 1 herein.
2. See chaps. 2, 3, and 4 herein.
3. See chaps. 3, 4, and 5 herein.
4. See chaps. 3 and 4 herein.
5. See chap. 6 herein.
6. Robert W. Righter, *Crucible for Conservation: The Struggle for Grand Teton National Park* (Boulder: Colorado Associated University Press, 1982), 43–125; see chaps. 5 and 6 herein.
7. Secretary of the Interior, Order No. 3342, *Identifying Opportunities for Cooperative and Collaborative Partnerships with Federally Recognized Indian Tribes in the Management of Federal Lands and Resources* (October 21, 2016); US Department of the Interior and US Department of Agriculture, Order No. 3404, *Joint Secretarial Order on Fulfilling the Trust Responsibility to Indian Tribes in the Stewardship of Federal Lands and Waters* (November 15, 2021); Anna V. Smith, "What Does the Nation's Commitment to Tribal Co-Stewardship Mean for Public Lands?," *High Country News*, February 1, 2023; Joseph L. Sax, "Do Communities Have Rights? The National Parks as a Laboratory for New Ideas," *University of Pittsburgh Law Review* 45 (1984): 499; see chaps. 2 and 4 herein.
8. See chap. 6 herein; Property and Environment Research Center, "Welcome to Paradise," *PERC Reports* (December 15, 2023), https://www.perc.org/2023/12/15/welcome-to-paradise/; Joseph L. Sax, "Ecosystems and Property Rights in Greater Yellowstone: The Legal System in Transition," in *The Greater Yellowstone Ecosystem: Redefining America's Wilderness Heritage*, ed. Robert B. Keiter and Mark S. Boyce (New Haven, CT: Yale University Press, 1991), 77–84.
9. Robert B. Keiter, "Wildlife Conservation, Climate Change, and Ecosystem Management," in *The Laws of Nature: Reflections on the Evolution of Ecosystem Management Law and Policy*, ed. Kalyani Robbins (Akron: University of Akron Press, 2013), 235–60; see chap. 7 herein.
10. See chap. 6 herein.
11. See chaps. 3, 4, 5, and 6 herein.
12. See chaps. 1 and 3 herein.

INDEX

Page numbers in italics refer to figures. Page numbers in boldface and italics refer to maps.

Absaroka-Beartooth Front, 141
Absaroka-Beartooth Wilderness Area, 144, 148, 151
ACECs (Areas of Critical Environmental Concern), 3, 121
adaptive management, 50, 103, 213
Adopt-a-Wildlife-Acre program, 173
Agricultural Conservation Easements Program, 176
Alaska National Interest Lands Conservation Act (ANILCA), 28
Albright, Horace, 8
Alkali Creek feedground, 113–14
amenity properties and buyers, 21, 170–71, 175
American Prairie reserve, 102, 223
America the Beautiful initiative, 31, 223–24
Anderson, Jack, 53
ANILCA (Alaska National Interest Lands Conservation Act), 28
Animal and Plant Health Inspection Services (APHIS), 96–97, 98

animal cruelty, 88–89
animal unit months (AUMs), 172
antelope. *See* pronghorn
Anthropocene era, 11, 31, 42, 190, 225, 234
antigovernment sentiment, 123, 176, 179, 196
Antiquities Act, 29
APHIS (Animal and Plant Health Inspection Services), 96–97, 98
Arapahoe tribe, 17
Areas of Critical Environmental Concern (ACECs), 3, 121
Audubon, John James, 80
AUMs (animal unit months), 80
automobiles, 59–60

Babbitt, Bruce, 83, *84*
bald eagles, 49; Bald and Golden Eagle Protection Act, 218
Barbee, Bob, 44, 45
Barrasso, John, 155
Bayer Company, 147–48

bears, 37, 59–60. *See also* black bears; grizzly bears
Bear 399, 65
Beattie, Molly, *84*
Beaverhead-Deerlodge forest plan, 132, 141–42
Beaverhead-Deerlodge National Forest, 130–32, 152–53, 157, 159, 199
Beaverhead-Deerlodge Partnership, 131, 152, 156
Biden administration, 29, 31, 43, 120, 223–24
"Big Burn" (fire), 44
bighorn sheep, 50–52, 161, 171
Big Sky Lumber, 129–30
Big Sky resort, 6, 20, 130, 160, 162–63, 168, 201
bikes and biking: allowed, 17, 158, 163; consequences of, 216; disagreement over, 58, 150, 153–54; prohibited, 17, 58, 159
bison, *54, 95, 99*; controversy over, 95–105, 230; diseased, 33, 34, 95–96, 219, 271n31; history of, 93–95; hunting of, 106, 111; migration of, 53–54, 172–73; and Native Americans, 39–40, 100–101, 210–11, 233–34; and park visitors, 59–60; as symbol, 92
bison management: consensus in, 97–98; controversy in, 96–97, 101–5; differences in, 193–94, 200; early, 37; and elk policy, 104, 109; feeding in, 111–13; improvement in, 98–100; jurisdictional issues in, 124–25, 206–7, 209; translocation, 100–103
black bears, *38*, 48–49, 67
Blackfeet tribe, 101
BLM (Bureau of Land Management): and conservation rule, 31; and ecosystem conservation, 196, 224, 237; and endangered species, 29; and energy development, 136, 137, 138–39, 194; and FLPMA, 27–28; and interagency coordination, 29, 206, 237; jurisdiction, 18–19, 117; and mining industry, 143, 147, 148; and multiple-use lands, 126–27; responsibilities of, 17; and wildlife management, 3, 113–14, 120–21, 123
Bonnie, Robert, 140
Bozeman, Montana, 20, 167, 177, 178
Bridger-Teton National Forest: and energy development, 136, 139–40, 199, 205, 209; feedgrounds in, 113; and Jackson Hole Mountain Resort, 162; and livestock, 173; and pronghorn, 3, 120–22, 182; recreation in, 158–59, 161–62; roadless area in, 157; and timber production, 129, 132, 198; and wilderness, 152, 155; wildfire management in, 134
brucellosis: about, 11, 95, 219; in bison, 95–96, 104; conflict over, 96–100; and domestic livestock, 112, 171, 172, 194; in elk, 104, 109–10; and feedgrounds, 111, 113, 114, 116, 210; lawsuits over, 97, 100, 208; policies on, 102–3, 200; spread of, 53–54, 92–93, 217–18, 230, 271n31; testing for, 101; and vaccination, 97, 98, 103, 111–12
buffalo. *See* bison
Buffalo Field Campaign, 104
Buffalo Ranch, 94–95
Bureau of Land Management (BLM). *See* BLM (Bureau of Land Management)
Bush I administration, 209
Bush II administration, 33, 55, 137, 157, 209, 224

Caldera National Monument, proposed, 156
Caldwell Canyon Mine, 148
California, 222–23
Caribou forest plan, 152
Caribou-Targhee National Forest, 143, 147–48, 155–56, 157, 198, 210
Carney Ranch, 182–83, 207, 232

Carson, Rachel, *Silent Spring*, 26
Carter, Jimmy, 159
Catlin, George, 93–94
cattle: and bison, 96; and brucellosis, 104, 111, 112, 219, 271n31; grazing regulation, 194, 230; and wildlife, 7, 40, 97–98, 110–11, 171–74, 234
cell-phone towers, 61–62
Centennial Mountain Range, 147, 156
checkerboard lands, 129–30, 160
Cheney, Liz, 155
chromium, and chromium mining, 143
chronic wasting disease (CWD). *See* CWD (chronic wasting disease)
Chronic Wasting Disease Research and Management Act, 116
cisco (fish), 50
Clarks Fork River, 17, 144, 199
Clean Air Act, 27–28
Clean Water Act, 27–28, 145, 148, 175
clearcutting, 6, 25, 32, 128, 130, 135
climate change: about, 7, 235; and bison, 101; consequences of, 34; and conservation, 9–10, 184–85, 191, 212–14; and cutthroat recovery, 50; and feedgrounds, 113; and grizzly bears, 73–74; in GYE, 11; and migration corridors, 196; and pika, 212, 214; presidential dealings with, 30–31, 43, 224; and science, 203; and water, 212–13; and wildfires, 46, 132–33, 135; and wildlife, 73–74, 151; and wolverines, 213, 214, 264n33
Clinton administration, 29, 32–33, 55, 137, 144–45, 209, 223–24
Cody, Buffalo Bill, 94
Cody, Wyoming, 20, 53, 69, 169
Collaborative Forest Restoration Program, 133
Colorado Plateau initiatives, 222
Columbia River Gorge Commission, and Scenic Area Act, 224
Committee of Scientists, 25
conservation biology, 9, 13, 24, 77

conservation easements: about, 187; alternatives to, 183–84; on amenity properties, 170–71; controversy over, 182–83, 232; and land trusts, 181; laws governing, 181–83, 207; and migration corridors, 3–4; on private lands, 30, 121, 174, 177; and ranchers, 22, 176; for recreation, 160
Conservation Strategy (grizzly bear), 72–73
Continental Divide Fire, *45*
Cooke City, Montana, 143
Court of Appeals. *See* DC Circuit Court of Appeals; Ninth Circuit Court of Appeals; Tenth Circuit Court of Appeals
COVID-19 pandemic, 20, 59, 167, 215
Craighead, Frank, 8–9, 32, 67–68
Craighead, John, 8–9, 32, 67–68
Crazy Mountains, 160
Crevice Mine, 146
Crown Manager's Partnership, 221–22
Crown of the Continent Ecosystem, 185, 204–5, 221–22
Custer Gallatin forest plan, 130, 134, 152, 153–54, 158, 217
Custer Gallatin National Forest: and Big Sky Resort, 162; and bison policy, 102; grazing in, 173; and mineral leases, 141; mining in, 143, 148, 284n75; planning process of, 153–54, 217; recreation in, 160; timber harvest in, 129–30; and wilderness designation, 153–54, 199; and wildfire management, 133–34
Custer National Forest, 1
cutthroat trout, 47, 48–50, 69
Cuyahoga River chemical fire, 25
CWD (chronic wasting disease), 11, 109–10, 111, 113–15, 116, 210, 217–18, 219–20
cyanide leaching, 143

DC Circuit Court of Appeals, 266n52
deer. *See* mule deer

Defenders of Wildlife, 82–83
Department of Agriculture. *See* USDA (US Department of Agriculture)
Department of Livestock (Montana), 102
Department of Livestock (Wyoming), 115
Desert Managers Group, 223
Director's Order 100, 43, 224
disease and disease transmission, 47, 50–51, 171. *See also* brucellosis; CWD (chronic wasting disease)
Dodge, Richard, 94
Driggs, Idaho, 20, 167, 178, 179, 180
drones, 61
Dubois, Wyoming, 20, 129

eagles, 49
Earth Day, 26
Earthjustice, 22
East Boulder Mine, 148, 149
East Rosebud Creek, 17, 199
Echo Canyon Dam, 150
ecological conservation, 6, 30–31, 196, 211, 221
ecological restoration: and conservation, 222, 225; for migrating wildlife, 124; in national forests, 126, 131, 194; in national parks, 38, 192; of predators, 65, 80, 205
ecological science, 9, 23–25, 26, 33, 192, 227, 229
ecosystem conservation: about, 13; challenges to, 19, 22, 33–34, 211–12, 214–16, 217–20; jurisdictional issues in, 165–66, 200–202, 208–11; laws affecting, 26–31, 204–8; science influencing, 23–25; shift toward, 15, 17, 31–32, 198–200, 202–4, 221–22, 224–26; support for, 19, 21–23; and wildlife management, 3, 4, 105
ecosystem management: about, 9, 228–29; changes in, 232–33; in conservation, 190, 195, 197–98, 202, 204–5, 220–21, 222–24; and grizzly bear policy, 65, 66, 69–70, 73; laws affecting, 28–29; shift toward, 11, 32–33, 35, 41; and wolf policy, 65, 90
EIS (Environmental Impact Statement). *See* Environmental Impact Statement (EIS)
elk, *108*, **118**; and bison policy, 103; development affecting, 168; diseased, 96, 104, 109–10, 219–20; feedgrounds, 110–16; and jurisdictional issues, 176, 194; management of, 17, 37, 92–93, 183–84, 210; migration of, 40, 116–19; and National Elk Refuge, 110–13; population of, 105–9; and wolf policy, 87, 109, 205
Elk Feedgrounds Management Plan, 115, 116
Emigrant Canyon, 146
Emigrant Mine (proposal), 146, 207
endangered species, 66, 73. *See* grizzly bears; wolverines; wolves
Endangered Species Act (ESA). *See* ESA (Endangered Species Act)
energy development: diminishing, 7; encouragement of, 136–39, 142, 209; future of, 214–15; physical location of, 135; protections from, 199; resistance to, 3–4, 139–40, 141, 142, 209–10; and wildlife management, 72, 136–39. *See also* extractive industries
Energy Policy Act, 136–37
Environmental Impact Statement (EIS), 82, 102–3, 136, 141, 149
Environmental Policy Act (Montana), 102
Environmental Quality Incentives Program, 176
Erickson, Mary, 153
ESA (Endangered Species Act): and bison policy, 104; and ecosystem management, 9; in environmental

reviews, 136; and jurisdictional issues, 193, 268n63; and land use, 174–75; in legal cases, 128–29, 134; regulations of, 70, 71, 73–74; revision of, 27; and wolverines, 264n33
ESA (Endangered Species Act), and grizzly bear policy: about, 65–66; and adoption of act, 68; and delisting, 74, 75, 77–78, 220, 261n19; factors affecting, 72–73; in legal cases, 69–77, 260n8; and livestock, 171–72; and population requirement, 5; significance of, 204–5
ESA (Endangered Species Act), and wolf policy: about, 65–66; and delisting, 85–86, 208; and experimental populations, 81–82; and grizzly bears, 88; jurisdictional issues in, 193, 205; and livestock, 171–72; and population requirement, 87; reintroduction of, 81–83; restoration supported by, 89–90; violations of, 1–2
Evanoff, Jim, 84
extractive industries: damage from, 27–28; diminishing, 6, 15, 19, 21, 216; and grizzly bear policy, 70; resistance to, 22; and wolf reintroduction, 82. See also energy development; mining; oil and gas industry

Farm Service Agency, 183
Farrell, Justin, 268n64
Faunal Survey (Wright), 37
federalism, 210, 232
Federal Land Policy and Management Act (FLPMA), 28
feedgrounds, for elk: and conflict with cattle, 110; disease affected by, 110, 113–16, 124, 210, 219–20; hunters encouraging, 93; interstate controversy caused by, 11, 201, 231; jurisdiction over, 119, 205–6

Fish and Wildlife Service (FWS), US. See FWS (Fish and Wildlife Service), US
fishing, 48, 213, 216
Flora, Gloria, 136, 139
Florida Everglades Restoration Initiative, 29, 223
FLPMA (Federal Land Policy and Management Act), 28
Flying D Ranch, 170
Forest Jobs and Recreation Act, 131
Forest Service, US: agencies formed by, 17; and bison policy, 97; changing policies of, 6, 23–24, 25, 32, 126–27, 203; and commodity production, 28; and domestic livestock, 172–73, 219; and ecosystem conservation, 199, 202, 205–6, 222, 224, 230, 236; and elk policy, 110, 113–14, 115; and endangered species, 29, 77; and energy development, 136, 140, 141, 142, 278n19; and grizzly bear policy, 68, 69–70, 71, 72–73; influences on, 33; and interagency coordination, 200; and mining industry, 143–44, 146, 147, 148, 149; and National Forest Management Act, 25, 28, 128, 195, 206; and private development, 167–68; and recreation, 20, 158–60, 161–62, 217; resource management by, 15, 30; responsibilities of, 26–27; and timber policy, 127–28, 129, 131; and wilderness policy, 150, 152–54, 156–57; and wildfire management, 130, 132–35, 213; and wildlife management, 3, 44–45, 120, 130, 194–95
Fort Belknap Indian Reservation, 100
Fort Laramie treaties, 39
Fort Peck Indian Reservation, 100–101
Fund for Animals, 54
FWS (Fish and Wildlife Service), US: and bison policy, 104, 111, 112–13; ecological management by, 30; and elk policy, 110, 111, 112–13, 206;

FWS (Fish and Wildlife Service), US (*continued*)
  and energy development, 136; and grizzly bear policy, 68, 70–75, 77–79, 212–13, 260n11, 261n19, 262n23, 264n33; and National Elk Refuge, 17, 106, 110–13; priorities of, 202; and ranching, 173; and science, 203, 206; and wolf policy, 2, 81–83, 84–86, 88, 90, 205, 266n46

"Gallatin Community Collaborative," 153
Gallatin County, Montana, 20, 167, 168, 177–78, 197
Gallatin Forest Partnership, 153, 156
Gallatin National Forest, 129–30, 136, 159, 205
Gallatin Range, 129–30, 154, 163
Gallatin River, 162
Gallatin Valley Sensitive Lands Protection Plan, 177–78
Gallatin Yellowstone Wilderness Alliance, 154
Gardiner, Montana, 21, 178
gas industry. *See* energy development; extractive industries; oil and gas industry
gateway communities, 19–21, 59–60, 197
General Mining Law, 143, 144, 208–9
Glacier National Park, 41, 81–82, 222
Good Neighbor Agreement (GNA), 148–49
Gordon, Mark, 121
Grand Canyon National Park, and Grand Canyon Protection Act, 222
Grand Targhee Resort, 20, 160–61, 168, 180, 201
Grand Teton National Park: disease in, 109, 114; expansion of, 6, 233; as GYE centerpiece, 14–15; inholdings in, 62–64; and Kelly Parcel, 63–64; maintenance costs in, 61; as nature conservation model, 35–36, 238; nonnative species in, 47, 50–52; and Path of the Pronghorn, 121; policy changes in, 191–92; recreational controversy in, 58; science employed by, 203; surroundings of, 151; technology in, 61–62; and visitation, 59–60; and wilderness designation, 152

Grant Village, 68–69
Grassland Conservation Reserve Program, 183
grazing, 6, 7, 18, 100, 171–74, 194
Great American Outdoors Act, 29–30, 61, 175
"Greater Yellowstone," term origins, 7–8
Greater Yellowstone Climate Assessment (Hostetler et al.), 213
Greater Yellowstone Coalition, 21–22, 32, 69, 128–29, 146, 211
Greater Yellowstone Coordinating Committee (GYCC). *See* GYCC (Greater Yellowstone Coordinating Committee)
Greater Yellowstone Ecosystem (GYE). *See* GYE (Greater Yellowstone Ecosystem)
Great Recession, 168, 179
Green Knoll fire, 133
Grizzly Bear Management Guidelines, 70
Grizzly Bear Recovery Plan, 70–71
Grizzly Bear Recovery Zone, 172
grizzly bears, 71, **76**; conflicts with, 40, 166, 171–74; and conservation, 229–30; decline of, 67–70, 260n8; as ecosystem indicator, 8–9, 32, 67–68, 204–5; food sources for, 48–50; and genetic diversity, 75–77, 185, 214, 220, 260n11; in history, 66–67; and hunting, 66, 67, 68, 72, 74–75, 77, 78, 90, 193, 218, 230; jurisdictional issues affecting, 218–19; litigation over, 70–77; management of, 34, 207, 209–10, 218; in national forests,

141; population of, 33; recovery of, 70–72, 90–91, 128–29, 193, 199, 200, 260n11; significance of, 65–66

grizzly bears, delisting of: attempted, 72–75, 77–79, 261n19, 262n23; consequences of, 218, 237; denied, 212–13; opposition to, 205; and state policies, 193

GYCC (Greater Yellowstone Coordinating Committee): about, 17–18; influence of, 198, 229, 237–38; membership, 17, 218, 229, 237; purpose of, 32, 42, 206; and Vision initiative, 18, 32–33, 70, 200, 206, 229; and wildfire management, 45–46; and wildlife management, 218

GYE (Greater Yellowstone Ecosystem), **16**; challenges to, 234–39; changes in, 32–34, 190–91, 228–32; conservation in, 232–34; controversy in, 5–7; description of, 14–23; ecosystem management in, 9–10, 13; historical roots of, 7–8; land conservation in, 165–66; migratory animals in, 92–93; multiple-use lands in, 126–27; nature conservation in, 5, 10–12; predators representing, 65–66; resource management in, 228; term, 7–9, 32, 67–68, 69, 197–98, 228; wildlife management in, 2–4

GYE states: anti-government sentiment in, 120; and climate change, 213; differences among, 201; and disease control, 98, 109–10, 219–20; and grizzly bear policy, 33, 34, 75, 91, 230, 236–37; land use in, 14–15, 174, 176–77, 181–82; political power of, 18–19, 52, 206, 209–10; population growth in, 167; ranching in, 169; timber industry in, 132; wilderness designations in, 199; wildlife management in, 117, 119, 207, 218; and wolf restoration, 82, 84–88, 91; and wolverine policy, 264n33. *See also* Idaho; Montana; Wyoming

Haaland, Deb, 218
Habitat Conservation Lease program, 183
habitat leases, 183–84
Hamilton, Kniffy, 121, 139–40
Harrison, Benjamin, 8
Hayden, Ferdinand, 36
HDC (High Divide Collaborative), 185, 187–88
Healthy Forests Restoration Act (HFRA), 133, 175
High Divide area, 123, 175, 185–87, **186**, 200, 232, 234
High Divide Collaborative (HDC), 185, 187–88
High Divide Collaborative Coordinating Committee, 185
High Noon Ranch development, 180
Hoffman, Paul, 57
Hough, Emerson, 8
housing issues, 21, 168, 178–80, 197
HPBH WSA (Hyalite-Porcupine-Buffalo Horn Wilderness Study Area), 153–54, 159
hunting: of bears, 74–75, 77–78, 218; of bison, 99; and disease threat, 104; and ecosystem conservation, 218; of elk, 106–9, 117; jurisdictional issues over, 63; and migrating animals, 92–93; by Native Americans, 39; and wildlife management, 193–94; of wolves, 86–88, 90, 218
Hyalite-Porcupine-Buffalo Horn Wilderness Study Area (HPBH WSA), 153–54, 159

IBMP (Interagency Bison Management Plan), 97–99, 101–3
Idaho: development in, 141–42, 157, 161; and grizzly bear policy, 73; mining industry in, 147–48; rental

Idaho (*continued*)
  policy in, 180; resort areas in, 168; and roadless area forest lands, 157; and wilderness, 155–56; and wildlife management, 123; and wolf policy, 1–2, 82, 89–90, 205. *See also* GYE states
Idaho Fish and Game, 128
IGBC (Interagency Grizzly Bear Committee). *See* Interagency Grizzly Bear Committee (IGBC)
immigrants, Hispanic, 178, 197
Indians. *See* Native Americans
Indigenous peoples. *See* Native Americans
Interagency Bison Management Plan (IBMP), 97–99, 101–3
Interagency Fire Management Policy Review, 45–46
Interagency Grizzly Bear Committee (IGBC): coordination by, 229; and government agencies, 33; and grizzly bear policy, 75, 78, 199, 204; guidelines of, 70; purpose of, 68; on sheep grazing, 172
Interior Columbia Basin Ecosystem Management Plan, 223
Interior Department, 57
International Snowmobile Industry Association, 53, 55

Jackson, Wyoming, 6, 19–20, 161–62, 167, 168
Jackson elk herd, 105, 110
Jackson Hole area, 184
Jackson Hole Mountain Resort (JHMR), 161–62, 168
Jardine mining district, 146
Jedediah Smith wilderness area, 155
Jewell, Sally, 146, 209
JHMR (Jackson Hole Mountain Report), 161–62, 168
John D. Rockefeller Jr. Memorial Parkway, 74
Jonah Energy, 138

kayaking, 58
Kelly Parcel, 63–64
keystone species, 82, 102

Lacey Act, 106
Lake Tahoe watershed, 224–25
lake trout, 42, 47, 48–50
Lamar Valley, *108*
Land and Water Conservation Act (LWCA), 27, 175
Land and Water Conservation Fund, 29–30, 185, 187
land development: and conservation, 189; impact of, 11; on multiple-use lands, 130, 161–63; on private lands, 176–81, 196, 207, 215; problems from, 166–69
land exchanges, 130, 160, 162
landscape conservation: and climate change, 184–85; in ecosystem management, 9–10, 223, 225; and ESA, 90; and grizzly bear policy, 204–5; and High Divide, 185–88; importance of, 184–85, 213–14; initiatives in, 10, 30–31, 43, 185, 187–88; and migration corridors, 214, 231, 232–33
Landscape Conservation Cooperatives, 30–31, 43, 224
land trusts, 181, 182–83, 196
land-use planning, 19, 174, 176–78, 207, 231, 236–37
Lane, Franklin, 36, 252n1
Leopold, A. Starker, 37–38, 81, 107
Leopold, Aldo, 81; *A Sand County Almanac*, 26
Leopold Report: ecological science promoted by, 192, 229; and grizzly bear policy, 67; influence of, 42–43; and park boundaries, 35; and resource management, 37–39, 40; and wildfire management, 44; and wildlife management, 24, 47, 81, 94, 107
Leopold Wolf Pack, 87

Lewis, Meriwether, 67, 93
Lewis and Clark Expedition, 66–67, 80, 93, 105–6
Lewis and Clark National Forest, 136
Lincoln County, Wyoming, 155, 201
Livingston, Montana, 20–21, 167
logging. *See* timber industry
Louisiana Pacific saw mill, 129–30
lumber mills, 128, 129, 132. *See also* timber industry
LWCA (Land and Water Conservation Act), 27, 175

Madison County, Montana, 170
Madison River, 168–69, 216
Madison Valley, 21
Magic Pack, 81–82
Management Policies (National Park Service), 30, 41–42, 46, 47–48, 52, 57, 60
migration: about, 92–93, 116–19; and climate change, 101; hunting enabled by, 107–8; management of, 98, 217–18; and park boundaries, 40, 53–54, 105, 193–94; scientific study of, 33
migration corridors, **118**, **122**; about, 3–4; awareness of, 7; and climate change, 84–85, 196, 212, 214; and landscape conservation, 231, 232–33; protections for, 105, 119–26, 172–73, 175–76, 188, 199–200, 220, 229–30; scientific study of, 185; severed, 106, 110; threats to, 142, 147, 166, 168–69, 195–96; and wilderness areas, 151. *See also* Path of the Pronghorn; Red Desert to Hoback (RD2H) migration corridor
military, as park caretakers, 39, 44, 67, 94
million board feet (MMBF), 128, 129, 130, 134
mineral leases, 135–36, 140–41, 233
Mineral Leasing Act, 136, 147

mining: decline of, 11, 143, 230; economic effect of, 284n75; and environment, 6, 144–47; in national forests, 127; New World Mine, 144–45; on private lands, 146–47; on public lands, 143–46, 147–49; resistance to, 33, 194, 198, 204, 207, 209; state involvement in, 146, 210
Mission 66 campaign, 68–69
Missouri Breaks, 102, 223
MMBF (million board feet), 128, 129, 130, 134
Molloy, Donald, 85, 86
Montage Big Sky luxury resort, 162
Montana: and bison policy, 96–97, 99–101, 102–5, 194; change in, 20–21; and disease control, 116; energy development in, 141–42; and grizzly bear policy, 73, 78–79; land use in, 130, 159, 184; mining industry in, 143, 146; wilderness designations in, 152–53; and wolf policy, 1–2, 90, 205. *See also* GYE states
Montana Department of Fish, Wildlife, and Parks, 119, 183
Montana Fish and Wildlife Commission, 88
Montana Supreme Court, 100, 177, 207
Montana Wilderness Study Act, 153
Moose-Wilson Road, 58
mountain bikes, 150, 158
mountain goats, 47, 50–52
Mud Volcano, *99*
Muir, John, 238
mule deer, 4, 117, **118**, 188
Multiple Use–Sustained Yield Act, 26–27
Murie, Adolf, 37, 81

National Academy of Sciences, 37–39, 101, 104, 109, 271n31
National Conservation Network Act (proposal), 225–26

National Elk Refuge (NER). *See* NER (National Elk Refuge)
National Environmental Policy Act (NEPA). *See* NEPA (National Environmental Policy Act)
National Forest Management Act (NFMA): and diversity, 25, 28, 195; and ecosystem management, 9; planning regulations of, 30, 206, 224; and resource management, 28; and timber production, 25, 128–29; and wildlife management, 195
national forests: about, 15, 17; changes in, 198–99, 229–30; ecosystem management for, 32–33; and energy development, 135–36; and grizzly bear policy, 70, 72; and livestock, 171–72; and mining industry, 146–48; and production management model, 6; recreation in, 158–63; resource management policy, 28, 163–64; roadless area rule, 156–57; role of, 194–95; and timber production, 126–27, 132–33, 135; as wilderness, 150, 154–57. *See also names of individual national forests*
National Landscape Conservation System, 29
national monuments, 29, 156, 222, 223, 224
National Park Service. *See* Park Service
National Parks Omnibus Management Act, 29
National Park system, 56, 59, 191–92
National Trails Act, 27
National Wild and Scenic Rivers System, 221
National Wildlife Federation (NWF), 172–73
National Wildlife Refuge System Administration Act, 27, 206
National Wildlife Refuge System Improvement Act, 29
Native Americans: and bison policy, 93, 98–99, 100–101, 103, 104–5, 194, 207, 210–11; buffalo slaughter affecting, 94; fire use of, 44; and grizzly bear policy, 75; involvement of, 22–23, 33, 210–11, 222, 233–34; and Yellowstone National Park, 39–40, 210
Native Fish Conservation Plan, 49–50
natural regulation policy, 40, 43–44, 45, 64, 107–8, 192
Natural Resources Conservation Service, 176
nature conservation: changing priorities in, 31, 33–34, 35–36; as controversial, 4; Forest Service promoting, 222–23; and GYE concept, 7, 9, 10–11, 14, 191, 201–2, 220–21, 226; and land use, 176–77, 181, 184; and Park Service, 52; and public lands, 29–30; and wildlife management, 2–3, 120–21, 172; Yellowstone National Park as symbol of, 4–5, 227–28
NEPA (National Environmental Policy Act): about, 27; and cell-phone towers, 61; and ecosystem conservation, 205–6, 208–9; and ecosystem management, 9, 27–29, 204; and elk feedgrounds, 113–14; and energy development, 136; and grizzly bear policy, 72; and migration corridors, 121, 123; and mining industry, 148; and multiple-use lands, 128–29, 134, 139, 147, 149, 162
NER (National Elk Refuge), 17, 106, 110, 111–13, 203, 206, 219
New West, 7, 21, 204
New World Mine, and New World Mining District, 143–46, 209
NFMA (National Forest Management Act). *See* National Forest Management Act (NFMA)
Ninth Circuit Court of Appeals, 2, 73, 148, 261n19
Nixon, Richard, 159
nonnative species, 36, 47–52, 229

Noranda, 144–45, 210
Normally Pressured Lance gas field, 121, 138–39
North American Model of Wildlife Management, 77, 86, 207
Northern Pacific Railroad, 59, 130
Northern Range elk herd, 105, 106–9
Northern Rockies Ecosystem Protection Act, 221
Northwest Forest Plan, 29, 222–23
NWF (National Wildlife Federation), 172–73

Obama administration, 9, 29, 30–31, 43, 56, 209, 224
off-road vehicles (ORVs), 16–17, 150, 153, 155, 158, 159
oil and gas industry: decline of, 163, 194; effects of, 20; encouragement of, 32, 119; management of, 17; in national forests, 127; and pronghorn, 120–21, 123; regulation of, 205; resistance to, 32–33. *See also* energy development; extractive industries; Pinedale Anticline Gas Field
Old Faithful Protection Act, 175
Olson, Sigurd, 81
One and Only Moonlight Basin development, 162–63
Organic Act (National Park Service): and bison policy, 97; influence of, 41–42; Park Service created by, 8, 94; and recreational policy, 52, 56, 57, 60, 206; science ignored in, 36; and wildfire management, 46; and wildlife management, 48
ORVs (off-road vehicles), 16–17, 150, 153, 155, 158, 159
ospreys, 49
otters, 49
Overthrust Belt, 6, 135
ozone, 138

PACs (protected area complexes), 225–26

Palisades Wilderness Study Area, 155–56, 158
palladium mining, 148
Paradise Valley, 21, 103, 168–69, 173
Paradise Valley project, 234
Park County, Montana, 180–81
Park County, Wyoming, 169
Park County Business Council (Montana), 146
Park County Resource Council v. US Dept. of Agriculture, 278n19
Parkman, Francis, 80
Park Service: agencies formed by, 17; and bison policy, 8, 94, 96, 97–98, 99–103, 105, 124–25; changing priorities of, 23–25, 32, 229; ecological science used by, 192, 202; and elk policy, 105, 106, 107–8, 110, 111; and grizzly bear policy, 67–69, 70, 79; and inholdings, 62–64; and interagency coordination, 200, 206; and Leopold Report, 24, 35, 38, 41–44, 47, 67, 81, 94, 107, 192, 226, 229; *Management Policies* controversy, 57–58; and mining industry, 144, 145–47; and Native Americans, 39–40, 234; and nonnative species, 47–48, 49–50, 51, 52; and recreation, 5, 52–58, 217; resource management policies, 5, 36–43; responsibilities of, 15, 28, 29, 33; science guiding, 37–39, 43–45, 192, 229; and visitation, 59–62; and wildfire management, 3, 24, 43–47, 81, 94–95, 107; and wolf policy, 2, 80–81, 91, 205
Path of the Pronghorn, **122**; in conservation easement, 182–83; establishment of, 120–21; politics affecting, 220; preservation of, 3; purpose of, 199–200; significance of, 123–24, 231; threats to, 3–4, 121, 123, 138–39, 169. *See also* migration corridors
philanthropy, 63, 142, 202, 233
Phillips, Mike, *84*

phosphate, and phosphate mining, 143, 147–48
Phosphate Patch, 147, 149
pikas, 214
Pinchot, Gifford, 23
pine beetles, 73, 128, 134, 212
Pinedale, Wyoming, 17, 20, 137–38
Pinedale Anticline gas field, 20, 120–21, 137, *138*, 142, 209
platinum mining, 148
poaching, 67, 94, 106
Primary Conservation Area (grizzly management), 72–73
private land: development on, 195–96, 236; factors affecting, 237; in GYE, 18; legislation affecting, 27, 174, 176–78, 207, 231, 236–37; migration corridors on, 120–21; mining on, 146; politics influencing, 180–81; recreation on, 160–61, 162–63; wildlife management on, 172–73; and zoning, 163, 176–77, 179–80
private land, conservation of: about, 165–66, 188–89, 231–34; evolving methods of, 184–85, 187–88; by federal government, 174–76, 183; with incentives, 119, 173–76, 215; by independent entities, 181–84; and land-use planning laws, 27, 174, 178–78, 207, 231, 236–37; by local governments, 177–78; by state government, 174, 175–77, 181–82, 183
pronghorn, 92, 116–17, 120–24, 138–39. *See also* Path of the Pronghorn
protected area complexes (PACs), 225–26
public land: conflict over, 26, 171; energy development on, 136–37, 138; evolving management of, 199–200; legislation affecting, 26–29, 32–33, 150–51; mining on, 143; multiple uses of, 154–55, 159–60; oversight of, 17; recreation on, 20–21

railroad, 59, 94, 129–30
ranchers: and bison policy, 95–98, 102; and conservation, 22, 176, 182–84, 234; and disease threat, 33, 95, 100, 104, 208; and elk policy, 104, 111; on park expansion, 8; and predator issues, 77, 82–83, 171–74, 187–88, 219
ranches, ownership changing, 18, 21, 169–71, 195
RD2H (Red Desert to Hoback) migration corridor, 4, 188, 220, 223
Reagan administration, 209
recreation: challenges caused by, 64; and climate change, 213; conflict over, 5, 158–60, 216–17; and conservation, 52, 56–58, 192, 203, 258n41; economic effect of, 6, 19, 20–21, 157–58, 284n75; and Forest Service, 127, 130; impact of, 11, 164–65, 216–17; increasing, 157–58, 160–63; legislation affecting, 26–27, 131; on multiple-use lands, 140–41; in national forests, 15, 17, 157–63, 195; in national parks, 52–59; regulation needed for, 235; resource management in, 163–64; and ski resorts, 159–63; in wilderness areas, 150–51, 153–55; in wilderness study areas, 158–59
Red Desert to Hoback (RD2H) migration corridor, 4, 188, 220, 233
Red Lodge, Montana, 21, 143, 167
Redwood Amendment, 28, 41
Redwood National Park, 41
resource management: in ecosystem management, 9–10; in GYE priorities, 69–70, 163, 191–92; influences on, 5, 23, 25–26, 43, 192; jurisdictional issues in, 201–2; legal issues in, 204; in nature conservation, 228–29; in wildfire management, 46–47
Revisiting Leopold, 42–43, 224
Rexburg, Idaho, 128, 129

roadless area rule, 30, 132, 150, 152, 156–57, 163, 199
roadless areas, 131–32, 154, 156–57, 195
Rockefeller family, 233
Roosevelt, Teddy, 80
Roosevelt fire, 133
Royal Teton Ranch, 172–73
Russell, Osborne, 106

sage grouse, 30–31, 121, 123, 138–39
Sand County Almanac, A (Leopold), 26
San Luis Valley Conservation Area, 223
Santa Barbara oil spill, 25
Sawyer, Hall, 4
selenium contamination, 147–48
sheep, domestic, 7, 171–72, 194, 230
Sheepeaters, 39
Sheridan, Philip, 8
Shoal Creek Wilderness Study Area, 155
Shoshone forest plan, 132, 133–34, 152
Shoshone National Forest, 141, 172, 278n19
Shoshone tribe, 17, 39
sideboards, in elk management, 115–16
Sierra Club, 41
Sierra Nevada Framework, 222–23
Silent Spring (Carson), 26
Simplot Company, 147–48
Simpson, Mike, 85–86
skiing: backcountry, 51; downhill, 6, 19–20, 127, 160–63, 168, 179, 213; helicopter, 158–59; resorts, 159–63
Smoky Canyon Mine, 148
Snake River Range, 51
snow coaches, 53, 54–55, 56
snowmobiling, 54; agreement on, 56; and animal cruelty, 89; controversy over, 5, 52–58, 217; impacts from, 53–54, 155; lessons from, 56–57; regulations for, 158, 159, 229; technological changes affecting, 158; in Yellowstone National Park, 52–58
South Plateau Area Landscape Treatment Project, 134
spotted owl, 28–29, 32–33, 222
sprawl, 20–21, 167, 176–81
St. Anthony, Idaho, 128, 129
Star Valley, Wyoming, 19, 168, 178
State of the Parks (National Park Service), 41
Stegner, Wallace, *Wilderness Letter*, 27
Stillwater mining complex, 143, 148–49, 284n75
Sublette County, Wyoming, 155, 168–69, 170, 201
Sublette mule deer corridor, 123
Sylvan Pass, 56, 60

Tahoe Regional Planning Agency, 224–25
tailings ponds, 144, 149
Targhee forest plan, 152, 155–56
Targhee National Forest, 6, 72, 127–29, 141
Tendoy drilling project, 141–42
Tenth Circuit Court of Appeals, 2, 123, 139, 157, 173, 278n19
Terrestrial Wildlife Movement and Habitat Strategy, 119
Tester, Jon, 85–86, 131–32, 146, 152
Teton County, Idaho, 161, 168, 179–80
Teton County, Wyoming: challenges in, 201; housing in, 178–79; immigrant population in, 197; land use in, 177; newcomers in, 168; real estate in, 63; recreational controversy in, 155, 161; wealth in, 19
Teton to Snake Fuels Management Project, 134
Teton Valley, Idaho, 19–20, 161, 168
Teton Village, Wyoming, 162
Theodore Roosevelt Conservation Partnership, 137
30x30 campaign, 9–10

timber industry: controversies in, 6, 130–34; decline of, 7, 11, 126–27, 130, 135, 194; encouragement of, 127–28; and endangered species, 28–29, 71–72, 222; increase of, 32; influences on, 33; legislation affecting, 25; mills, 128, 129, 132; resistance to, 128–30, 152; restrictions on, 30; and wildfire management, 163, 222–23

tourism: and cell-towers, 61–62; economic effect of, 60, 159, 203, 216; and elk policy, 110, 117; and gateway communities, 15, 21, 59; in GYE states, 19; in winter, 53, 162; and wolf policy, 83, 89. *See also* visitation

Townsend Mill, 132

trapping, 87–88

trophic cascades, 49, 108–19

trout. *See* cutthroat trout; lake trout

Trout Unlimited, 50

Trump administration, 31, 33, 43, 75, 120, 146, 224

Turner, Ted, 100, 170

Udall, Stewart, 24, 37, 107

ungulates, 3, 7, 116–17, 124, 193–94, 195, 214. *See also* bison; elk; mule deer; pronghorn

Uniform Conservation Easement Act, 181–82

Union Pass Road, 129

United Nations, 145

United Property Owners of Montana (UPOM), 102

United Steelworkers Union, 148

University of Wyoming, 119

UPOM (United Property Owners of Montana), 102

Upper Green River, 77, 120, 142, 173, 194

US Constitution, 208

USDA (US Department of Agriculture), 96, 140, 175–76, 183, 215

US Fish and Wildlife Service. *See* FWS (Fish and Wildlife Service), US

US Forest Service. *See* Forest Service, US

US Geological Survey, 116, 120

Victor, Idaho, 20, 179, 180

Vision initiative, 18, 32–33, 70, 200, 206, 229

visitation: challenges caused by, 33, 59–62, 64, 216–17; and conservation, 15, 216–17; and gateway communities, 20–21, 35–36, 59; in national forests, 158; in national parks, 59–62, 192–93; in ski resorts, 161; winter, 53; wolves affecting, 83

Washburn expedition, 36

watersheds, 130, 145, 213–14

Wenk, Dan, 74–75

Western Governors' Association, 120

West Yellowstone, Montana, 20, 53, 55, 56

whitebark pine tree, 73–74, 212

wild and scenic rivers, 17, 144, 199, 208, 221–22

Wild and Scenic Rivers Act, 27

Wilderness Act, 6, 27, 29, 150–51, 157, 206, 208

wilderness designation: areas recommended for, 155–56; difficulty in achieving, 11; disagreement over, 131, 150–54, 163, 199; in national parks, 15; and recreation, 158–59; restrictions in areas of, 156–57

Wilderness Letter (Stegner), 27

wilderness study areas (WSAs). *See* WSAs (wilderness study areas)

wildfires, 45; changing policies on, 5, 7, 23–24, 32, 40–41; and climate change, 132–33, 213; factors affecting, 132–35; management of, 43–47, 132–35, 192; and new homes, 167–68, 169, 175–76. *See also* wildland-urban interface zone

wildland-urban interface zone, 133–35, 169, 175, 213, 236
wildlife conservation: and agricultural lands, 175–76; encouragement of, 22, 173; and energy development, 3–4, 135–42; federal programs, 174–76; and feedgrounds, 110–16; growing importance of, 11, 194–96; and GYCC, 218; and migration corridors, 116–24; and North American Model, 77, 207; and recreation issues, 154; for sage grouse, 30–31, 139, 142; and science, 13, 23–25, 36–42; and state law, 207
Wildlife Corridors Conservation Act, 120
wildlife habitat: and amenity buyers, 170–71; and conservation easements, 181, 182, 184; conservation of, 196; in GYE, 17; and habitat leases, 184; humans affecting, 64; in island settings, 24; in multiple-use lands, 137, 141, 142, 145–46, 155, 159; in national forests, 130, 173–74; on private lands, 18, 166, 171, 195; protection for, 175–76, 178, 188; threats to, 6, 34, 185
wildlife management: aerial gunning in, 51–52; and bison policy, 97, 200; changing focus of, 81, 94, 107; and climate change, 213–14; conflict in, 66, 217, 230–31, 235, 236–37; and elk policy, 113; in inholdings, 62–63; jurisdictional issues in, 192–93, 207, 232; and predators, 218; science influencing, 37, 42; and state law, 11, 15, 105, 207–8, 220, 231; and wolf policy, 88, 218
wildlife refuges, 17, 27, 29, 92, 94, 151, 223. *See also* NER (National Elk Refuge)
Wild Migrations Atlas (Kauffman et al.), 119
Wind River Indian Reservation, 17, 20, 22, 211

Wind River Range wilderness area, 138
Winegar Hole wilderness area, 155
Wolf Compensation Fund, 82–83
wolverines, 214, 264n33
wolves, 84, 87; and elk, 106–7, 114–15; extermination of, 23, 80–81; killing of, 1, 86–89, 175; research on, 37, 83, 86; significance of, 65–66, 79–80
wolves, restoration of: about, 1–3, 81–86, 90–91, 205; conflicts from, 166; debate, 40, 81–83; and delisting, 84–87, 89, 90–91, 208; and elk, 105, 108–9; as experimental population, 81–86; genetic exchange in, 266n47, 266n52; jurisdictional issues in, 192–93, 201, 207, 209–10, 218–19, 230, 266n46, 268n63; lessons from, 89–90, 205; and livestock, 171–72; opposition to, 82; reactions to, 5, 86–90, 268n64; success of, 33, 83–85, 229; under state management, 86–89
Working Lands for Wildlife initiative, 176
World Heritage Sites, 145
WPI (Wyoming Public Lands Initiative), 154–56, 199
Wright, George Melendez, 36–37; *Faunal Survey*, 37
WSAs (wilderness study areas): in GYE national forests, 150, 195, 199; and motorized recreation, 158–59, 206, 217; protection of, 158–59, 199, 201; and wilderness designation, 150, 199
WUI (Wildland Urban Interface) zone. *See* wildland-urban interface zone
Wyoming: and elk policy, 110–11, 112, 113–16, 231; energy development in, 135, 139–40, 141, 142; and feedgrounds, 110–11, 113–16; and Grand Teton National Park, 51–52, 62–64; grizzly bear policy, 69, 77–78, 173; and housing density,

Wyoming (*continued*)
178–79; and land ownership issues, 62–63; and migration corridors, 3–4, 121, 123, 138, 183, 188, 220; and mining industry, 144, 147; park proximity benefiting, 60; and roadless area rule, 157; and snowmobile policy, 55–56; timber industry in, 129; wilderness opposed by, 155–56; and wolf policy, 2, 88–89, 90, 266n46, 266n52. *See also* GYE states

Wyoming Board of Land Commissioners, 63, 123

Wyoming County Commissioners Association, 154

Wyoming Farm Bureau, 83

Wyoming Game and Fish Department, 110, 115–16, 119, 123, 137, 139, 161, 188, 207–8

Wyoming Highway Department, 121, 123

Wyoming Land Board, 123

Wyoming Migration Initiative, 119

Wyoming Public Lands Initiative, 154–56, 199

Wyoming Range, 140, 142, 198, 209–10

Wyoming Range Legacy Act, 140, 199

Wyoming Supreme Court, 177, 182, 208

Wyoming Wilderness Act, 151, 156, 158, 208

Yellowstone (television series), 169

Yellowstone Bison Management Plan, 102–3, 105, 219

Yellowstone Club, 20, 160, 162

Yellowstone Ecosystem Grizzly Bear Subcommittee (YES), 70, 204

Yellowstone Forever, 50

Yellowstone Gateway Protection Act, 146, 209

Yellowstone Grizzly Bear Coordinating Committee, 75

Yellowstone Lake, 47; and lake trout, 47–50

Yellowstone National Park: history, 4–5, 8, 36–41, 44–45, 67–69, 94–95, 106–8; as GYE centerpiece, 14–15; and mining industry, 144–47; and Native Americans, 39–40, 100, 104–5, 194, 210–11, 233–34; as nature conservation model, 4–5, 35–36, 226, 227–28, 238; policy changes in, 36–41, 191–92; recreational controversies in, 58; and science, 36–43, 203; snowmobiling in, 53–57; surroundings of, 151; technology in, 61–62; as testing ground, 31–32; visitation increasing in, 59–61, 216; and wilderness designation, 152; and wildfire, 44–47; wildlife management in, 37–42, 82–83, 92, 106–9, 192–94

Yellowstone National Park Bison Management Plan, 194

Yellowstone Timber Land and Reserve, 8

Yellowstone to Yukon (Y2Y), 221

YES (Yellowstone Ecosystem Grizzly Bear Subcommittee), 70, 204

Y2Y (Yellowstone to Yukon), 221

Zinke, Ryan, 120, 146, 209

zoning, 163, 176–77, 179–80